C0-ATX-370

Advances in Nuclear Dynamics 5

Advances in Nuclear Dynamics 5

Edited by

Wolfgang Bauer
Michigan State University
East Lansing, Michigan

and

Gary D. Westfall
Michigan State University
East Lansing, Michigan

Kluwer Academic / Plenum Publishers
New York, Boston, Dordrecht, London, Moscow

Proceedings of the 15th Winter Workshop on Nuclear Dynamics, held January 9–16, 1999, in Park City, Utah

ISBN: 0-306-46280-X

233 Spring Street, New York, N.Y. 10013

10 9 8 7 6 5 4 3 2 1

A C.I.P. record for this book is available from the Library of Congress

Printed in the United States of America

Preface

This workshop was the 15th in a series that addresses the subject of the dynamics of nuclear reactions. These workshops are dedicated to the concept that bringing together scientists from diverse areas of nuclear reactions promotes the vibrant exchange of ideas. This workshop hosted presentations from experimentalists and theorists, intermediate energy to ultrarelativistic energies, and final results to recent speculations. Many of these scientists would not normally be exposed to the work done in other subfields. Thus the Winter Workshop on Nuclear Dynamics plays a unique role in information exchange and the stimulation of new ides.

The field of nuclear dynamics has a bright future. New accelerators are being planned and completed around the world. New detectors are being constructed. New models and theories are being developed to describe these phenomena. The Winter Workshop on Nuclear Dynamics will continue to promote this lively and compelling field of research.

WOLFGANG BAUER AND GARY D. WESTFALL

Previous Workshops

The following table contains a list of the dates and locations of the previous Winter Workshops on Nuclear Dynamics as well as the members of the organizing committees. The chairpersons of the conferences are underlined.

1. Granlibakken, California, 17-21 March 1980
   W.D. Myers, J. Randrup, G.D. Westfall

2. Granlibakken, California, 22-26 April 1982
   W.D. Myers, J.J. Griffin, J.R. Huizenga, J.R. Nix, F. Plasil, V.E. Viola

3. Copper Mountain, Colorado, 5-9 March 1984
   W.D. Myers, C.K. Gelbke, J.J. Griffin, J.R. Huizenga, J.R. Nix, F. Plasil, V.E. Viola

4. Copper Mountain, Colorado, 24-28 February 1986
   J.J. Griffin, J.R. Huizenga, J.R. Nix, F. Plasil, J. Randrup, V.E. Viola

5. Sun Valley, Idaho, 22-26 February 1988
   J.R. Huizenga, J.I. Kapusta, J.R. Nix, J. Randrup, V.E. Viola, G.D. Westfall

6. Jackson Hole, Wyoming, 17-24 February 1990
   B.B. Back, J.R. Huizenga, J.I. Kapusta, J.R. Nix, J. Randrup, V.E. Viola, G.D. Westfall

7. Key West, Florida, 26 January - 2 February 1991
   B.B. Back, W. Bauer, J.R. Huizenga, J.I. Kapusta, J.R. Nix, J. Randrup

8. Jackson Hole, Wyoming, 18-25 January 1992
   B.B. Back, W. Bauer, J.R. Huizenga, J.I. Kapusta, J.R. Nix, J. Randrup

9. Key West, Florida, 30 January - 6 February 1993
   B.B. Back, W. Bauer, J. Harris, J.I. Kapusta, A. Mignerey, J.R. Nix, G.D. Westfall

10. Snowbird, Utah, 16-22 January 1994
    B.B. Back, W. Bauer, J. Harris, A. Mignerey, J.R. Nix,
    G.D. Westfall
11. Key West, Florida, 11-18 February 1995
    W. Bauer, J. Harris, A. Mignerey, S. Steadman, G.D. Westfall
12. Snowbird, Utah, 3-10 February 1996
    W. Bauer, J. Harris, A. Mignerey, S. Steadman,
    G.D. Westfall
13. Marathon, Florida, 1-8 February 1997
    W. Bauer, J. Harris, A. Mignerey, H.G. Ritter, E. Shuryak,
    S. Steadman, G.D. Westfall
14. Snowbird, Utah, 31 Januray - 7 February 1998
    W. Bauer, J. Harris, A. Mignerey, H.G. Ritter, E. Shuryak,
    G.D. Westfall
15. Park City, Utah, 9-16 January 1999
    W. Bauer, J. Harris, A. Mignerey, H.G. Ritter, E. Shuryak,
    G.D. Westfall

Contents

Advances in Nuclear Dynamics 5

Chapter 1

HEATING OF NUCLEAR MATTER AND MULTIFRAGMENTATION: ANTIPROTONS VS PIONS

L. Beaulieu[1,*], T. Lefort[1], W.c-Hsi[1], K. Kwiatkowski[1], V.E. Viola[1], L.Pienkowski[2], R.G. Korteling[3], R. Laforest[4], E. Martin[4], E. Ramakrishnan[4], D. Rowland[4], A. Ruangma[4], E. Winchester[4], S.J. Yennello[4], H. Breuer[5], S. Gushue[6], L.P. Remsberg[6], and B. Back[7]

[1] *Department of Chemistry and IUCF, Indiana University, Bloomington IN 47405*

[2] *Heavy Ion Laboratory, Warsaw University, Warsaw, Poland*

[3] *Department of Chemistry, Simon Fraser University, Burnaby, B.C., V5A 1S6, Canada*

[4] *Department of Chemistry and Cyclotron Laboratory, Texas A&M University, College Station, TX 77843*

[5] *Department of Physics, University of Maryland, College Park, MD 20742*

[6] *Physics Division, Brookhaven National Laboratory, Upton, NY 11973*

[7] *Physics Division, Argonne National Laboratory, Argonne, IL 60439*

* email: LBeaulieu@iucf.indiana.edu
url: http://www.iucf.indiana.edu/~lbeaulieu/

Abstract Heating of nuclear matter with 8 GeV/c $\bar{p}$ and π^- beams has been investigated in an experiment conducted at BNL AGS accelerator. All charged particles from protons to $Z \simeq 16$ were detected using the Indiana Silicon Sphere 4π array. Significant enhancement of energy deposition in high multiplicity events is observed for antiprotons compared to other hadron beams. The experimental trends are qualitatively consistent with predictions from an intranuclear cascade code.

Keywords: 8 GeV/c π^-, $\bar{p}$+Au, 4π detector, multifragmentation, heating of nuclear matter, event-by-event excitation energy, cascade calculations.

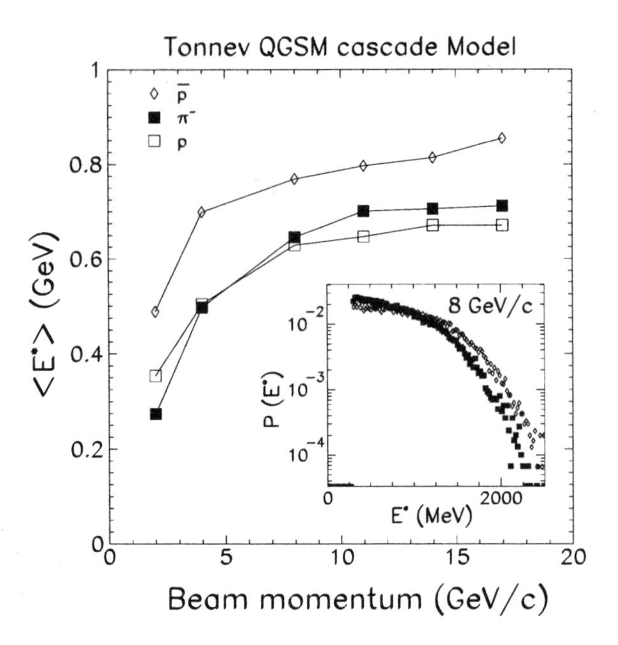

Figure 1.1 Average excitation energy deposited by $\bar{p}$, π^- and p beams as a function of beam momentum as predicted by a cascade code [12]. The insert shows the predicted excitation energy distributions for $\bar{p}$ and π^- beams at 8 GeV/c.

1. INTRODUCTION

The "quest" for the liquid gas phase transition in nuclear matter using heavy ion collisions at intermediate energies has been punctuated by heated debates in recent years, ranging from the very small cross section for fusion of the two partners [1, 2] to the dominance of dynamical effects [3-8] over thermal equilibrium [9].

In the midst of such complexities, the use of GeV hadron projectiles impinging on nuclei offer many unique advantages in producing highly excited nuclear matter. Hard N-N scattering leads to a very efficient and fast ($\leq$ 30-40 fm/c) heating of the target nuclei via excitation of $\Delta(N^*)$ resonances and pion reabsorption [10, 11]. Moreover, hadron beams impart little compression and angular momentum to the excited nuclei. Therefore, any futher decay of the (single!) excited source should be dominated by thermal effects.

Of all the hadron projectiles, the antiproton beams are expected to offer a significant enhancement of the excitation energy deposition relative to other hadrons, while retaining the same simplicity of the reaction dynamics described previously. This enhancement is related to the probability for reabsorption of the large number of pions created by the annihilation process ($n_\pi \simeq 5$). This effect is illutrated in Fig. 1.1, using Toneev's QGSM cascade code [12], which shows the average excitation energy imparted to the nuclei as function of beam momentum for protons, pions and antiprotons. The enhancement of excitation energy predicted for antiproton beams is on the order of 30%.

For all beams, a saturation of energy deposition is seen at high beam momentum due to a decreasing probability for nuclear stopping. The saturation of energy deposition has been seen experimentally in ^{3}He-induced reactions at around 5 GeV kinetic energy [13], and also in pion- and proton-induced reactions from 5 GeV/c to 14.6 GeV/c [14]. Futhermore, Hsi et al. [14] have observed very regular behavior of various experimental distributions and averages, leading to conclusion that 5 GeV/c pion-induced reactions produce a thermal source with essentially the same characteristics as 14.6 GeV/c proton-induced reactions on the same target. This constitutes an experimental verification that pion- and proton-induced reactions have the same heating "efficiency" as shown on Fig. 1.1.

Based of previous results and the cascade predictions, the antiproton-nucleus reactions would be expected to yield a data set that spans the complete rise of multifragmentation and extend into the vaporization regime (fall of multifragmentation) [15]. In the following, a comparison of the heating efficiency of pions versus antiprotons at the same beam momentum will be presented.

2. EXPERIMENT

Experiment E900a was performed at the Brookhaven National Laboratory AGS accelerator. Negative secondary beams consisting of pions, kaons and antiprotons were tagged with a time-of-flight/Cerenkov counter system. The time-of-flight consisted of a 12 mm thick Bicron 480 scintillator start detector followed, 64 meters downstream, by a 5mm thick Bicron 418 scintillator stop detector. Clean separation between $\bar{p}$ and π^- projectiles was achieved with a timing resolution of 200 ps (standard deviation). Negative pions overlapping with $\bar{p}$ were identified and vetoed using a 7 m CO_2 Cerenkov counter operated at atmospheric pressure. The purity of the beam at the target is 98% π^-, $\sim$ 1% $\bar{p}$ and $\sim$ 1% K^-. The identification of K^- remains a difficult task even after veto of negative pions.

Beams consisting of $\approx 4\times10^6$ particle/spill (4.5 s spill time, $\approx$ 2.2 s flat top) were incident on a 2×2 cm^2 self-supporting 2mg/cm^2 thick ^{197}Au target. The target was suspended on two 50 μm tungsten wires to reduce halo reactions. Charged particles from the π^-, $\bar{p}$+Au reactions

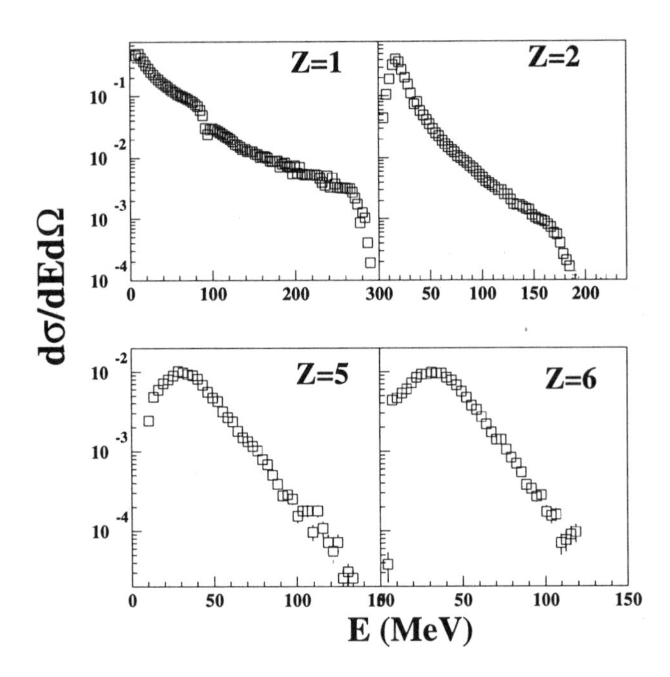

Figure 1.2 Kinetic energy spectra in the laboratory frame for Z=1, 2, 5 and 6 measured between 128° and 147°.

were measured with the ISiS 4π array consisting of 162 triple telescope detectors covering 74% of the 4π solid angle between 14° and 166° [16]. Each telescope is made of an ion chamber (IC), 500 μm Si and 2.8 cm CsI(Tl) crystal read by a photodiode. The first stage (IC-Si) provides elemental charge resolution up to $Z \simeq 16$ for kinetic energies between 0.7 to 8 MeV/nucleon. Mass and charge resolution is achieved in the second stage (Si-CsI) for hydrogen, helium and lithium with kinetic energies between 8A MeV to 92A MeV. Unidentified ("grey") charged particles, mainly protons, up to about 300MeV are also detected.

The trigger for this experiment was a fast signal in at least three Si detectors. Acquisition was permitted only during the flat top period of the spill. This is reflected in our software requirement of having 3 or more detected charged particles. Additional software cuts required the kinetic energy of Z=1 to 5 fragments to be greater than 1 MeV/nucleon, and at least one fragment ($Z \geq 3$) or one helium must be detected in the IC-Si stage. These last cuts were made to reduce noise. The final event sample is made of 25 000 $\bar{p}$ and 2 500 000 π^-.

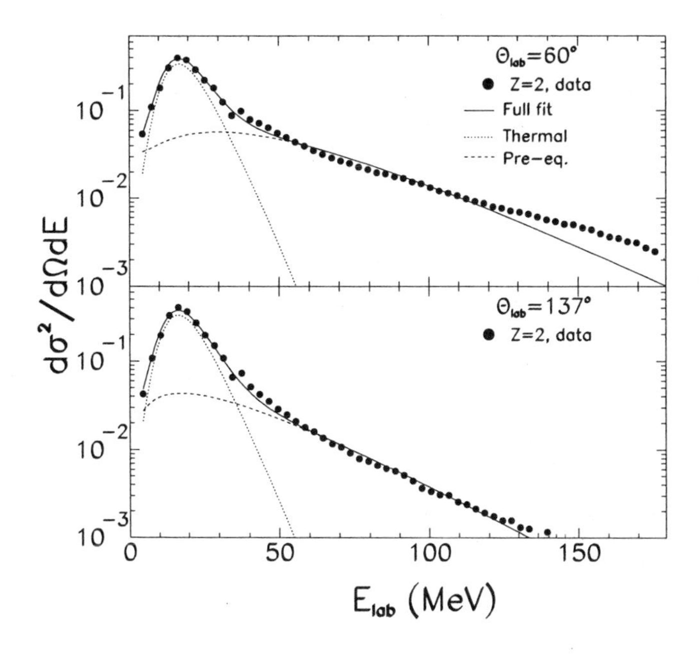

Figure 1.3 Moving source fits on the helium kinetic energy spectra at forward angles (top panel) and backward angles (bottom panel). The dotted line corresponds to the thermal source fit, and the dashed line to the pre-equilibrium component. The full curve is the sum of the thermal and pre-equilibrium fits

3. "UNPROCESSED" RESULTS

Laboratory energy spectra for hydrogen, helium, boron and carbon fragemnts measured between 128° and 147° are shown in Fig. 1.2. The spectra for the heavier fragments peak roughly at the Coulomb barrier (or slighly lower) and can be characterized by a single slope, reminiscent of emission from a single thermalized source. Light charged particles have a dominant thermal component at low kinetic energies but clear deviation is seen at higher energies. The contribution of high energy non-equilibrium particles is larger at forward angles, and is also present for light fragments (Z=3-4).

Two moving source fits, using Moretto's formalism [17], yield very good reproduction of the helium energy spectra at backward angles (Fig. 1.3, bottom panel). At forward angles (Fig. 1.3, top panel), the thermal source is well described but not the high energy tails, which require a much flatter slope beyond 100 MeV of kinetic energy. These particles probably come from the early time of the reactions and are related to the initial knock-out ("splash") of nucleons by the primary projectile. They constitute a measure "centrality", as defined in ref. [18]. The parameters of the thermal source, deduced from the helium energy spectra, are: T_{th}=4.8 MeV and V_{th}=0.0012c. Notice the very small source velocity for the thermal component.

The separation of the thermal source from the preequilibrium contributions is achieved by selecting the particles based on their kinetic energy in the source frame (E_{cm}), assuming that the source velocity of 0.0012c is along the beam axis. This selection makes use of the experimental systematic analysis of the ^{3}He-induced reactions between 1.8 GeV and 4.8 GeV [19, 20]:

$$E_{cm} \leq 9.0Z + 40MeV \quad ; Z \geq 2 \tag{1.1}$$

$$E_{cm} \leq 30MeV \quad ; Z = 1 \tag{1.2}$$

The effectiveness of the method was tested by constructing the angular distribution of the selected thermal particles, and verifying that it has the expected feature of an isotropic source [21].

Our hability to isolate a single thermal component combined with very small collective effects for this reaction, open the possibility of extracting the source temperature using Maxwell-Boltzmann fits to the energy spectra [22]. Comparison of the kinetic thermometer in such a simple scenario to the isotope ratios thermometer [23, 24] should bring new insights to the temperature measurement problems [25].

3.1 ANTIPROTONS VS PIONS

Any enhancement of energy deposition using $\bar{p}$ instead of π^- projectiles should be seen in "raw" global variables to be of significant interest. This first point was tested by looking at the probability distributions of the observed multiplicity of charged particles (N_c), the observed multiplicity of fragments (N_{imf}) and the transverse energy of all charged particles, defined by $E_t = \sum_{i=1}^{N_c} E_i sin^2\Theta_i$. These probability distributions, shown on the top panels of Fig. 1.4, exhibit a strong enhancement in favor of the $\bar{p}$ projectiles for the last 15-20% of the observed cross-section (from N_c and E_t). These events are expected to correspond to high excitation energy events, and are possibly the best candidates for observing phase transition-like behavior [20].

The enhancement is quantified in the bottom panels of Fig. 1.4 by taking the ratios of the above probabilities, P($\bar{p}$)/P(π^-). As stated before, most of the emitted IMFs come from the thermal source, and should be

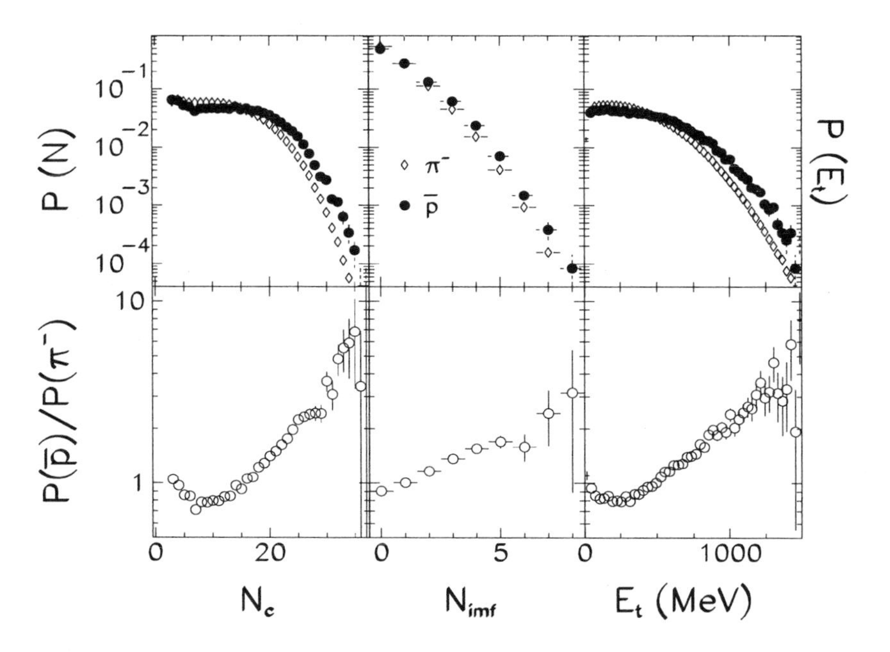

Figure 1.4 Top panels: Probability distributions for the observed charged-particle (left) and IMF multiplicities (middle), and for the total transverse energy (right). The open and black symbols are for pion and antiproton beams, respectively. Bottom panels: Ratio of the $\bar{p}$ to π^- probability distributions of the corresponding global variable from the top panels.

examined at first. The increase in probability is about 80% for 5 IMF events. On the other hand, the ratio for N_c is even stronger going up to a factor of 8 for the largest multiplicities. The probability ratio for E_t is somewhat between these two extremes showing a four-fold increase. All three observables show signs of a significant enhancement of energy deposition with $\bar{p}$ projectiles. The difference between N_c and N_{imf} could be interpreted as a signature that the decay mechanism is favoring more the light charged particles than the IMFs. This would mean that the hot source has enter the vaporization regime. However, these are observed multiplicities, and therefore include fast particles. A separation of the multiplicity in term of fast and thermal gives an increase of about 300-400% for the fast particles but only a 100% for the thermal source. Therefore, the increase of thermal charged particle multiplicity follows closely that of the IMF to first order (cf. Fig. 1.4). The immediate implication is that the observed increases are partially due to fast/non-

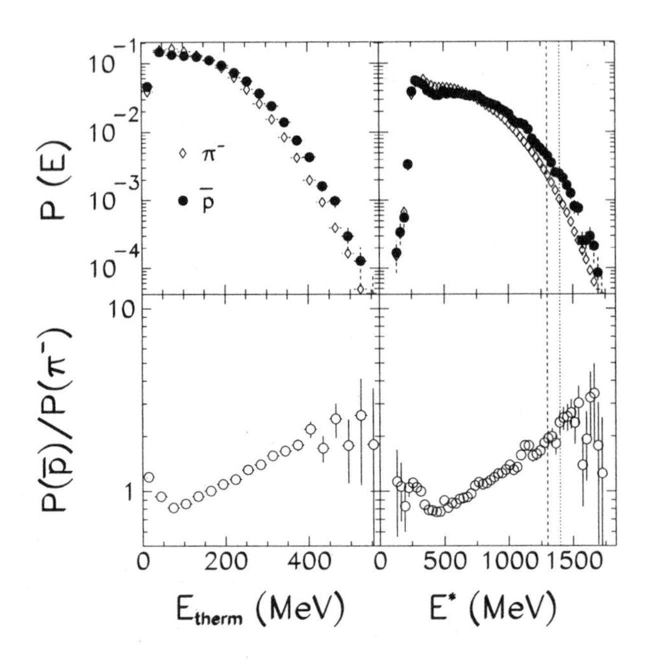

Figure 1.5 Thermal kinetic energy (top left) and thermal excitation energy (top right panel) distributions for pions and antiprotons beams. Bottom panels: Ratio of the $\bar{p}$ to π^- probability distributions of the corresponding global variable from the top panels. The dashed and dotted vertical lines represent the last 1% of the excitation energy distribution for pions and antiprotons respectively.

equilibrium emission of light charged particles $Z \leq 2$. This gives a rough idea of how the annihilation energy is divided among the thermal and non-thermal components, and invites comparisons to models in order to extract the details. Since E_t is constructed from all the observed charged particles, the same interpretation as for N_c applies. To be more quantitative, it is necessary to construct observable sensitive only to the thermal excitation energy.

4. "MASSAGE" RESULTS

The kinetic energy of all thermal particles, selected according to Eq. 1.1-1.2, are used to construct the total thermal energy, E_{therm}. The probability distribution of E_{therm} in Fig. 1.5 indicates an increase in cross-section at larger thermal energies for $\bar{p}$ compared to π^- beams. The enhancement is about a factor of 2 at the highest value of E_{therm}.

Going from thermal energy to excitation energy requires several assumptions [20]. Each event is separated into two groups: fast and thermal particles. The excited residue (primary source after non-equilibrium emission) mass and charge are obtained by subtracting all the fast particles from the target mass and charge

$$A_{res} = A_{tgt} - \sum A^{fast} - \sum M_n^{fast} \tag{1.3}$$

$$Z_{res} = Z_{tgt} - \sum Z^{fast} \tag{1.4}$$

Corrections for geometrical efficiency are taken into account. The multiplicity of fast neutrons, M_n^{fast} is taken to be $1.93 \times M_p^{fast}$, where M_p^{fast} corresponds to the efficiency-corrected experimental multiplicity of fast protons ($E_{cm} > 30$ MeV). This procedure to estimate the fast neutrons is intermediate between the experimental low-energy systematics of Polster et al. [26] and that expected from INC calculations [27].

For thermal particles, the kinetic energy of each particle in the source frame (K_i) is computed. The multiplicity of thermal neutrons, M_n, is obtained using the measured multiplicity of thermal charged particles M_c according to the experimental work of Goldenbaum and co-workers [28]. Again, corrections for geometrical efficiency was made on all observables related to the detected charged particles. The excitation energies E^* were assigned on an event-by-event basis according to the following prescription

$$E^* = \sum_{i=1}^{M_c} K_i + M_n \langle K_n \rangle + Q + E_\gamma \tag{1.5}$$

E_γ is taken to be $1 \times (M_c + M_n)$ MeV. The reconstructed event is used to determine the mass difference Q. $\langle K_n \rangle$ is assigned a value of $3T/2$ where $T = \sqrt{E^*/a}$ and $a = A_{res}/11$ MeV^{-1}, and then iterated to obtain a self-consistent value. It should be stressed that this procudere was appplied the same way to both the $\bar{p}$ and the π^- beams.

The excitation energy distributions are presented in Fig. 1.5 (right panels). The increase of excitation energy imparted to the target nuclei using $\bar{p}$ projectiles is consistent with that of the thermal energy on the left panels with almost a factor 2 more cross-sections at high excitation energies. The dashed and dotted lines correspond to the last 1% of the observed cross-sections (roughly 10-20 mb) for π^- and $\bar{p}$, respectively. This translates in an increase of the maximum excitation energy of 1.3 MeV/nucleon using $\bar{p}$ beams (last 1% is 9.0 MeV/nucleon for π^- and 10.3 MeV/nucleon for $\bar{p}$). This is in qualitative agreement with QGSM code as shown in **Fig.** 1.1 (see insert).

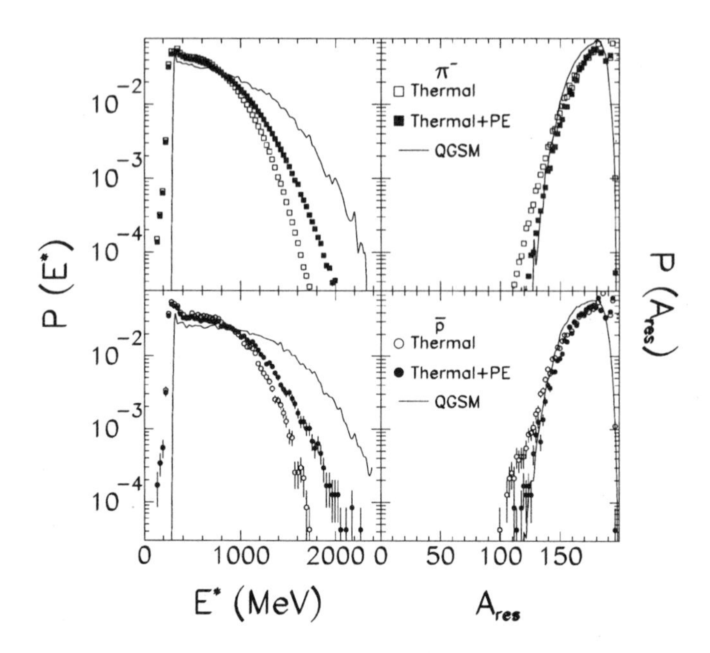

Figure 1.6 Excitation energy and residue mass distributions for pions (top panels) and antiprotons (bottom panels) obtained using Eq. 1.1-1.2 (open symbols) and an extension to 30 MeV/nucleon of kinetic energy for thermal particles (black symbols). See text for details. The curve correspond to the prediction of Toneev's QGSM cascade code [12].

The excitation energy and residue mass distributions are compared directly to the QGSM cascade code in Fig. 1.6, using the default values for the various parameters. The excitation energy distributions predicted by QGSM reach higher values than the data for both beams while A_{res} tend to be slightly bigger. A possible explanation may be the use of a too high cut-off energy for nucleons to escape during the cascade, trapping more particles and increasing E^* and A_{res}. To verify this hypothesis, the excitation energies were re-evaluated by relaxing the definition of thermal particles to include all particles with kinetic energies smaller than 30 MeV/nucleon, thus including some non-equilibrium particles. The result of this exercise is shown in Fig. 1.6 as the black symbols. A_{res} probability distributions are now in good agreement with Toneev's QGSM but the code still predicts more excitation energy than the data. Therefore, a more important effect than the nucleon escape energy cut-off might be the pion reabsorption cross-section or other cross-sections related to pions and resonances used in the code.

5. SUMMARY

Heating of nuclear matter has been investigated using $\bar{p}$ and π^- projectiles at 8 GeV/c beam momentum. The raw multiplicity distributions, N_c and N_{imf}, the transverse and the thermal energy distributions show significant enhancement of energy deposition with $\bar{p}$ beams relative to other hadrons. Separation of thermally emitted particles from preequilibrium ones was performed on the kinetic energy spectra. The primary residue mass and excitation energy distributions were reconstructed event-by-event for both beams under consideration. The increase in cross-section for high excitation energy events reaches 100% with $\bar{p}$. The excitation energy distribution for antiprotons reaches higher values (1.3 MeV/nucleon higher) than for pions as measured by the last 1% of their respective distributions. The results are in qualitative agreement with a cascade code but the code overpredicts the absolute values. The enhancement of thermal energy deposition is also accompanied by a stronger increase of fast proton and light cluster emission, and should be related to the details of the annihilation process itself. Certainly, an in-depth comparison to models is needed to shed light on the energy deposition in this beam momentum region.

Acknowledgments

The authors wish to thank John Vanderwerp, Bill Lozowski and Dick Yoder at IUCF and Phil Pile, Joe Scaduto, Larry Toler, John Gould and Jerry Bunce at AGS for their assistance with the experiement. This work was supported by the US Department of Energy and the National Science Foundation, the National Sciences and Engineering Research Council of Canada and the Robert A. Welch Foundation.

References

[1] J. Péter *et al.* (1995) Nucl. Phys. **A593**, 95.
[2] L. Beaulieu *et al.* (1996) Phys. Rev. Lett. **77**, 462.
[3] C.P. Montoya *et al.* (1994) Phys. Rev. Lett. **73**, 3070.
[4] J. Lukasik *et al.*, (1997) Phys. Rev. C **55**, 1906.
[5] Y. Larochelle *et al.*, (1997) Phys. Rev. C **55**, 1869.
[6] J. Toke *et al.*, (1995) Phys. Rev. Lett. **75**, 2920.
[7] J.F. Lecolley *et al.*, (1995) Phys. Lett. B **354**, 202 .
[8] J.F. Dempsey *et al.*, (1996) Phys. Rev. C **54**, 1710.
[9] L.G. Moretto, *et al.*, Phys. Rep. **287**, 249 (1997), and references therein.

[10] G.Wang *et al.,* (1996) Phys. Rev. C **53**, 1811.

[11] J. Cugnon *et al.*, (1987) Nucl. Phys. **A470**, 558, (1989) Ann Phys. (Paris) **14**, 49.

[12] V. Toneev *et al.*, (1990) Nucl. Phys. **A519**, 463.

[13] K.B. Morley *et al.*, (1995) Phys. Lett. **B 355**, 52.

[14] W-c. Hsi, *et al.*, (1997) Phys. Rev. Lett. **79**, 817.

[15] C.A. Ogilvie, *et al.*, (1991) Phys. Rev. Lett. **67**, 1214.

[16] K. Kwiatkowski *et al.*, (1995) Nucl. Instr. Meth. **A360**, 571.

[17] L.G. Moretto, (1975) Nucl. Phys. **A247**, 211.

[18] R. Soltz *et al.*, Contribution to these proceedings; I. Chemakin *et al.*, (1999) submitted to Phys. Rev. C and nucl-ex 9902003.

[19] D.S. Bracken, (1996) Ph.D. thesis, Indiana University.

[20] K. Kwiatkowski *et al.*, (1998) Phys. Lett.**B 423**, 21.

[21] T. Lefort *et al.*, (1999) Proceedings of the 37th International Winter Meeting on Nuclear Physics, Bormio, Italy, Jan 25-30 1999.

[22] A. Siwek *et al.*, (1998) Phys. Rev. C **57**, 2507.

[23] J. Albergo *et al.*, (1985) Nuovo Cim. **89A**, 1.

[24] J. Pochodzalla *et al.*, Phys. Rev. Lett. **75**, 1040 (1995).

[25] V.E. Viola, K. Kwiatkowski and W.A Friedman, (1999) accepted for publication in Phys. Rev. C.

[26] D. Polster *et al.*, (1995) Phys. Rev. C **51**, 1167.

[27] I.A. Pshenichnov *et al.*, (1995) Phys. Rev. C **52**, 947.

[28] F. Goldenbaum *et al.*, (1996) Phys. Rev. Lett. **77** 1230; L. Pienkowski *et al.*, (1994) Phys. Lett. B **B336**, 147.

Chapter 2

IN-MEDIUM MODIFICATION OF THE $\Delta(1232)$ RESONANCE AT SIS ENERGIES

D. Pelte

Physics Institute, University of Heidelberg

Philosophenweg 12

*D-69120 Heidelberg, Germany**

Dietrich.Pelte@mpi-hd.mpg.de

*Partial funding provided by grants 06 HD 525 I(3) and HD Pel K.

Abstract From the pion production rates n_π/A_{part} and the pion charge ratios $R_\pi = n_{\pi^-}/n_{\pi^+}$ and $S_\pi = n_{\pi^0}/(n_{\pi^-} + n_{\pi^+})$, measured in p + A and A + A reactions, it is concluded that the mass of the $\Delta(1232)$ resonance is reduced in nuclear matter. This conclusion is supported by the $\Delta(1232)$ mass distribution extracted from correlated (p,$\pi^\pm$) pairs and the $\pi^\pm$ transverse momentum distributions.

Keywords: relativistic H.I. reactions, baryon resonances, in-medium effects

1. INTRODUCTION

The experimental verification that the properties of hadrons are modified in the dense nuclear medium produced by relativistic nucleus-nucleus collisions is of great interest since the partial restoration of chiral symmetry would lead to such modifications, in particular it predicts the reduction of the K^- mass [1]. In this contribution I will survey the experimental information with regard to pions, produced in p + A and A + A reactions at SIS energies. I will show that these data can be interpreted as the result of a $\Delta(1232)$ mass reduction. This interpretation is based on two simple models, the thermal and the isobar models, and the conclusion is confirmed by the direct measurement of the $\Delta(1232)$ mass distribution.

2. THE THERMAL MODEL

In order to obtain the charge states of hadrons from the thermal model one has to introduce the isospin chemical potential μ_I, besides the baryon chemical potential μ_B and the temperature T. Defining $x = e^{-\mu_I/T}$ one obtains for the charges states of nucleons and pions:

$$n_p = \frac{n_N}{2}x^{+1} \quad , \quad n_n = \frac{n_N}{2}x^{-1} \tag{2.1}$$

$$n_{\pi^+} = \frac{n_\pi}{3}x^{+2} \quad , \quad n_{\pi^0} = \frac{n_\pi}{3} \quad , \quad n_{\pi^-} = \frac{n_\pi}{3}x^{-2} \tag{2.2}$$

$$n_{\Delta^{++}} = \frac{n_\Delta}{4}x^{+3}, n_{\Delta^+} = \frac{n_\Delta}{4}x^{+1}, n_{\Delta^0} = \frac{n_\Delta}{4}x^{-1}, n_{\Delta^-} = \frac{n_\Delta}{4}x^{-3}, \tag{2.3}$$

where n_N, n_π, and n_Δ are determined by μ_B and T. Since charge and baryon number are conserved the following relations between x, the pion production rate n_π/A_{part}, and the N/Z ratio ζ_{part} of the participants should hold:

$$1 = \frac{\zeta_{part}+1}{2}x + \frac{n_\pi}{A_{part}}\frac{\zeta_{part}+1}{3}\left(x^2 - x^{-2}\right) \tag{2.4}$$

$$1 = \frac{\zeta_{part}+1}{2}x + \frac{n_\pi}{A_{part}}\frac{\zeta_{part}+1}{4}\left(2x^3 - x - x^{-3}\right) \quad . \tag{2.5}$$

The first equation is valid in case the participant region only contains nucleons and pions, the second if it contains only nucleons and $\Delta(1232)$ resonances, where the $\Delta(1232)$ after freeze-out decay into pions. The importance of these equations is due to the fact that they only depend on known (ζ_{part}) and measured (n_π/A_{part}) quantities, and that x can be deduced from the measured pion charge ratios $R_\pi = n_{\pi^-}/n_{\pi^+}$. Note that n_π/A_{part} requires to know the production rates n_{π^0}/A_{part} of neutral pions, these rates have been measured for a number of A + A reactions by the TAPS collaboration [2], but they are unknown for the p + A reactions considered here. In this latter case I have assumed $n_{\pi^0} = n_{\pi^-} + n_{\pi^+}$. This assumption has only little consequences for the present discussion, but its experimental confirmation would add support to the interpretation of the existing data advocated in this report.

The data have been measured for Ni + Ni [3], Au + Au [4], and p + C/Nb/Pb [5] reactions in the energy range between 1 to 2 AGeV and at several impact parameters b, I have chosen central collisions by extrapolating to $b \to 0$. The extrapolated results, which include the production rates of the $\Delta(1232)$ resonance [6] [7], are in conflict with the equations 2.4 2.5, if one assumes for each reaction a unique temperature and baryon chemical potential. In order to be consistent with charge

Table 2.1 Participant temperatures T extracted from different observables assuming a participant density $\rho = 0.3\rho_0$. Numbers in brackets: incident energy in AGeV.

reaction	$T(R_\pi)$	$T(n_\pi)$	$T(n_\Delta)$	$T(slope)$
Au+Au(1.06)	56 ± 5	55 ± 2	66 ± 5	81 ± 24
Ni+Ni(1.06)	< 50	62 ± 2	75 ± 3	79 ± 10
Ni+Ni(1.45)	< 50	68 ± 2	84 ± 4	84 ± 10
Ni+Ni(1.93)	< 55	73 ± 2	93 ± 3	92 ± 12

and baryon number conservation the different observables would require in general different temperatures which are shown in table 2.1. In this table I have also included the temperature values which were deduced from the slopes of the energy spectra of various particles emitted from the participant region [8] [9]. One finds that the slope temperatures are within errors consistent with $T(n_\Delta)$, but that in general

$$T(R_\pi) < T(n_\pi) < T(n_\Delta). \tag{2.6}$$

In case of p + A reactions the data would not allow to obtain any meaningful solution to the equations 2.4 2.5. This implies that within this very simple framework the thermal model is unable to explain the measured data, and the problem can be traced in case of the A + A reactions to the measured number of pions which is too small to accommodate the measured number of $\Delta(1232)$ resonances and the calculated number of thermally produced pions. In addition the R_π ratios suggest that their values largely depend on the type of charge exchange reactions that occur between the pions and the nucleons of the participant region.

3. THE ISOBAR MODEL

The isobar model allows to calculate the charge ratios of pions and nucleons with the assumption that pions are exclusively produced by the decay of baryon resonances B, which were excited in a first step by the reaction $N + N \to N + B$. Isospin conservation yields for this first step in the cases in which B stands for the Δ resonance

$$\begin{aligned} &n_{\pi^-} : n_{\pi^0} : n_{\pi^+} = \\ &(\zeta_P + \zeta_T + 10\zeta_P\zeta_T) : (2 + 4\zeta_P + 4\zeta_T + 2\zeta_P\zeta_T) : (10 + \zeta_P + \zeta_T), \end{aligned} \tag{2.7}$$

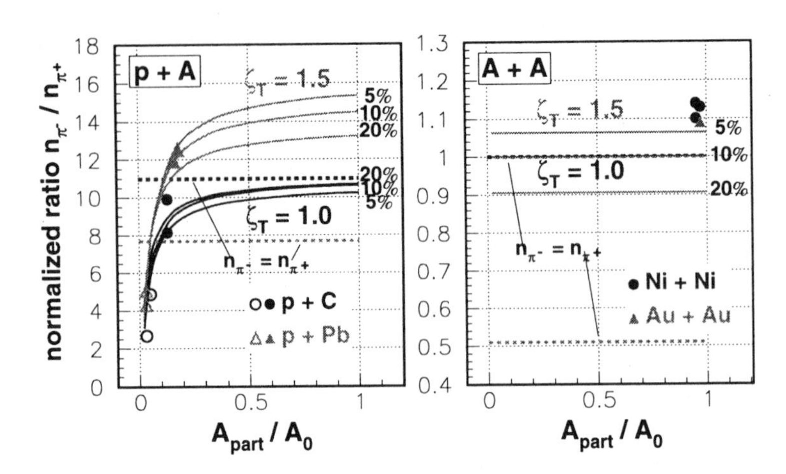

Figure 2.1 Left panel: The A_{part}/A_0 dependence of the normalized π^- to π^+ ratio $R_\pi^{(0)}$ of p + A reactions for two different values ζ_T of the target neutron to proton ratio. The percentage numbers give the probability to excite the $\Delta(1232)$ resonance in the first step, the open symbols correspond to peripheral, the full symbols to central data from p + ^{12}C (black) or p + ^{208}Pb (grey) reactions. In case of identical symbols the larger one corresponds to the higher incident energy. Right panel: The same for symmetric A + A reactions as specified in table 2.1.

where ζ_P, ζ_T are the N/Z ratios of the projectile respectively the target. These relative numbers are modified when the emitted pions are reabsorbed in participant matter and form new Δ resonances which subsequently decay by pion emission. The final pion numbers depend on the number n_{loop} of such $\pi N \Delta$ loops since two successive loops i and $i+1$ ($i < n_{loop}$) are coupled by the equations:

$$\begin{aligned} n_{\pi^+}^{(i+1)} &= \left(n_p^{(i)} (n_{\pi^+}^{(i)} + \frac{1}{3} n_{\pi^0}^{(i)}) + n_n^{(i)} \frac{1}{3} n_{\pi^+}^{(i)} \right) / A_{part} \\ n_{\pi^0}^{(i+1)} &= \left(n_p^{(i)} (\frac{2}{3} n_{\pi^0}^{(i)} + \frac{2}{3} n_{\pi^-}^{(i)}) + n_n^{(i)} (\frac{2}{3} n_{\pi^+}^{(i)} + \frac{2}{3} n_{\pi^0}^{(i)}) \right) / A_{part} \\ n_{\pi^-}^{(i+1)} &= \left(n_p^{(i)} \frac{1}{3} n_{\pi^-}^{(i)} + n_n^{(i)} (\frac{1}{3} n_{\pi^0}^{(i)} + n_{\pi^-}^{(i)}) \right) / A_{part}. \end{aligned} \tag{2.8}$$

The n_π numbers do not change anymore once $n_{loop} > 10$ in the reactions considered here. For the R_π ratios these saturation values are shown in Fig. 2.1, and for the ratios $S_\pi = n_{\pi^0}/(n_{\pi^-} + n_{\pi^+})$ in Fig. 2.2. Note that I have normalized the R_π and S_π values by the predictions derived from equation 2.7, i.e. $R_\pi^{(0)} = S_\pi^{(0)} = 1$ would imply that the value of n_{loop} is sufficiently small not to modify the ratios of the first step. The saturation values of $R_\pi^{(0)}$ and $S_\pi^{(0)}$ depend in case of p + A reactions on

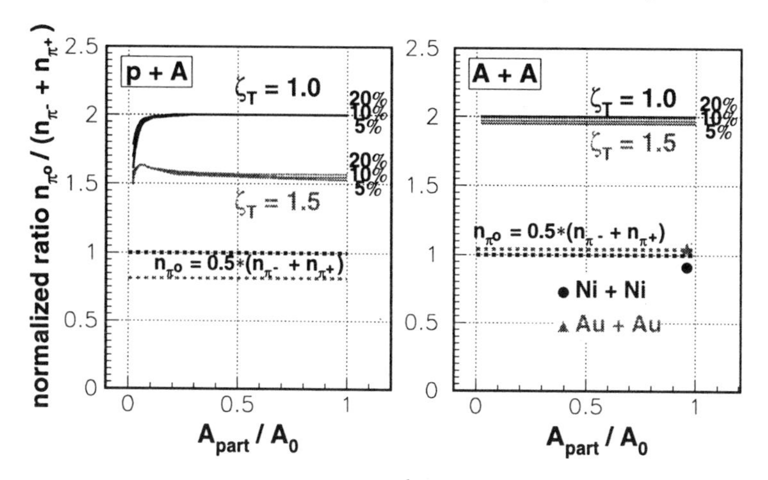

Figure 2.2 As in Fig.4 but for the normalized neutral to charged pion ratio $S_\pi^{(0)}$. Experimental data (symbols) only exist for A + A reactions.

the size A_{part} of the participant, but not in case of A + A reactions. The dependence on the excitation probability of the Δ resonance, listed in Fig. 2.1 and 2.2 by the percentages, is only weak. The comparison with the data shown by symbols indicates that to explain the p + A results one has to assume the presence of sufficiently many $\pi N \Delta$ loops, but that in the case of A + A reactions the number of such loops is much smaller since for both ratios one observes $R_\pi^{(0)} \approx S_\pi^{(0)} \approx 1$.

There exist 2 possible explanations for this difference between the behavior of p + A and A + A reactions:

- The decay probability of $\Delta(1232)$ resonances decreases in participant matter which reduces the number of $\pi N \Delta$ loops,
- the absorption probability of $\Delta(1232)$ resonances via the $\Delta + N \rightarrow N + N$ reaction increases in participant matter which also reduces the number of $\pi N \Delta$ loops.

Of these two alternatives the second explanation is favored by the experimental observation that the pion production rate depends on the total system mass $A_0 = A_P + A_T$. The measured n_π / A_{part} values, integrated over the impact parameter, may be fitted for incident energies from 1 to 2 AGeV by the relation [10]

$$\frac{\langle n_\pi \rangle}{\langle A_{part} \rangle} = a_\pi \cdot (E_{kin} - 0.11), \tag{2.9}$$

where the cm energy E_{kin} is given in AGeV. The production coefficient a_π is shown in Fig. 2.3, it displays a linear decrease with rising A_0.

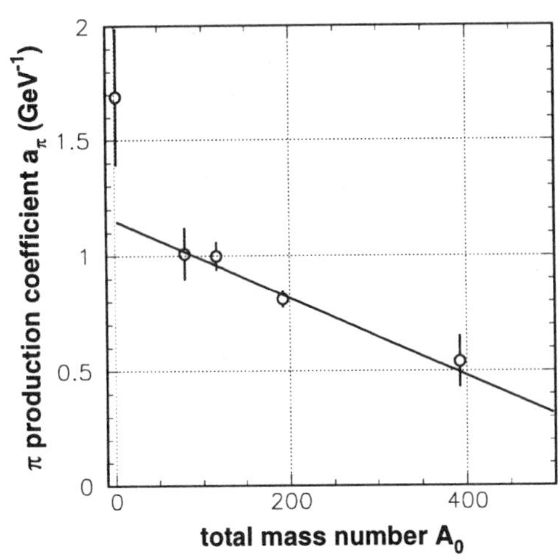

Figure 2.3 The measured pion production coefficients a_π as function of the system mass A_0. The line is a linear fit through all points except the one at $A_0 = 2$.

In addition the n_π rate from p + p reactions is noticeably enhanced compared to A + A reactions [11]. It is evident that compared to p + p reactions the pion production rates are suppressed in A + A reactions, and the suppression is the stronger the heavier the system is, or the larger the participant density is which can be attained in nucleus-nucleus collisions. More insight into the mechanism which might cause the rise of the absorption cross section is gained by studying the $\Delta(1232)$ mass distribution in participant matter.

4. THE $\Delta(1232)$ MASS DISTRIBUTION

The mass distribution of the $\Delta(1232)$ resonance in participant matter can be reconstructed from the transverse momentum spectra dn/dp_t of pions, or from the invariant mass of (p,π) pairs. The former technique was applied in A + A reactions [7], the latter also in p + A reactions [12]. I display the results for central Ni + Ni and Au + Au reactions in Fig. 2.4 lower panels, where the dark points correspond to the dn/dp_t technique and the stars to the (p,π) technique. For the p + C reaction the equivalent data are displayed in Fig. 2.4 upper panel. In general the mass distributions are shifted from the mass distribution of the free $\Delta(1232)$ resonance (dashed curves) towards smaller masses. In case of the p + C reaction this shift can be explained in the framework of the thermal model by the finite temperature $T = 65$ MeV of the participants, the deviations from the expected mass distribution at higher masses are most likely due to first-chance N + N collisions. Contrary to these findings, in Ni + Ni and Au + Au reactions the observed mass shifts are only partly reproduced by the participant temperatures $T(n_\pi)$ of table refPeltatab1 (dotted curves). To obtain a fit of the experimental mass distributions additional shifts of ≈ -100 MeV/c^2 in case of Au +

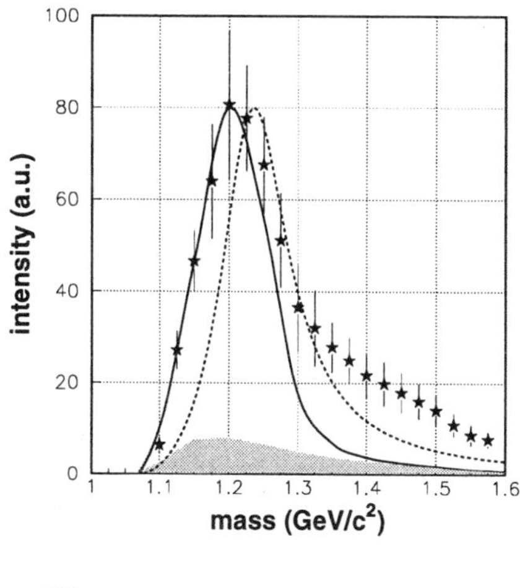

Figure 2.4 The invariant mass spectrum of the $\Delta(1232)$ resonance excited in the p + C reaction at 1.6 GeV incident energy (top), the Au + Au reaction at 1.06 AGeV (bottom left), and Ni + Ni reaction at 1.93 AGeV (bottom right). The shaded areas correspond to the contributions from higher baryon resonances. For the symbols and different curves see text.

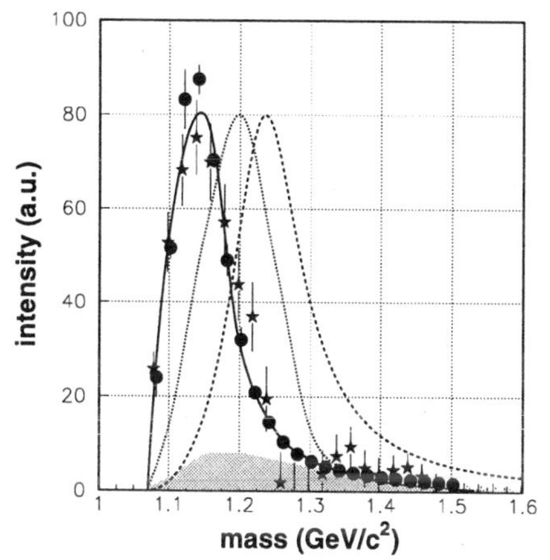

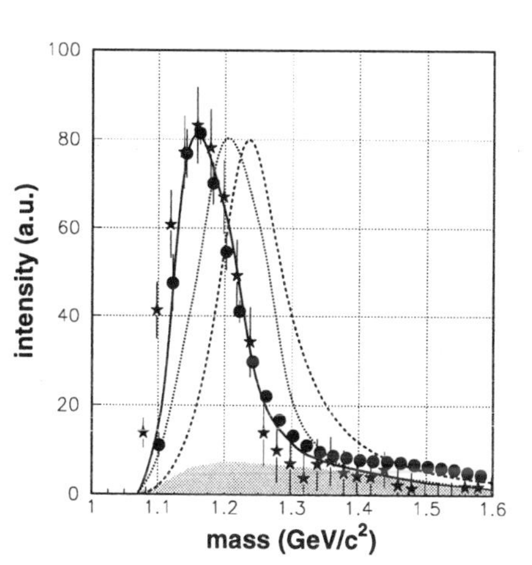

Au, and ≈ -50 MeV/c^2 in case of Ni + Ni are necessary. The data do not allow to obtain a more quantitative result, in particular they do not allow to disentangle the medium effects onto the mass and the width of the distribution. The size of the in-medium modification depends on the system mass, and the result for Ni + Ni is in fair agreement with similar data for the Ni + Cu reaction published in [13].

5. DISCUSSION

The analyses of the pion data from p + A and A + A reactions at SIS energies have yielded convincing evidence that

1. the $\Delta(1232)$ absorption increases,
2. the $\Delta(1232)$ average mass is reduced

as function of the average participant density which increases with the system mass. These two phenomena are indeed related by the "extended principle of detailed balance" via

$$\sigma_{n\Delta^{++}\to pp} = \frac{1}{4}\frac{p_N^2}{p_\Delta^2}\sigma_{pp\to n\Delta^{++}} \frac{1}{\int_{(m_N+m_\pi)^2}^{(\sqrt{s}-m_N)^2} f_\Delta(m^2)\, dm^2} , \tag{2.10}$$

where $\sigma_{n\Delta^{++}\to pp}$ is the partial absorption cross section for the $\Delta(1232)$ resonance in the reaction $\Delta + N \to N + N$, and $f_\Delta(m^2)$ is the in-medium mass distribution of the $\Delta(1232)$ resonance. The equation 2.10 has been studied in [14] and [15], the differences in these studies have to be resolved before a detailed comparison with the experimental results can be accomplished. It is of interest that in [13] the mass shift was found to become weaker with rising impact parameter. In terms of the presently advocated interpretation this implies a reduction of the pion absorption. Experimentally the pion production rate n_π/A_{part} depends on A_{part}, however the dependence is different for π^- and π^+ [3] [4]. Therefore Coulomb effects have to play an important role in the pion dynamics.

Acknowledgments

The author wishes to thank the FOPI and the TAPS collaborations which supplied most of the data used in this study.

References

[1] Cassing, W., and F. L. Bratkovskaya (1999) Phys. Rep. **308**, 66.
[2] Metag, V. (1998) private communication.
[3] Pelte, D., et al. (1997) Z. Phys. A **359**, 55.
[4] Pelte, D., et al. (1997) Z. Phys. A **357**, 215.
[5] Lemaire, M. C., et al. (1988) Phys. Rev. c **43**, 2711.
[6] Hong, B., et al. (1997) Phys. Lett. B **407**, 115.
[7] Eskef, M., et al. (1998) Eur. Phys. J. A **3**, 335.
[8] Lisa, M. A., et al. (1995) Phys. Rev. Lett. **75**, 2662.
[9] Hong, B., et al. (1998) Phys. Rev. C **57**, 244.
[10] Harris, J.W., et al. (1987) Phys. Rev. Lett. **58**, 463.
[11] Gazdzicki, M., et al. (1998) Eur. Phys. J. C **5**, 129.
[12] Trzaska, M., et al. (1991) Z. Phys. A **340**, 325.
[13] Hjort, E. L., et al. (1997) Phys. Rev. Lett. **79**, 4345.
[14] Danielewicz, P., and G. F. Bertsch (1991) Nucl. Phys. A **533**, 712.
[15] Wolf, G., et al. (1993) Nucl. Phys. A **552**, 549.

Chapter 3

PARTICLE PRODUCTION IN 158·A GEV ^{208}PB+^{208}PB COLLISIONS

Terry C. Awes[7]
for the WA98 Collaboration
http://www.cern.ch/WA98/Welcome.html

[1] *University of Panjab, Chandigarh 160014, India*
[2] *University of Rajasthan, Jaipur 302004, Rajasthan, India*
[3] *Variable Energy Cyclotron Centre, Calcutta 700 064, India*
[4] *University of Geneva, CH-1211 Geneva 4,Switzerland*
[5] *RRC "Kurchatov Institute", RU-123182 Moscow, Russia*
[6] *Joint Institute for Nuclear Research, RU-141980 Dubna, Russia*
[7] *Oak Ridge National Laboratory, Oak Ridge, Tennessee 37831-6372, USA*
[8] *University of Jammu, Jammu 180001, India*
[9] *University of Münster, D-48149 Münster, Germany*
[10] *SUBATECH, Ecole des Mines, Nantes, France*
[11] *Gesellschaft für Schwerionenforschung (GSI), D-64220 Darmstadt, Germany*
[12] *Universiteit Utrecht/NIKHEF, NL-3508 TA Utrecht, The Netherlands*
[13] *Lund University, SE-221 00 Lund, Sweden*
[14] *University of Tsukuba, Ibaraki 305, Japan*
[15] *Nuclear Physics Institute, CZ-250 68 Rez, Czech Rep.*
[16] *KVI, University of Groningen, NL-9747 AA Groningen, The Netherlands*
[17] *Institute for Nuclear Studies, 00-681 Warsaw, Poland*
[18] *MIT Cambridge, MA 02139, USA*
[19] *Institute of Physics, 751-005 Bhubaneswar, India*
[20] *University of Tennessee, Knoxville, Tennessee 37966, USA*

Abstract The production of neutral pions in 158·A GeV ^{208}Pb+^{208}Pb collisions has been studied in the WA98 experiment. The centrality dependence of the neutral pion production is investigated. An invariance of the spectral shape and a simple scaling of the yield with the number of participating nucleons is observed for centralities with more than about 50 participants. The transverse mass spectrum is analyzed in terms of a

Advances in Nuclear Dynamics, 5,
Edited by Bauer and Westfall, Kluwer Academic / Plenum Publishers, New York, 1999.

thermal model with hydrodynamic expansion. The high accuracy and large kinematic coverage of the measurement constrains the extracted freeze-out parameters, and provides information on the freeze-out velocity profile.

Keywords: Quark Gluon Plasma, Heavy-ion collisions, Neutral pions, Particle production, Nuclear dependence, Thermalization, Freeze-out.

1. INTRODUCTION

Ultra-relativistic heavy-ion collisions provide the means to produce dense matter which at sufficiently high energy densities may undergo a phase transition from normal hadronic matter to a deconfined phase of quarks and gluons creating a Quark-Gluon Plasma (QGP). A primary goal of the ultra-relativistic heavy-ion programme is thus to identify and characterize the QGP phase transition. The classic means to search for a phase transition in condensed matter physics it to measure the heat capacity of the matter as a function of temperature, and to search for a discontinuous change in the heat capacity indicating a sudden change in the number of available degrees of freedom available to the system in the new phase. In the case of heavy-ion collisions one would like to perform the analagous measurement, which is to measure the temperature of the system as a function of the deposited energy, or energy density. Information about the temperature can be extracted from the spectral distributions of the emitted particles, while the energy density attained by the system is reflected in the amount of transverse energy or particle production. Indeed, one of the earliest signatures of QGP formation, proposed by Van Hove [1], was the observation of a saturation of the average transverse momentum with increasing energy (or entropy) density for systems excited just above the critical energy density. With increasing energy density, the initial temperature would not rise above the critical temperature until all of the latent heat of the QGP phase transition had been supplied.

For these reasons it is of interest to study the centrality dependence of the particle production. It is generally believed that the initial energy density increases with increasing centrality, due to the many overlapping interactions. Also, the volume of the excited matter increases with centrality, as well as the amount of rescattering. Since rescattering is the feature which distinguishes AA collisions non-trivially from pp collisions, and since significant rescattering is a prerequisite for thermalization, it is imperative to demonstrate an understanding of the centrality dependence of the AA results in order to understand the effects of rescattering. While those effects may be minor on extensive observables, like the particle multiplicity or transverse energy, they should be most evident on the momentum distribution of the produced particles.

The transverse momentum spectra of produced pions can provide information on both the initial and final state properties of the hot matter. The low p_T pion production would dominantly reflect the temperature of the hadronic system at the freeze-out stage occurring late in the reaction. It is strongly influenced by rescattering among the final state hadrons. The high p_T pion production is expected to be dominated by hard scattering of the partons. In pA collisions, the high p_T region is known to be enhanced (Cronin effect [2]) due to initial state scattering of the incident partons leading to a broadening of their incoming p_T. In AA collisions, many of the scattered partons must traverse the excited matter to escape and therefore may undergo additional rescatterings and energy loss [3]. In the case of significant parton rescattering, the parton distributions may approach thermal distributions with a temperature reflecting the initial state of the excited matter. The intermediate p_T region of the pion spectrum might then reflect this initial temperature. In this paper we investigate in detail the centrality dependence of the neutral pion spectral shape and yields for 158·A GeV ^{208}Pb+^{208}Pb collisions as measured by the WA98 experiment and discuss the implications for a thermal description. Some of the results presented have been published elsewhere [4, 5].

2. THE WA98 EXPERIMENT

The CERN experiment WA98 [6] consists of large acceptance photon and hadron spectrometers together with several other large acceptance devices which allow various global measurements on an event-by-event basis. Neutral pions are reconstructed via their $\gamma\gamma$ decay branch using the WA98 lead-glass photon detector, LEDA, which consisted of 10,080 individual modules with photomultiplier readout. The detector was located at a distance of 21.5 m from the target and covered the pseudorapidity interval $2.35 < \eta < 2.95$. The minimum bias distribution ($\sigma_{min.bias} \approx 6300$ mb) is divided into various centrality classes using the transverse energy E_T measured in the MIRAC calorimeter. The impact parameter is extracted by the assumption of a monotonic relation between impact parameter and transverse energy and using the resulting correspondence between measured cross section and impact parameter. The average number of participants, N_{part}, is calculated from nuclear geometry using the extracted impact parameter. The results presented here were obtained from an analysis of the data taken with Pb beams in 1995 and 1996. In total, $\approx 9.6 \cdot 10^6$ reactions have been analyzed.

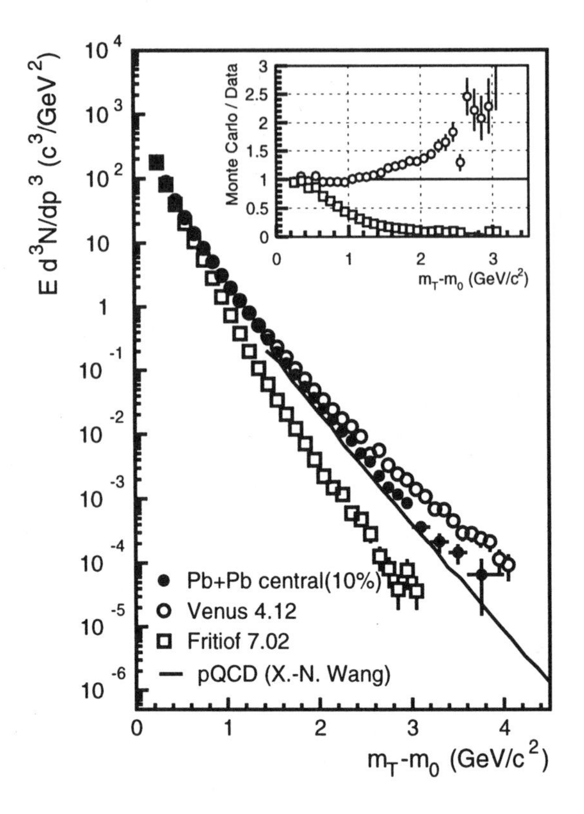

Figure 3.1 Transverse mass spectra of neutral pions in central collisions of 158 *A*GeV Pb+Pb. Invariant yields per event are compared to calculations using the FRITIOF 7.02 [8] and VENUS 4.12 [9] Monte Carlo programs. Predictions of a pQCD calculation [10] are included as a solid line. The inset shows the ratios of the results of the Monte Carlo codes to the experimental data.

3. π^0 PRODUCTION

The general π^0 analysis procedure is similar to that used in the WA80 experiment and described in [7]. Hits in the lead-glass detector are combined in pairs to provide distributions of pair mass vs. pair transverse momentum (or transverse mass) for all possible combinations. Subtraction of the combinatorial background is performed using mixed event distributions. The resulting momentum distributions are corrected for geometrical acceptance and reconstruction efficiency. The efficiency depends on the particle occupancy in the detector and therefore has been calculated independently for each centrality bin. The systematic error of the pion yields is mainly due to errors in the reconstruction efficiency for central collisions and to corrections for non-target interactions for peripheral collisions. The systematic error on the absolute yield is $\approx 10\%$ and increases sharply below $p_T = 0.4\,\mathrm{GeV}/c$. An additional systematic error originates from the uncertainty of the momentum scale of 1%. The influence of this rises slowly for higher p_T and leads to an error of 15% at $p_T = 4\,\mathrm{GeV}/c$.

The measured neutral pion spectrum from central Pb+Pb reactions (10% of the minimum bias cross section) as a function of $m_T - m_0$ is shown in Fig. 3.1. The data are compared to predictions of the string model Monte Carlo generators FRITIOF 7.02 [8] and VENUS 4.12 [9]. As already observed in S+Au reactions [7], both generators fail to describe the data well at large m_T. The FRITIOF prediction is more than an order of magnitude lower at high m_T while VENUS significantly overpredicts the data. Alternatively, it has recently been shown that perturbative QCD calculations, including initial state multiple scattering and intrinsic p_T [10], are able to describe the preliminary WA98 data at intermediate and high p_T. This prediction is included in Fig. 3.1 as a solid line. The pQCD calculation shows a very good agreement in the high m_T region. This surprising agreement has been interpreted as an indication for unexpectedly small effects of parton energy loss [10]. On the other hand, the parton cascade Monte Carlo code, VNI, which provides a more detailed pQCD description, overpredicts the measured WA98 result by more than a factor of ten at large p_T [11].

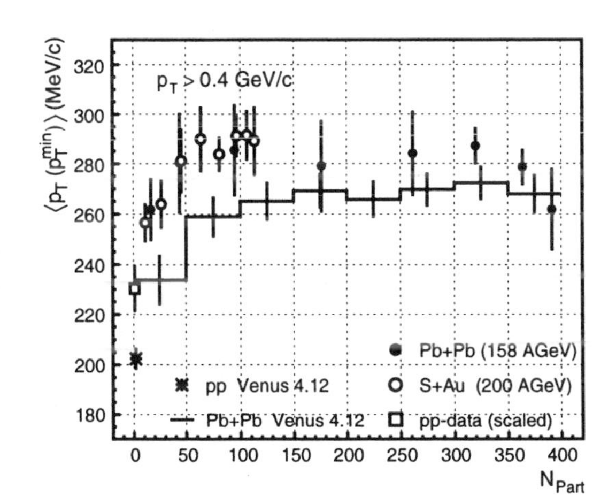

Figure 3.2 Truncated mean transverse momentum $\langle p_T(p_T^{min})\rangle$ of π^0 mesons as defined by Eq. 3.1 plotted as a function of the average number of participants N_{part}. The solid circles correspond to the 8 E_T based centrality selections for Pb+Pb. The open square shows $\langle p_T(p_T^{min})\rangle$ extracted from a parametrization of pp data scaled to the same cms-energy [12], the open circles the results for S+Au collisions at 200 AGeV [7].

In view of the above discussion and the difficulty to describe the details of the neutral pion spectrum, it is apparent that the theoretical description of ultra-relativistic nucleus-nucleus collisions remains uncertain. In order to demonstrate a consistent description of nuclear effects it is important to investigate the details of the pion production as a function of the system size. To study the centrality dependence of the spectral shape in a manner which is independent of model or fit function we have used the truncated mean transverse momentum $\langle p_T(p_T^{min})\rangle$, where

$$\langle p_T(p_T^{min})\rangle = \left(\int_{p_T^{min}}^{\infty} p_T \frac{dN}{dp_T} dp_T \Big/ \int_{p_T^{min}}^{\infty} \frac{dN}{dp_T} dp_T \right) - p_T^{min}. \qquad (3.1)$$

The lower cutoff $p_T^{min} = 0.4$ GeV/c is introduced to avoid systematic errors from extrapolation to low p_T and has been chosen according to the lowest p_T of the present data where systematic uncertainties imposed by the necessary corrections are still small.

Figure 3.2 shows $\langle p_T(p_T^{min})\rangle$ as a function of the average number of participants N_{part} for 158·A GeV ^{208}Pb+Pb collisions. For comparison, $\langle p_T(p_T^{min})\rangle$ values for 200·A GeV S+Au [7] and from a parametrization of pp data [12] are also included. Together these data show the general trend of a rapid increase of $\langle p_T(p_T^{min})\rangle$ compared to pp results for small system sizes. For N_{part} greater than about 50 the mean transverse momentum appears to attain a limiting value of $\approx 280\,\mathrm{MeV}/c$. VENUS 4.12 [9] calculations show a qualitatively similar behaviour, although the values of $\langle p_T(p_T^{min})\rangle$ are somewhat lower than the experimental data.

Initial state multiple scattering, as suggested as explanation for the Cronin effect [2], would imply a continuing increase of $\langle p_T(p_T^{min})\rangle$ for more central collisions. Here, however, the surprising observation is that additional multiple scattering, implied by increasing N_{part}, does not alter the pion distributions. This is most easily understood as a consequence of final state rescattering and is, of course, the behaviour expected for a thermalized system.

More detailed information about the centrality dependence of the pion spectral shape and yield is shown in Fig. 3.3 where the neutral pion yield per event has been parameterized as $Ed^3N/dp^3 \propto N_{part}^{\alpha(p_T)} \cdot \sigma_0(p_T)$. The results for $N_{part} > 30$ are well described by this scaling with an exponent $\alpha(p_T) \approx 1.3$, independent of p_T. Consistent with the previous discussion, the results indicate a constant spectral shape over the entire interval of measurement from $0.5 < p_T < 3$ GeV/c. The observed $N_{part}^{4/3}$ scaling for symmetric systems implies a scaling with the number of nucleon collisions, as confirmed by a similar analysis. However, this scaling does not extrapolate from the pp results. On the contrary, when comparing semi-peripheral Pb+Pb collisions with pp the exponent α varies over the entire p_T interval, confirming the very different spectral shapes.

4. THERMAL FREEZEOUT

A successful thermal interpretation of relativistic heavy ion collisions must provide an accurate description of the pion spectra since pions

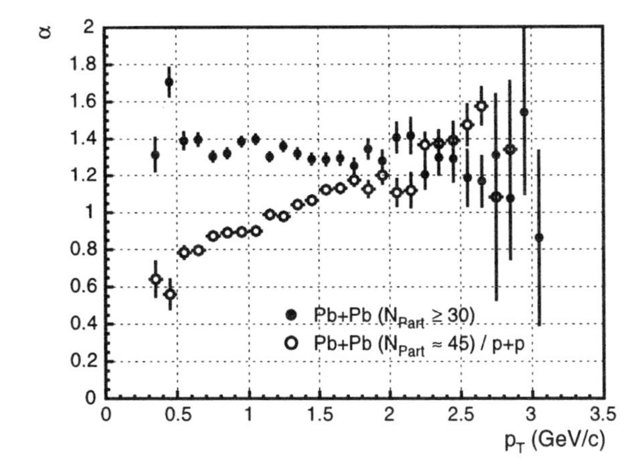

Figure 3.3 The exponent $\alpha(p_T)$ of the dependence of the π^0 yield on the average number of participants N_{part} plotted as a function of the transverse momentum for 158 *A*GeV Pb+Pb. The solid circles are calculated based on a fit to the centrality selections with $N_{part} \geq 30$. The open circles are calculated based on the ratio of the semi-peripheral data ($N_{part} \approx 45$) to a parameterization of pp data.

provide the "thermal bath" of the late stages the collision. The WA98 π^0 data provide important constraints due to their accuracy and coverage in transverse mass.

The measured neutral pion cross section from central Pb+Pb reactions as a function of $m_T - m_0$ is shown in Fig. 3.1. The data have been fit with a hydrodynamical model [13] which includes transverse flow and resonance decays. This computer program calculates the direct production and the contributions from the most important resonances having two- or three-body decays including pions (ρ, K^0_S, $K^\star$, Δ, $\Sigma + \Lambda$, η, ω, η'). The code, originally intended for charged pions, has been adapted to predict neutral pion production. The model uses a gaussian transverse spatial density profile truncated at 4σ. The transverse flow rapidity is assumed to be a linear function of the radius. For all results presented here, a baryonic chemical potential of $\mu_B = 200\,\text{MeV}$ has been used. The results are not very sensitive, however, to the choice of μ_B for the $m_T - m_0$ region considered here.

This model provides an excellent description of the neutral pion spectra with a temperature $T = 185\,\text{MeV}$ and an average flow velocity of $\langle\beta_T\rangle = 0.213$. These values are very similar to the parameters obtained with similar fits to neutral pion spectra in central reactions of ^{32}S+Au [7]. The 2σ lower limit on the temperature is $T^{low} = 171\,\text{MeV}$ and the corresponding upper limit on the flow velocity is $\langle\beta_T^{upp}\rangle = 0.253$.

The high statistical accuracy and large transverse mass coverage of the present π^0 measurement reveals the concave curvature of the π^0

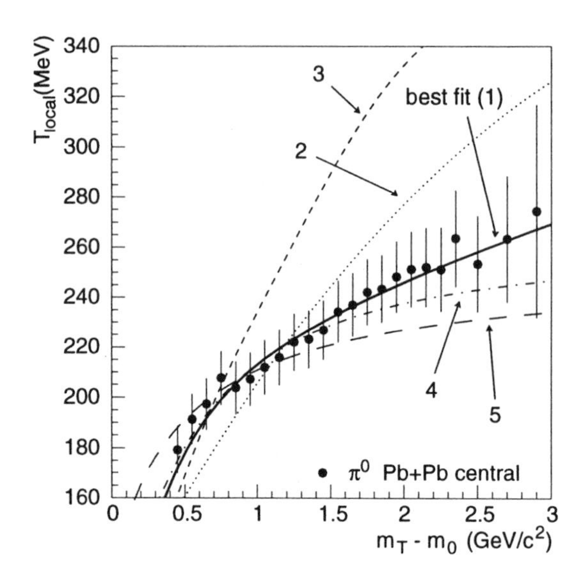

Figure 3.4 The local inverse slope of the transverse mass spectrum of neutral pions in central collisions of 158 *A*GeV Pb+Pb. The measured results (solid points) are compared to the hydrodynamical model best fit result (solid line; T = 185 MeV and $\langle\beta_T\rangle$ = 0.213) and to the other results given in table 3.1.

spectrum over a large m_T range, which constrains the parameters significantly. This is further demonstrated by studying the local slope at each m_T. The local (inverse) slope is given by

$$T_{local}^{-1} = -\left(E\frac{d^3\sigma}{dp^3}\right)^{-1} \frac{d}{dm_T}\left(E\frac{d^3\sigma}{dp^3}\right). \tag{3.2}$$

The local slope results are plotted in Fig. 3.4. Each individual value of T_{local} has been extracted from 3 adjacent data points of Fig. 3.1. The data are compared to the hydrodynamical model best fit results, as well as fits in which the transverse flow velocities have been fixed to larger values comparable to those obtained by Refs. [14] and NA49 [15] (sets 2 and 3). The corresponding fit parameters are given in Table 3.1. The comparison demonstrates that while the large transverse flow velocity fits can provide a reasonable description of the data up to transverse masses of about 1 GeV, they significantly overpredict the local slopes at large transverse mass. While application of the hydrodynamical model at large transverse mass is questionable, the model cannot overpredict the measured yield. The observed overprediction therefore rules out the assumption of large transverse flow velocities, or points to a deficiency in the model assumptions used in these fits.

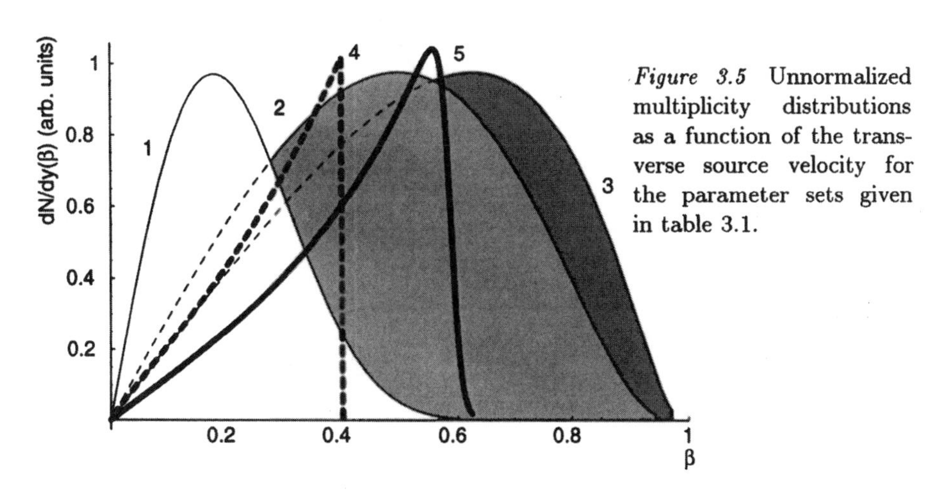

Figure 3.5 **Unnormalized multiplicity distributions as a function of the transverse source velocity for the parameter sets given in table 3.1.**

The curvature in the π^0 spectrum at large transverse mass is a result of the distribution of transverse velocities. Although the spectrum is not directly sensitive to the spatial distribution of particle emission, within this model it is dependent indirectly on the spatial distribution due to the assumption that the transverse rapidity increases linearly with radius. The large curvature at large transverse mass is due to high velocity contributions which result from the tail of the assumed gaussian density profile [16]. Figure 3.5 shows the transverse source velocity distributions $dN/dy(\beta)$ for the different parameter sets. The curves labelled 1-3 correspond to the calculations in figure 3.4 using a gaussian spatial profile. In addition, velocity profiles are shown for a uniform density profile (set 4) and for a Woods-Saxon distribution:

$$\rho(r) = \frac{1}{1 + \exp\left[(r - r_0)/\Delta\right]} \tag{3.3}$$

with $\Delta/r_0 = 0.02$ (set 5). These are included in figures 3.4 and 3.5. It is seen that the uniform density assumption truncates the high velocity tail resulting in less curvature in the pion spectrum, while the Woods-Saxon has a more diffuse edge at high β.

While the gaussian and uniform density assumptions have very different velocity profiles, it is interesting that both can provide acceptable fits to the pion spectrum with best fit results with similar $\langle\beta_T\rangle$ and T parameters, which give similar effective temperatures, and which have similar velocity widths, β_{RMS}, as shown in Table 3.1. Compared to the gaussian profile result, the best fit result using the uniform profile gives a lower temperature of 178 MeV and would lead to weaker limits of

Table 3.1 Parameters for different hydrodynamical model fits to the neutral pion spectrum shown in figures 3.4 and 3.5. The temperature T, average and RMS transverse flow velocity $\langle\beta_T\rangle$ and β_{RMS} are given together with the effective temperature $T_{eff} = T/\sqrt{(1-\langle\beta_T\rangle)/(1+\langle\beta_T\rangle)}$.

Set	spatial profile	T (MeV)		$\langle\beta_T\rangle$		β_{RMS}	T_{eff} (MeV)	χ^2/dof
1	Gauss	185	± 4	0.213	± 0.020	0.107	230	25.9/18
2	Gauss	75	± 1	0.469		0.199	125	386/19
3	Gauss	49	± 1	0.527		0.213	88	578/19
4	Uniform	178	± 13	0.274	± 0.046	0.093	235	33.3/18
5	WS	146	$^{+21}_{-16}$	0.365	$^{+0.056}_{-0.069}$	0.137	214	26.7/18

$\langle\beta_T^{upp}\rangle = 0.42$ and $T^{low} = 134\,\text{MeV}$. However, using the Woods-Saxon profile similar limits cannot be obtained. The best fit using data for $m_T - m_0 > 2\,\text{GeV}/c^2$ only as upper limits is obtained with $T = 129\,\text{MeV}$ and $\langle\beta_T\rangle = 0.42$.

The data presented here can be well described with high thermal freeze-out temperatures similar to temperatures which have been extracted for chemical freeze-out [17] and small transverse flow velocities. On the other hand, if the larger velocities obtained in other analyses which have considered limited particle spectra together with HBT results [14, 15, 18] persist, the present data obviously provide important information on the shape of the freeze-out velocity distribution.

5. SUMMARY

We have analyzed the centrality dependence of high precision transverse momentum spectra of neutral pions from 158AGeV Pb+Pb collisions. The neutral pion spectra are observed to show increasing deviation from pp results with increasing centrality, indicating the importance of multiple scattering effects. However, for centralities with more than about 50 participating nucleons, the shape of the transverse momentum spectrum becomes invariant over the interval $0.5 < p_T < 3$ GeV/c. In this interval the pion yield scales like $N_{part}^{1.3}$, or like the number of nucleon collisions, for this range of centralities. Since the amount of rescattering increases with centrality, the invariance of the spectral shape with respect to the number of rescatterings, most naturally suggests a dominantly thermal emission process.

We have argued that hydrodynamical models which attempt to extract the thermal freeze-out parameters of relativistic heavy ion collisions must provide an accurate description of the pion spectra, since pions most directly reflect the thermal evironment in the late stage of the collision. In particular, models, or parameter sets, which overpredict the observed pion yields, even at large transverse mass, can immediately be ruled out. We have demonstrated that the high accuracy neutral pion spectra with large transverse mass coverage can constrain the thermal freeze-out parameters and model assumptions. Within the context of the hydrodynamical model of Ref. [13], only special choices of the velocity profile allow freeze-out parameters similar to those extracted from other recent analyses which consider also HBT results [14, 15, 18]. Other profiles favor large thermal freeze-out temperatures consistent with chemical freeze-out temperatures determined for the same system [17].

Acknowledgments

We wish to thank Urs Wiedemann for assistance with the model calculations and valuable discussions. ORNL is managed by Lockheed Martin Energy Research Corporation under contract DE-AC05-96OR22464 with the U.S. Department of Energy.

References

[1] L. Van Hove, Phys. Lett. B **118**, 138 (1982).

[2] D. Antreasyan, et al. Phys. Rev. D **19**, 764 (1979).

[3] X.-N. Wang and M. Gyulassy. Phys. Rev. Lett. **68**, 1480 (1992).

[4] WA98 Collaboration, M.M. Aggarwal et al., Phys. Rev. Lett. **81** (1998) 4087.

[5] WA98 Collaboration, M.M. Aggarwal et al., nucl-ex/9901009.

[6] WA98 Collaboration, *Proposal for a large acceptance hadron and photon spectrometer*, 1991, Preprint CERN/SPSLC 91-17, SPSLC/P260.

[7] WA80 Collaboration, R. Albrecht et al., Eur. Phys. J. C **5** (1998) 255.

[8] B. Andersson, G. Gustafson, and H. Pi, Z. Phys. C **57**, 485 (1993).

[9] K. Werner, Phys. Rep. **232**, 87 (1993).

[10] X.-N. Wang, 1998, preprint hep-ph/9804384 and private communication.

[11] D.K. Srivastava and K. Geiger, Phys. Rev. C **56**, 2718 (1997).

[12] C. Blume, doctoral thesis, University of Münster, Germany (1998).

[13] U.A. Wiedemann and U. Heinz, Phys. Rev. C **56**, 3265 (1997).

[14] J.R. Nix et al.,nucl-th/9801045.

[15] NA49 Collaboration, H. Appelshäuser et al., Eur. Phys. J. C **2** (1998) 661–670.

[16] U. Heinz and U.A. Wiedemann, private communication.

[17] F. Becattini, M. Gaździcki, and J. Sollfrank, Eur. Phys. J. C **5** (1998) 143.

[18] J.R. Nix, Phys. Rev. C **58** (1998) 2303.

Chapter 4

HADRON FREEZE-OUT AND QGP HADRONIZATION

J. Rafelski and J. Letessier
*Department of Physics, University of Arizona, Tucson, AZ 85721, USA**
Laboratoire de Physique Théorique et Hautes Energies†
Université Paris 7, 2 place Jussieu, F-75251 Cedex 05.

*Supported in part by a grant from the U.S. Department of Energy, DE-FG03-95ER40937.
†LPTHE, Univ. Paris 6 et 7 is: Unité mixte de Recherche du CNRS, UMR 7589.

Abstract Abundances and $m_\perp$-spectra of strange and other hadronic particles emanating from central 158–200 A GeV reactions between nuclei are found to originate from a thermally equilibrated, deconfined source in chemical non-equilibrium. Physical freeze-out parameters and physical conditions (pressure, specific energy, entropy, and strangeness) are evaluated. Properties of the source we obtain are as expected for direct hadron emission from a deconfined quark-gluon plasma phase.

Keywords: QGP, SPS, quarks, gluons, strangeness, hadrons

1. INTRODUCTION

Quark-gluon plasma (QGP) is, by the meaning of these words, a thermally equilibrated state consisting of mobile, color charged quarks and gluons. Thermal equilibrium is established faster than the particle abundance (chemical) equilibrium and thus in general quark and gluon abundances in QGP can differ from their equilibrium Stefan-Boltzmann limit. This in turn impacts the hadronic particle production yields, and as we shall see, the chemical non-equilibrium is a key ingredient in the successful data analysis of experiments performed at the CERN-SPS in the past decade. We address here results of 200 A GeV Sulphur (S) beam interactions with laboratory stationary targets, such as Gold (Au), Tungsten

(W) or Lead (Pb) nuclei and, Pb–Pb collisions at 158 A GeV. In these interactions the nominal center of momentum (CM) available energy is 8.6–9.2 GeV per participating nucleon.

Considerable refinement of data analysis has occurred since last comprehensive review of the field [1]. Our present work includes in particular the following:

1. We considered aside of strange also light quark ($q = u, d$) chemical non-equilibrium abundance [2] and introduce along with the statistical strangeness non-equilibrium parameter γ_s, its light quark analogue γ_q.
2. Coulomb distortion of the strange quark phase space has been understood [3].

Incorporating these developments, we accurately describe abundances of strange as well as non-strange hadrons, both at central rapidity and in 4π-acceptance. We are thus combining in the present analysis the strangeness diagnostic tools of dense matter with the notion of entropy enhancement in deconfined phase [4].

As particle emerge from the QGP hadronization, not only their abundance but also their spectral shape is of interest. Our analysis considers the impact of explosive radial flow on the spectra of particles at high $m_\perp$. This contributes significant information about the fireball dynamics and the possibly deconfined nature of the hadron source.

2. COULOMB EFFECT IN QGP

The diverse statistical chemical parameters that we need to consider will in a self-explanatory way be now introduced, considering the concept of strangeness balance: since strangeness can only be made and destroyed by hadronic interactions in pairs, the net abundance in the hadronic fireball must vanish. We consider a hot gas of free quarks, and evaluate the difference between strange and anti-strange quark numbers (net strangeness). The Coulomb potential originating in the initial proton abundance distorts slightly the Fermi particle distributions: strange quarks (negative charge) are attracted, while anti-strange quarks repelled. Allowing for this slowly changing potential akin to the relativistic Thomas-Fermi phase space occupancy model at finite temperature, the energy of a quark depends on both the momentum and the Coulomb potential V:

$$E_p = \sqrt{m^2 + p^2} - \frac{1}{3}V \,. \tag{4.1}$$

It is helpful to see here in first instance the potential V as a square well, Eq. (4.1) within the volume of interest. Counting the states in the fireball we obtain:

$$\langle N_s - N_{\bar{s}} \rangle = \int_{R_\mathrm{f}} g_s \frac{d^3r d^3p}{(2\pi)^3} \left[\frac{1}{1+\gamma_s^{-1}\lambda_s^{-1} e^{(\sqrt{m^2+p^2}-\frac{1}{3}V)/T}} - \frac{1}{1+\gamma_s^{-1}\lambda_s e^{(\sqrt{m^2+p^2}+\frac{1}{3}V)/T}} \right]. \tag{4.2}$$

In Eq. (4.2) the subscript R_f on the spatial integral reminds us that only the classically allowed region within the fireball is covered in the integration over the level density, $g_s(=6)$ is the quantum degeneracy of strange quarks. The magnitude of the charge of strange quarks $(Q_s/|e| = 1/3)$ is shown explicitly, the Coulomb potential refers to a negative integer probe charge.

The fugacity λ_s of strange quarks enters particles and antiparticles with opposite power, while the occupancy parameter γ_s enters both term with same power. For $\gamma_s < 1$ one obtains a rather precise result for the range of parameters of interest to us (see below) considering the Boltzmann approximation:

$$\langle N_s - N_{\bar{s}} \rangle \simeq \frac{\int_{R_\mathrm{f}} d^3r \left[\lambda_s e^{\frac{V}{3T}} - \lambda_s^{-1} e^{-\frac{V}{3T}}\right]}{\int_{R_\mathrm{f}} d^3r} g_s \int \frac{d^3p\, d^3x}{(2\pi)^3} \gamma_s e^{-\sqrt{p^2+m^2}/T}. \tag{4.3}$$

The meaning of the different factors is now evident. γ_s controls overall abundance of strange quark pairs, multiplying the usual Boltzmann thermal factor while λ_s controls the difference between the number of strange and non-strange quarks. Since strangeness is produced as $s, \bar{s}$-pair, the value of λ_s fulfills the constraint

$$\int_{R_\mathrm{f}} d^3r \left[\lambda_s e^{\frac{V}{3T}} - \lambda_s^{-1} e^{-\frac{V}{3T}}\right] = 0, \tag{4.4}$$

which is satisfied exactly both in the Boltzmann limit Eq. (4.3), and for the exact quantum distribution Eq. (4.2), when:

$$\tilde{\lambda}_s \equiv \lambda_s \lambda_\mathrm{Q}^{1/3} = 1, \qquad \lambda_\mathrm{Q} \equiv \frac{\int_{R_\mathrm{f}} d^3r e^{\frac{V}{T}}}{\int_{R_\mathrm{f}} d^3r}. \tag{4.5}$$

λ_Q is not a fugacity that can be adjusted to satisfy a chemical condition, its value is defined by the applicable Coulomb potential V. More generally, in order to account for the Coulomb effect, the quark fugacities within a deconfined region should be renamed as follows in order to absorb the Coulomb potential effect:

$$\lambda_s \to \tilde{\lambda}_s \equiv \lambda_s \lambda_Q^{1/3}, \qquad \lambda_d \to \tilde{\lambda}_d \equiv \lambda_d \lambda_Q^{1/3},$$
$$\lambda_u \to \tilde{\lambda}_u \equiv \lambda_u \lambda_Q^{-2/3}, \qquad \lambda_q \to \tilde{\lambda}_q \equiv \sqrt{\tilde{\lambda}_u \tilde{\lambda}_d} = \lambda_q \lambda_Q^{-1/6}. \quad (4.6)$$

Since $Q_d = Q_s = -1/3$, the first line is quite evident after the above strangeness discussion, the second follows with $Q_u = +2/3$. Note that for a negatively charged strange quark the tilded fugacity Eq. (4.5) contains a factor with positive power 1/3 but the potential that enters the quantity λ_Q is negative, and thus $\tilde{\lambda}_s < \lambda_s$. Because Coulomb-effect acts in opposite way on u and d quarks, its net impact on λ_q is smaller than on λ_s, and it also acts in the opposite way with $\tilde{\lambda}_q > \lambda_q$. To see the relevance of the tilde fugacities for light quarks, note that in order to obtain baryon density in QGP one needs to use the tilde-quark fugacity to account for the Coulomb potential influence on the phase space.

It is somewhat unexpected that for the Pb–Pb fireball the Coulomb effect is at all relevant. Recall that for a uniform charge distribution within a radius $R_{\rm f}$ of charge $Z_{\rm f}$:

$$V = \begin{cases} -\dfrac{3}{2}\dfrac{Z_{\rm f}e^2}{R_{\rm f}}\left[1-\dfrac{1}{3}\left(\dfrac{r}{R_{\rm f}}\right)^2\right], & \text{for} \quad r < R_{\rm f}\,; \\ -\dfrac{Z_{\rm f}e^2}{r}, & \text{for} \quad r > R_{\rm f}\,. \end{cases} \quad (4.7)$$

Choosing $R_{\rm f} = 8\,\text{fm}$, $T = 140\,\text{MeV}$, $m_s = 200\,\text{MeV}$ (value of γ_s is practically irrelevant) we find as solution of Eq. (4.2) for $\langle N_s - N_{\bar{s}} \rangle = 0$ for $Z_{\rm f} = 150$ the value $\lambda_s = 1.10$ (precisely: 1.0983; $\lambda_s = 1.10$ corresponds to $R_{\rm f} = 7.87\,\text{fm}$). We will see that both this values within the experimental precision arise from study of particle abundances. For the S–W/Pb reactions this Coulomb effect is practically negligible.

In the past we (and others) have disregarded in the description of the hadronic final state abundances the electrical charges and interactions of the produced hadrons. This is a correct way to analyze the chemical properties since, as already mentioned, the quantity λ_Q is not a new fugacity: conservation of flavor already assures charge conservation in chemical hadronic reactions, and use of λ_i, $i = u, d, s$ exhausts all available chemical balance conditions for the abundances of hadronic particles. As shown here, the Coulomb deformation of the phase space in the QGP fireball makes it necessary to rethink the implications that the final state particle measured fugacities have on the fireball properties.

3. FREEZE-OUT OF HADRONS

The production of hadrons from a QGP fireball occurs mainly by way of quark coalescence and gluon fragmentation, and there can be some quark fragmentation as well. We will explicitly consider the recombination of quarks, but implicitly the gluon fragmentation is accounted for by our allowance for chemical nonequilibrium. The relative number of primary particles freezing out from a source is obtained noting that the fugacity and phase space occupancy of a composite hadronic particle can be expressed by its constituents and that the probability to find all j-components contained within the i-th emitted particle is:

$$N_i \propto \prod_{j\in i} \gamma_j \lambda_j e^{-E_j/T}\,, \qquad \lambda_i = \prod_{j\in i} \lambda_j\,, \qquad \gamma_i = \prod_{j\in i} \gamma_j\,. \tag{4.8}$$

Experimental data with full phase space coverage, or central rapidity region $|y - y_{\rm CM}| < 0.5$, for $m_\perp > 1.5$ GeV are considered; recall that the energy of a hadron 'i' is expressed by the spectral parameters $m_\perp$ and y as follows,

$$E_i = \sum E_j\,, \qquad E_i = \sqrt{m_i^2 + p^2} = \sqrt{m_i^2 + p_\perp^2}\cosh(y - y_{\rm CM})\,,$$

where $y_{\rm CM}$ is the center of momentum rapidity of the fireball formed by the colliding nuclei.

The yield of particles is controlled by the freeze-out temperature $T_{\rm f}$. This freeze-out temperature is different from the $m_\perp$-spectral temperature $T_\perp$, which also comprises the effect of collective matter flows originating in the explosive disintegration driven by the internal pressure of compressed hadronic matter. In order to model the flow and freeze-out of the fireball surface, one in general needs several new implicit and/or explicit parameters. We therefore will make an effort to choose experimental variables (compatible particle ratios) which are largely flow independent. This approach also diminishes the influence of heavy resonance population — we include in Eq.(4.8) hadronic states up to $M = 2$ GeV, and also include quantum statistical corrections, allowing for first Bose and Fermi distribution corrections in the phase space content. It is hard to check if indeed we succeeded in eliminating the uncertainty about high mass hadron populations. As we shall see comparing descriptions which exclude flow with those that include it, our approach is indeed largely flow-insensitive.

We consider here a simple radial flow model, with freeze-out in CM frame at constant laboratory time, implying that causally disconnected domains of the dense matter fireball are synchronized at the instant of the collision. Within this approach [5], the spectra and thus also multiplicities of particles emitted are obtained replacing the Boltzmann exponential factor in Eq.(4.8),

$$e^{-E_j/T} \to \frac{1}{2\pi}\int d\Omega_v \gamma_v (1+\vec{v}_c\cdot\vec{p}_j/E_j) e^{-\frac{\gamma_v E_j}{T}(1+\vec{v}_c\cdot\vec{p}_j/E_j)}, \tag{4.9}$$

where as usual $\gamma_v = 1/\sqrt{1-\vec{v}_c^{\,2}}$. Eq. (4.9) can be intuitively obtained by a Lorentz transformation between an observer on the surface of the fireball, and one at rest in the general CM (laboratory) frame. One common feature of all flow scenarios is that, at sufficiently high $m_\perp$, the spectral temperature (inverse slope) $T_\perp$ can be derived from the freeze-out temperature $T_{\rm f}$ with the help of the Doppler formula:

$$T_\perp = T_{\rm f}\gamma_v(1+v_c)\,. \tag{4.10}$$

In actual numerical work, we proceed as follows to account for the Doppler effect: for a given pair of values $T_{\rm f}$ and $v_{\rm c}$, the resulting $m_\perp$ particle spectrum is obtained and analyzed using the spectral shape and procedure employed for the particular collision system by the experimental groups, yielding the theoretical inverse slope 'temperature' $T_\perp^j$.

Once the parameters $T_{\rm f}$, λ_q, λ_s, γ_q, γ_s and v_c have been determined studying available particle yields, and $m_\perp$ slopes, the entire phase space of particles produced is fully characterized within our elaborate statistical model. Our model is in fact just an elaboration of the original Fermi model [6], in fact all we do is to allow hadronic particles to be produced in the manner dictated by the phase space size of valance quarks. With the full understanding of the phase space of all hadrons, we can evaluate the physical properties of the system at freeze-out, such as, *e.g.*, energy and entropy per baryon, strangeness content.

4. RESULTS OF DATA ANALYSIS

As noted our analysis requires that we form particle abundance ratios between what we call compatible hadrons. We considered for S–Au/W/Pb reactions 18 data points listed in table 4.1 (of which three comprise results with Ω's). For Pb–Pb we address here 15 presently available particle yield ratios listed in table 4.2 (of which four comprise the Ω's). We believe to have included in our discussion most if not all particle multiplicity results available presently.

The theoretical particle yield results shown columns in tables 4.1 and 4.2 are obtained looking for a set of physical parameters which will minimize the difference between theory and experiment. The resulting total

Table 4.1 Particle ratios studied in our analysis for S–W/Pb/Au reactions: experimental results with references and kinematic cuts are given, followed by columns showing results for the different strategies of analysis B–F. Asterisk * means a predicted result (corresponding data is not fitted). Subscript s implies forced strangeness conservation, subscript v implies inclusion of collective flow. The experimental results considered are from:

[1] S. Abatzis *et al.*, WA85 Collaboration, *Heavy Ion Physics* **4**, 79 (1996).
[2] S. Abatzis *et al.*, WA85 Collaboration, *Phys. Lett.* B **347**, 158 (1995).
[3] S. Abatzis *et al.*, WA85 Collaboration, *Phys. Lett.* B **376**, 251 (1996).
[4] I.G. Bearden *et al.*, NA44 Collaboration, *Phys. Rev.* C **57**, 837 (1998).
[5] D. Röhrich for the NA35 Collaboration, *Heavy Ion Physics* **4**, 71 (1996).
[6] S–Ag value adopted: T. Alber *et al.*, NA35 Collaboration, *Eur. Phys. J.* C **2**, 643 (1998).
[7] A. Iyono *et al.*, EMU05 Collaboration, *Nucl. Phys.* A **544**, 455c (1992) and Y. Takahashi *et al.*, EMU05 Collaboration, private communication.

Ratios	Ref.	Cuts [GeV]	Exp.Data	B	C	D	D$_s$	F	D$_v$	F$_v$
Ξ/Λ	1	$1.2 < p_\perp < 3$	0.097 ± 0.006	0.16	0.11	0.099	0.11	0.10	0.11	0.11
$\overline{\Xi}/\bar{\Lambda}$	1	$1.2 < p_\perp < 3$	0.23 ± 0.02	0.38	0.23	0.22	0.18	0.22	0.23	0.22
$\bar{\Lambda}/\Lambda$	1	$1.2 < p_\perp < 3$	0.196 ± 0.011	0.20	0.20	0.203	0.20	0.20	0.20	0.20
$\overline{\Xi}/\Xi$	1	$1.2 < p_\perp < 3$	0.47 ± 0.06	0.48	0.44	0.45	0.33	0.44	0.44	0.43
$\overline{\Omega}/\Omega$	2	$p_\perp > 1.6$	0.57 ± 0.41	1.18*	0.96*	1.01*	0.55*	0.98	1.09*	1.05*
$\frac{\Omega+\overline{\Omega}}{\Xi+\overline{\Xi}}$	2	$p_\perp > 1.6$	0.80 ± 0.40	0.27*	0.17*	0.16*	0.16*	0.16	0.13*	0.13*
K^+/K^-	1	$p_\perp > 0.9$	1.67 ± 0.15	2.06	1.78	1.82	1.43	1.80	1.77	1.75
$K^0_{\rm s}/\Lambda$	3	$p_\perp > 1$	1.43 ± 0.10	1.56	1.64	1.41	1.25	1.41	1.38	1.39
$K^0_{\rm s}/\bar{\Lambda}$	3	$p_\perp > 1$	6.45 ± 0.61	7.79	8.02	6.96	6.18	6.96	6.81	6.86
$K^0_{\rm s}/\Lambda$	1	$m_\perp > 1.9$	0.22 ± 0.02	0.26	0.28	0.24	0.24	0.24	0.24	0.24
$K^0_{\rm s}/\bar{\Lambda}$	1	$m_\perp > 1.9$	0.87 ± 0.09	1.30	1.38	1.15	1.20	1.16	1.18	1.17
Ξ/Λ	1	$m_\perp > 1.9$	0.17 ± 0.01	0.27	0.18	0.17	0.18	0.17	0.16	0.17
$\overline{\Xi}/\bar{\Lambda}$	1	$m_\perp > 1.9$	0.38 ± 0.04	0.64	0.38	0.38	0.30	0.37	0.35	0.35
$\frac{\Omega+\overline{\Omega}}{\Xi+\overline{\Xi}}$	1	$m_\perp > 2.3$	1.7 ± 0.9	0.98*	0.59*	0.58*	0.52*	0.58	0.72*	0.75*
$p/\bar{p}$	4	Mid-rapidity	11 ± 2	11.2	10.1	10.6	7.96	10.5	10.6	10.4
$\bar{\Lambda}/\bar{p}$	5	$4\,\pi$	1.2 ± 0.3	2.50	1.47	1.44	1.15	1.43	1.58	1.66
$\frac{h^-}{p-\bar{p}}$	6	$4\,\pi$	4.3 ± 0.3	4.4	4.2	4.1	3.6	4.1	4.2	4.2
$\frac{h^+-h^-}{h^++h^-}$	7	$4\,\pi$	0.124 ± 0.014	0.11	0.10	0.103	0.09	0.10	0.12	0.12
		$\chi^2_{\rm T}$		264	30	6.5	38	12	6.2	11

error of the ratios R is shown at the bottom of tables 4.1 and 4.2:

$$\chi^2_{\rm T} = \sum_j \left(\frac{R^j_{\rm th} - R^j_{\rm exp}}{\Delta R^j_{\rm exp}} \right)^2 . \tag{4.11}$$

It is a non-trivial matter to determine the confidence level that goes with the different data analysis approaches since some of the results considered are partially redundant, and a few data points can be obtained from others by algebraic relations arising in terms of their theoretical definitions; there are two types of such relations:

$$\frac{\Omega+\overline{\Omega}}{\Xi+\overline{\Xi}} = \frac{\overline{\Omega}}{\overline{\Xi}} \cdot \frac{1+\overline{\Omega}/\Omega}{1+\overline{\Xi}/\Xi}, \qquad \frac{\overline{\Lambda}}{\Lambda} = \frac{\overline{\Lambda}}{\overline{\Xi}} \cdot \frac{\overline{\Xi}}{\Xi} \cdot \frac{\Xi}{\Lambda} . \tag{4.12}$$

Table 4.2 Particle ratios studied in our analysis for Pb–Pb reactions: experimental results with references and kinematic cuts are given, followed by columns showing results for the different strategies of analysis B–F. Asterisk * means a predicted result (corresponding data is not fitted or not available). Subscript s implies forced strangeness conservation, subscript v implies inclusion of collective flow. The experimental results considered are from:

[1] I. Králik, for the WA97 Collaboration, *Nucl. Phys.* A **638**,115, (1998).
[2] G.J. Odyniec, for the NA49 Collaboration, *J. Phys.* G **23**, 1827 (1997).
[3] P.G. Jones, for the NA49 Collaboration, *Nucl. Phys.* A **610**, 188c (1996).
[4] F. Pühlhofer, for the NA49 Collaboration, *Nucl. Phys.* A **638**, 431,(1998).
[5] C. Bormann, for the NA49 Collaboration, *J. Phys.* G **23**, 1817 (1997).
[6] G.J. Odyniec, *Nucl. Phys.* A **638**, 135, (1998).
[7] D. Röhrig, for the NA49 Collaboration, *Recent results from NA49 experiment on Pb-Pb collisions at 158 A GeV*, see Fig. 4, EPS-HEP Conference, Jerusalem, Aug. 1997, p. 19.
[8] A.K. Holme, for the WA97 Collaboration, *J. Phys.* G **23**, 1851 (1997).

Ratios	Ref.	Cuts [GeV]	Exp.Data	B	C	D	D_s	F	D_v	F_v
Ξ/Λ	1	$p_\perp > 0.7$	0.099 ± 0.008	0.138	0.093	0.095	0.098	0.107	0.102	0.110
$\bar{\Xi}/\bar{\Lambda}$	1	$p_\perp > 0.7$	0.203 ± 0.024	0.322	0.198	0.206	0.215	0.216	0.210	0.195
$\bar{\Lambda}/\Lambda$	1	$p_\perp > 0.7$	0.124 ± 0.013	0.100	0.121	0.120	0.119	0.121	0.123	0.128
$\bar{\Xi}/\Xi$	1	$p_\perp > 0.7$	0.255 ± 0.025	0.232	0.258	0.260	0.263	0.246	0.252	0.225
$\frac{(\Xi+\bar{\Xi})}{(\Lambda+\bar{\Lambda})}$	2	$p_\perp > 1.$	0.13 ± 0.03	0.169	0.114	0.118	0.122	0.120	0.123	0.121
K^0_s/ϕ	3,4		11.9 ± 1.5	6.3	10.4	9.89	9.69	16.1	12.9	15.1
K^+/K^-	5		1.80 ± 0.10	1.96	1.75	1.76	1.73	1.62	1.87	1.56
$p/\bar{p}$	6		$18.1 \pm 4.$	22.0	17.1	17.3	17.9	16.7	17.4	15.3
$\bar{\Lambda}/\bar{p}$	7		$3. \pm 1.$	3.02	2.91	2.68	3.45	0.65	2.02	1.29
K^0_s/B	3		0.183 ± 0.027	0.305	0.224	0.194	0.167	0.242	0.201	0.281
h^-/B	3		1.83 ± 0.2	1.47	1.59	1.80	1.86	1.27	1.83	1.55
Ω/Ξ	1	$p_\perp > 0.7$	0.192 ± 0.024	0.119*	0.080*	0.078*	0.080*	0.192	0.077*	0.190
$\bar{\Omega}/\bar{\Xi}$	8	$p_\perp > 0.7$	0.27 ± 0.06	0.28*	0.17*	0.17*	0.18*	0.40	0.18*	0.40
$\bar{\Omega}/\Omega$	1	$p_\perp > 0.7$	0.38 ± 0.10	0.55*	0.56*	0.57*	0.59*	0.51	0.60*	0.47
$\frac{(\Omega+\bar{\Omega})}{(\Xi+\bar{\Xi})}$	8	$p_\perp > 0.7$	0.20 ± 0.03	0.15*	0.10*	0.10*	0.10*	0.23	0.10*	0.23
		χ^2_{T}		88	24	1.6	2.7	19	1.5	18

However, due to smallness of the total error found for some of the approaches it is clear without detailed analysis that only these are statistically significant.

In addition to the abundance data, we also explored the transverse mass $m_\perp$-spectra when the collective flow velocity was allowed in the description, and the bottom line of tables 4.1 and 4.2 in columns D_v, F_v includes in these cases the error found in the inverse slope parameter of the spectra. The procedure we used is as follows: since within the error the high $m_\perp$ strange (anti)baryon inverse slopes are within error, overlapping we decided to consider just one 'mean' experimental value $T_\perp = 235 \pm 10$ for S–induced reactions and $T_\perp = 265 \pm 15$ for Pb–induced reactions. Thus we add one experimental value and one parameter, without changing the number of degrees of freedom. Once we find values of T_{f} and v_{c}, we evaluate the slopes of the theoretical spectra. The resulting theoretical T^j_{th} values are in remarkable agreement with experimental $T^j_\perp$, well beyond what we expected, as is shown

Table 4.3 Particle spectra inverse slopes: theoretical values $T_{\rm th}$ are obtained imitating the experimental procedure from the T_f, v_c-parameters. Top portion: S–W experimental $T_\perp$ from experiment WA85 for kaons, lambdas and cascades; bottom portion: experimental Pb–Pb $T_\perp$ from experiment NA49 for kaons and from experiment WA97 for baryons. The experimental results are from:

D. Evans, for the WA85-collaboration, APH N.S., Heavy Ion Physics **4**, 79 (1996).
E. Andersen *et al.*, WA97-collaboration, *Phys. Lett.* B **433**, 209, (1998).
S. Margetis, for the NA49-collaboration, J. Physics G, Nucl. and Part. Phys. **25**, 189 (1999).

	$T^{\rm K^0}$	T^{Λ}	$T^{\overline{\Lambda}}$	T^{Ξ}	$T^{\overline{\Xi}}$	$T^{\Omega+\overline{\Omega}}$
$T_\perp$ [MeV]	219 ± 5	233 ± 3	232 ± 7	244 ± 12	238 ± 16	—
$T_{\rm th}$ [MeV]	215	236	236	246	246	260
$T_\perp$ [MeV]	223 ± 13	291 ± 18	280 ± 20	289 ± 12	269 ± 22	237 ± 24
$T_{\rm th}$ [MeV]	241	280	280	298	298	335

in table 4.3. An exception is the fully strange $\Omega + \overline{\Omega}$ spectrum. We note in passing that when v_c was introduced we found little additional correlation between now 6 theoretical parameters. The collective flow velocity is a truly new degree of freedom and it helps to attain a more consistent description of the experimental data available.

Although it is clear that one should be using a full-fledged model such as D_v, we address also cases B and C. The reason for this arises from our desire to demonstrate the empirical need for chemical non-equilibrium: in the approach B, complete chemical equilibrium $\gamma_i = 1$ is assumed. As we see in tables 4.1 and 4.2 this approach has rather large error. Despite this the results in column B in tables 4.1 and 4.2 are often compared favorably with experiment, indeed this result can be presented quite convincingly on a logarithmic scale. Yet as we see the disagreement between theory and experiment is quite forbidding. With this remark we wish to demonstrate the need for comprehensive and precisely measured hadron abundance data sample, including abundances of multi-strange antibaryons, which were already 20 years ago identified as the hadronic signals of QGP phenomena [7]. The strange antibaryon enhancement reported by the experiment WA97 fully confirms the role played by these particles [8].

In the approach C, we introduce strangeness chemical non-equilibrium [9], *i.e.*, we also vary γ_s, keeping $\gamma_q = 1$. A nearly valid experimental data description is now possible, and indeed, when the error bars were smaller this was a satisfactory approach adapted in many studies [1]. However, only the possibility of light quark non-equilibrium in fit D produces a statistically significant data description.

It is interesting to note that a significant degradation of $\chi^2_{\rm T}$ occurs, especially in the Pb–Pb data, when we require in column F that the particle ratios comprising Ω and $\overline{\Omega}$-particles are also described. We

Table 4.4 Statistical parameters which best describe the experimental S–Au/W/Pb results shown in table 4.1. Asterisk (*) means a fixed (input) value or result of a constraint. In approaches B to D, particle abundance ratios comprising Ω are not considered. In case D_s strangeness conservation in the particle yields was enforced. In case F the three data-points with Ω are considered. Lower index v implies that radial collective flow velocity has been considered.

S–W	T_f [MeV]	λ_q	λ_s	γ_s/γ_q	γ_q	v_c	χ^2_T
B	144 ± 2	1.53 ± 0.02	0.97 ± 0.02	1*	1*	0*	264
C	147 ± 2	1.49 ± 0.02	1.01 ± 0.02	0.62 ± 0.02	1*	0*	30
D	143 ± 3	1.50 ± 0.02	1.00 ± 0.02	0.60 ± 0.02	1.22 ± 0.06	0*	6.5
D_s	153 ± 3	1.42 ± 0.02	1.10*± 0.02	0.56 ± 0.02	1.26 ± 0.06	0*	38
F	144 ± 3	1.49 ± 0.02	1.00 ± 0.02	0.60 ± 0.02	1.22 ± 0.06	0*	12
D_v	144 ± 2	1.51 ± 0.02	1.00 ± 0.02	0.69 ± 0.03	1.41 ± 0.08	0.49 ± 0.02	6.2
F_v	145± 2	1.50 ± 0.02	0.99 ± 0.02	0.69 ± 0.03	1.43 ± 0.08	0.50 ± 0.02	11

Table 4.5 Same as Fig. 4.4 for the experimental Pb–Pb results shown in table 4.2.

Pb–Pb	$T_f[MeV]$	λ_q	λ_s	γ_s/γ_q	γ_q	v_c	χ^2_T
B	142 ± 3	1.70 ± 0.03	1.10 ± 0.02	1*	1*	0*	88
C	144 ± 4	1.62 ± 0.03	1.10 ± 0.02	0.63 ± 0.04	1*	0*	24
D	134 ± 3	1.62 ± 0.03	1.10 ± 0.02	0.69 ± 0.08	1.84 ± 0.30	0*	1.6
D_s	133 ± 3	1.63 ± 0.03	1.09*± 0.02	0.72 ± 0.12	2.75 ± 0.35	0*	2.7
F	334 ± 18	1.61 ± 0.03	1.12 ± 0.02	0.50 ± 0.01	0.18 ± 0.02	0*	19
D_v	144 ± 2	1.60 ± 0.02	1.10 ± 0.02	0.86 ± 0.03	1.72 ± 0.08	0.58 ± 0.02	1.5
F_v	328 ± 17	1.59 ± 0.03	1.13 ± 0.02	0.51 ± 0.03	0.34 ± 0.13	0.38 ± 0.31	18

thus conclude that a large fraction of these particles must be made in processes that are not considered in the present model.

Another notable study case is shown in column D_s, with strangeness conservation enforced. Remarkably, the S–W data, table 4.1, do not like this constraint. A possible explanation is that for S-induced reactions, the particle abundances are obtained at relatively high $p_\perp$. Thus only a small fraction of all strange particles is actually observed, and therefore the overall strangeness is hard to balance. Similar conclusion results also when radial flow is explicitly allowed for, with a significant unbalanced strangeness fractions remaining, as we shall discuss below. On the other hand, this constraint is relatively easily satisfied for the Pb–Pb collision results, table 4.2, where a much greater proportion of all strange particles is actually experimentally detected.

The statistical parameters associated with the particle abundances described above are shown in the table 4.4 for S-W/Pb and in table 4.5 for Pb–Pb reactions. The errors of the statistical parameters shown are those provided by the program MINUIT96.03 from CERN program library. When the theory describes the data well, this is a one standard deviation error in theoretical parameters arising from the experimental measurement error. In tables 4.4 and 4.5 in cases in which $\gamma_q \neq 1$ we present the ratio γ_s/γ_q, which corresponds approximately to the parameter γ_s in the data studies in which $\gamma_q = 1$ has been assumed. It is notable that whenever we allow phase space occupancy to vary from the equilibrium, a significant deviation is found. In the S–W case, table 4.4, there is a 25% excess in the light quark occupancy, while strange quarks are 25% below equilibrium. In Pb–Pb case, table 4.5, the ratio of the nonequilibrium parameters $\gamma_s/\gamma_q \simeq 0.7$ also varies little (excluding the failing cases F with $\Omega, \overline{\Omega}$ data), though the individual values γ_s, γ_q can change significantly, even between the high confidence cases.

We note that in the Pb–Pb reaction, table 4.5, $\gamma_s > 1$. This is an important finding, since an explanation of this effect involves formation prior to freeze-out in the matter at high density of near chemical equilibrium, $\gamma_s(t < t_f) \simeq 1$. The ongoing rapid expansion (note that the collective velocity at freeze-out is found to be $1/\sqrt{3}$) preserves this high strangeness yield, and thus we find the result $\gamma_s > 1$. In other words the strangeness production freeze-out temperature $T_s > T_f$. Thus the strangeness equilibration time is proven implicitly to be of magnitude expected in earlier studies of the QGP processes [10]. It is hard, if not really impossible, to arrive at this result without the QGP hypothesis. Moreover, inspecting figure 38 in [10] we see that the yield of strangeness we expect from the kinetic theory in QGP is at the level of 0.75 per baryon, the level we indeed will determine below.

Another notable results is $\tilde{\lambda}_s \simeq \lambda_s \simeq 1.0$ in the S–Au/W/Pb case, see table 4.4, and $\lambda_s \simeq 1.1$ in the Pb–Pb case, see table 4.5, implying here $\tilde{\lambda}_s = 1$, see section 2.. We see clearly for both S- and Pb-induced reactions a value of λ_s, characteristic for a source of freely movable strange quarks with balancing strangeness.

5. PHYSICAL PROPERTIES AT FREEZE-OUT

Given the precise statistical information about the properties of the hadron phase space, we can determine the physical properties of the hadronic particles at the chemical freeze-out, see tables 4.6 and 4.7. We show for the same study cases B–F, along with their temperature, the specific energy and entropy content per baryon, and specific anti-strangeness content, along with specific strangeness asymmetry, and finally pressure at freeze-out. We note that it is improper in general to refer to these properties as those of a 'hadronic gas' formed in nuclear collisions, as the particles considered may be emitted in sequence from a deconfined source, and thus there may never be a evolution stage corresponding to a hadron gas phase. However, the properties presented are those carried away by the emitted particles, and thus characterize the properties of their source. The energy per baryon seen in the emitted

Table 4.6 T_f and physical properties (specific energy, entropy, anti-strangeness, net strangeness, pressure and volume) of the full hadron phase space characterized by the statistical parameters given in table 4.4 for the reactions S–Au/W/Pb. Asterisk * means fixed input.

S–W	T_f [MeV]	E_f/B	S_f/B	$\bar{s}_f/B$	$(\bar{s}_f - s_f)/B$	P_f [GeV/fm^3]
B	144 ± 2	8.9 ± 0.5	50± 3	1.66 ± 0.06	0.44 ± 0.02	0.056 ± 0.005
C	147 ± 2	9.3 ± 0.5	49± 3	1.05 ± 0.05	0.23 ± 0.02	0.059 ± 0.005
D	143 ± 3	9.1 ± 0.5	48± 3	0.91 ± 0.04	0.20 ± 0.02	0.082 ± 0.006
D_s	153 ± 3	8.9 ± 0.5	45 ± 3	0.76 ± 0.04	0*	0.133 ± 0.008
F	144 ± 2	9.1 ± 0.5	48 ± 3	0.91 ± 0.05	0.20 ± 0.02	0.082 ± 0.006
D_v	144 ± 2	8.2 ± 0.5	44 ± 3	0.72 ± 0.04	0.18 ± 0.02	0.124 ± 0.007
F_v	145 ± 2	8.2 ± 0.5	44 ± 3	0.73 ± 0.05	0.17 ± 0.02	0.123 ± 0.007

Table 4.7 Same as Fig. 4.6 for the reactions Pb–Pb with the statistical parameters given in table 4.5.

Pb–Pb	T_f [MeV]	E_f/B	S_f/B	$\bar{s}_f/B$	$(\bar{s}_f - s_f)/B$	P_f [GeV/fm^3]
B	142 ± 3	7.1 ± 0.5	41 ± 3	1.02 ± 0.05	0.21 ± 0.02	0.053 ± 0.005
C	144 ± 4	7.7 ± 0.5	42 ± 3	0.70 ± 0.05	0.14 ± 0.02	0.053 ± 0.005
D	134 ± 3	8.3 ± 0.5	47 ± 3	0.61 ± 0.04	0.08 ± 0.01	0.185 ± 0.012
D_s	133 ± 3	8.7 ± 0.5	48 ± 3	0.51 ± 0.04	0*	0.687 ± 0.030
F	334 ± 18	9.8 ± 0.5	24 ± 2	0.78 ± 0.05	0.06 ± 0.01	1.64 ± 0.06
D_v	144 ± 2	7.0 ± 0.5	38 ± 3	0.78 ± 0.04	0.01 ± 0.01	0.247 ± 0.007
F_v	328 ± 17	11.2 ± 1.5	28 ± 3	0.90 ± 0.05	0.09 ± 0.02	1.40 ± 0.06

hadrons is, within error, equal to the available specific energy of the collision (8.6 GeV for Pb–Pb and about 8.8–8.9 GeV for S–Au/W/Pb). This implies that the fraction of energy left in the central fireball must be the same as the fraction of baryon number. We further note that hadrons emitted at freeze-out carry away a specific $\bar{s}$ content which is determined to be 0.72 ± 0.04 in S–Au/W/Pb case and 0.78 ± 0.04 for the Pb–Pb collisions (cases D_v). Here we see the most significant impact of flow, as without it the specific strangeness content seemed to diminish as we moved to the larger collision system. We have already alluded repeatedly to the fact that for S–Au/W/Pb case the balance of strangeness is not seen in the particles observed experimentally. The asymmetry is 18%, with the excess emission of $\bar{s}$ containing hadrons at high $p_\perp > 1$ GeV. In the Pb–Pb data this effect disappears, perhaps since the $p_\perp$ lower cut-off is smaller. One could also imagine that longitudinal flow which is stronger in S–Au/W/Pb is responsible for this effect.

The small reduction of the specific entropy in Pb–Pb compared to the lighter S–Au/W/Pb is driven by the greater baryon stopping in the larger system, also seen in the smaller energy per baryon content. Both systems freeze out at $E/S = 0.185$ GeV (energy per unit of entropy). Aside of T_f, this is a second universality feature at hadronization of both systems. The overall high specific entropy content agrees well with the entropy content evaluation made earlier [4] for the S–W case. This is so because the strange particle data are indeed fully consistent within the chemical-nonequilibrium description with the 4π total particle multiplicity results.

6. CURRENT STATUS AND CONCLUSIONS

We have presented detailed analysis of hadron abundances observed in central S–Au/W/Pb 200 A GeV and Pb–Pb 158 A GeV interactions within thermal equilibrium and chemical non-equilibrium phase space model of strange and non-strange hadronic particles. In the analysis of the freeze-out the structure of the particle source was irrelevant. However, the results that we found for the statistical parameters point rather clearly towards a deconfined QGP source, with quark abundances near but not at chemical equilibrium: statistical parameters obtained characterize a strange particle source which, both for S–Au/W/Pb and for Pb–Pb case, when allowing for Coulomb deformation of the strange and anti-strange quarks, is exactly symmetric between s and $\bar{s}$ quark carriers, as is natural for a deconfined state.

There are similarities in freeze-out properties seen comparing results presented in tables 4.6 and 4.7 which suggest universality in the extensive physical properties of the two freeze-out systems [3]. Despite considerably varying statistical parameters we obtain similar physical conditions such as T_f and E/S corresponding possibly to the common physical properties of QGP at its breakup into hadrons. The precision of the data description we have reached, see tables 4.1–4.3 strongly suggests that despite its simplicity the model we developed to analyze experimental data provides a reliable image of the hadron production processes. When flow is allowed for the freeze-out temperature is identical in both physical systems considered, even though, *e.g.*, the baryochemical potential $\mu_B = 3T \ln \lambda_q$ and other physical parameters found are slightly different. It is worth to restate the values we obtained, case D_v (6 parameter description, no Ω, $\overline{\Omega}$):

$$T_f = 144 \pm 2\,\mathrm{MeV}, \quad \mu_B = 178 \pm 5\mathrm{MeV}\,, \quad \text{for S–Pb/W}\,,$$
$$T_f = 144 \pm 2\,\mathrm{MeV}, \quad \mu_B = 203 \pm 5\mathrm{MeV}\,, \quad \text{for Pb–Pb}\,,$$

Even though there is still considerable uncertainty about other freeze-out flow effects, such as longitudinal flow (memory of the collision axis), the level of consistency and quality of agreement between a wide range of experimental data and our chemical non-equilibrium, thermal equilibrium statistical model suggests that, for the observables considered, these effects do not matter.

The key results we found analyzing experimental data are:
$\tilde{\lambda}_s = 1$ for both S and Pb collisions ;
$\gamma_s^{\rm Pb} > 1$, $\gamma_q > 1$; $v_c^{\rm Pb} = 1/\sqrt{3}$; $S/B \simeq 40$ and $\bar{s}/B \simeq 0.75$.
This results are in remarkable agreement with the properties of a deconfined QGP source hadronizing without chemical reequilibration. The only natural interpretation of our findings is thus that hadronic particles seen at 158–200 GeV A nuclear collisions at CERN-SPS are emerging directly from a deconfined QGP state and do not undergo a chemical re-equilibration after they have been produced.

References

[1] J. Sollfrank, *J. Phys.* G **23**, 1903 (1997), and references therein.

[2] J. Letessier and J. Rafelski, *Phys. Rev.* C **59**, 947 (1999).

[3] J. Letessier and J. Rafelski, *Acta Phys. Pol.*; **B30**, 153 (1999), and *J. Phys.* Part. Nuc. **G25**, 295, (1999).

[4] J. Rafelski, J. Letessier and A. Tounsi, Dallas–ICHEP (1992) p. 983 (QCD161:H51:1992); [hep-ph/9711350];
J. Letessier, A. Tounsi, U. Heinz, J. Sollfrank and J. Rafelski *Phys. Rev. Lett.* **70**, 3530 (1993); and *Phys. Rev.* D **51**, 3408 (1995).

[5] E. Schnedermann, J. Sollfrank and U. Heinz, pp175–206 in *Particle Production in Highly Excited Matter*, NATO-ASI Series B303, H.H. Gutbrod and J. Rafelski, Eds., (Plenum, New York, 1993)

[6] E. Fermi, *Progr. theor. Phys.* **5** 570 (1950).

[7] J. Rafelski, GSI Report 81-6, pp.282–324, Darmstadt, May 1981; Proceedings of the Workshop on *Future Relativistic Heavy Ion Experiments*, held at GSI, Darmstadt, Germany, October 1980, R. Bock and R. Stock, Eds., (see in particular pp. 316–320); see also: J. Rafelski, pp. 619–632 in *New Flavor and Hadron Spectroscopy*, Ed. J. Tran Thanh Van (Editions Frontiers 1981), Proceedings of XVIth Rencontre de Moriond — Second Session, Les Arcs, March 1981;
J. Rafelski, *Nucl. Physics* A **374**, 489c (1982) — Proceedings of ICHEPNC held 6–10 July 1981 in Versailles, France.

[8] E. Andersen *et al.*, WA97-collaboration, *Phys. Lett.* B **433**, 209, (1998);
Strangeness enhancement at mid-rapidity in Pb-Pb collisions at 158

A GeV/c, E. Andersen *et al.*, WA97-collaboration, *Phys. Lett.* B *in press*, CERN-EP preprint, January 5, 1999.

[9] J. Rafelski, *Phys. Lett.* B **262**, 333 (1991); *Nucl. Phys.* A **544**, 279c (1992).

[10] J. Rafelski, J. Letessier and A. Tounsi, *Acta Phys. Pol.* B **27**, 1035 (1996), and references therein.

[11] J. Letessier, A. Tounsi and J. Rafelski, *Phys. Rev.* C **50**, 406 (1994); J. Rafelski, J. Letessier and A. Tounsi, *Acta Phys. Pol.* A **85**, 699 (1994).

Chapter 5

NUCLEAR COLLECTIVE EXCITATIONS AT FINITE TEMPERATURE

S. Ayik[a], D. Lacroix[b] and Ph. Chomaz[b]
[a] *Tennessee Technological University, Cookeville TN38505, USA*
[b] *G.A.N.I.L., B.P. 5027, F-14021 Caen Cedex, France*

Abstract

The nuclear collective response at finite temperature is investigated in the quantum framework of the small amplitude limit of the extended TDHF approach, including a non-Markovian collision term. By employing a Skyrme force, the isoscalar monopole, isovector dipole and isoscalar quadrupole excitations in ^{40}Ca are calculated. The collisional damping due to decay into incoherent 2p-2h states is small at low temperatures but increases rapidly at higher temperatures.

Keywords: extended TDHF, linear response, giant resonance

1. INTRODUCTION

During last several years, a systematic effort has been made to investigate the properties of giant dipole resonance built on excited states. These investigations indicate that as a function of the excitation energy or temperature, the mean resonance energy does not change much but the resonance becomes broader as observed in giant dipole resonance in ^{120}Sn [1, 2]and ^{208}Pb nuclei [3], and it appears to saturates at large excitations. On theoretical side much, work has been done to understand the properties of collective vibrations at finite temperature [4, 5]. These calculations are based on different damping mechanisms, such as the mechanism due to coupling with the surface modes [6, 7], the damping due to incoherent 2p-2h doorway states which is usually referred to as the collisional damping [8]. These calculations have been partially successful for explaining the broadening of the giant dipole resonance with increasing temperature, but the saturation is still an open problem.

The small amplitude limit of the extended TDHF provides an appropriate basis for investigating collective response, in which damping due the incoherent 2p-2h decay is included in the form of a non-Markovian collision term [9, 10, 11]. Based on this approach, the incoherent contribution to damping at finite temperature has been calculated in Thomas-Fermi approximation in refs. [12, 13]. In this work, we carry out a quantal investigation of the nuclear collective response at zero and finite temperature on the extended TDHF framework in small amplitude limit, which may be referred as an extended RPA approach [14]. In contrast to the semi-classical treatments, the shell effects are incorporated into the strength distributions as well as the collisional damping widths. We present calculations for the isoscalar monopole, isocalar quadrupole and isovector dipole strength distributions in ^{40}Ca at finite temperature by employing an effective Skyrme force.

2. COLLECTIVE RESPONSE AT FINITE TEMPERATURE

The formal basis of the extended TDHF theory has been developed some years ago [15], and it provides an effective quantal transport description for the evolution of the single particle density matrix $\rho(t)$ including the mean-field and the residual interactions in terms of a collision term. The equation of motion of the single particle density matrix determined by the first equation of the BBGKY hierarchy,

$$i\hbar\frac{\partial}{\partial t}\rho - [h(\rho), \rho] = K(\rho). \tag{5.1}$$

Here $h(\rho)$ is an effective mean-field Hamiltonian and the right hand side represents a non-Markovian collision term determined by the correlated part of the two-particle density matrix, $K(\rho) = Tr_2[v, C_{12}]$. The two-body correlations C_{12} is determined by the second equation of the BBGKY hierarchy which involves three-body correlations. In the extended TDHF, the hierarchy is truncated at the second level by neglecting the higher order correlations, and two-body correlations are calculated in terms of effective residual interactions according to

$$i\hbar\frac{\partial}{\partial t}C_{12} - [h(\rho), C_{12}] = F_{12}(\rho) \tag{5.2}$$

where the source term is given by

$$F_{12}(\rho) = (1-\rho_1)(1-\rho_2)v\widetilde{\rho_1\rho_2} - \widetilde{\rho_1\rho_2}v(1-\rho_1)(1-\rho_2). \tag{5.3}$$

We obtain a description for small density fluctuations, $\delta\rho(t) = \rho(t) - \rho_0$, by linearizing the extended TDHF theory around a finite temperature equilibrium state ρ_0 ,

$$i\hbar\frac{\partial}{\partial t}\delta\rho - [h_0, \delta\rho] - [\delta U + F, \rho_0] = Tr_2[v, \delta C_{12}] \tag{5.4}$$

where $\delta U = (\partial U/\partial\rho)_0 \cdot \delta\rho$ represents small deviations in the effective mean-field potential. The small deviation of two-body correlations $\delta C_{12}(t) = C_{12}(t) - C^0_{12}$ is specified by the second equation of the hierarchy,

$$i\hbar\frac{\partial}{\partial t}\delta C_{12} - [h_0, \delta C_{12}] - [\delta U + F, C^0_{12}] = \delta F_{12} \tag{5.5}$$

where δF_{12} is the small deviations of the source term and C^0_{12} denotes the equilibrium correlation function. In order to study the collective response of the system, we include an external harmonic perturbation of the form, $F(\mathbf{r}, t) = F(\mathbf{r})\exp(-i\omega t) + h.c.$ into the equation of motion. The two-body correlations can be determined by solving eq.(5) with the help of one-sided Fourier transform, and under a weak-damping approximation, the solution can be expressed as [13],

$$\delta C_{12}(t) = \int^t dt' \rho^0_1 \rho^0_2 e^{-ih_0(t-t')}[Q(t'), v]e^{+ih_0(t-t')}(1-\rho^0_1)(1-\rho^0_2) \tag{5.6}$$

where $Q(t)$ denotes the distortion function associated with the single particle density matrix, $\delta\rho(t) \approx [Q(t), \rho_0]$. Then, an expression for the linearized collision term can be obtained by substituting the correlation function to the right hand side of the transport eq.(4).

We analyze the linear response of the system to an external perturbation F by expanding the small deviation $\delta\rho(t)$ in terms of finite temperature RPA modes $O^\dagger_\lambda$ and O_λ,

$$\delta\rho(t) = \sum_{\lambda>0} z_\lambda(t)[O^\dagger_\lambda, \rho_0] - z^*_\lambda(t)[O_\lambda, \rho_0] \tag{5.7}$$

where $z_\lambda(t)$ and $z^*_\lambda(t)$ denote the amplitudes associated with the RPA modes, $O^\dagger_\lambda$ and O_λ, which are determined by the finite temperature RPA equation. In the Hartree-Fock basis the finite temperature RPA equation reads [16],

$$\begin{aligned}(\hbar\omega_\lambda - \epsilon_i + \epsilon_j) \quad & < i|O^\dagger_\lambda|j> + \\ & \sum_{l\neq k} < ik|v|jl>_A (n_l - n_k) < l|O^\dagger_\lambda|k> = 0\end{aligned} \tag{5.8}$$

where $v = (\partial U/\partial\rho)_0$, the indices $i, j, ..$ represent all single particle quantum numbers including spin-isospin, and $n_k = 1/[1+\exp(\epsilon_k - \epsilon_F)/T]$

denotes the finite temperature Fermi-Dirac occupation numbers of the Hartree-Fock states. At zero temperature, occupation numbers are zero or one, so that the RPA functions $O_\lambda^\dagger$, O_λ have only particle-hole and hole-particle matrix elements. At finite temperatures the RPA functions involve more configurations including particle-particle and hole-hole states. Substituting the expansion (7) into eq.(4) and projecting by O_λ, we find that the amplitudes of the RPA modes execute forced harmonic motion,

$$\frac{d}{dt}z_\lambda + i\omega_\lambda z_\lambda - \frac{i}{\hbar} < [O_\lambda, F] >_0 = -\frac{1}{2\hbar}\int^t dt'\bar{\Gamma}_\lambda(t-t')z_\lambda(t') \qquad (5.9)$$

where the deriving term is determined by the external perturbation $< [O_\lambda, F] >_0 = Tr[O_\lambda, F]\rho_0$ and the right hand side describe a non-Markovian damping term. Fourier transform of $\bar{\Gamma}_\lambda(t)$ is the collisional damping width due to mixing with the incoherent 2p-2h states. In the Hartree-Fock representation, it is given by [13],

$$\Gamma_\lambda(\omega) = \frac{1}{2}Im\sum\frac{|< ij|[O_\lambda, v]|kl >_A|^2}{\omega - \Delta\epsilon - i\eta}[n_k n_l \bar{n}_i \bar{n}_j - n_i n_j \bar{n}_k \bar{n}_l] \qquad (5.10)$$

where $\Delta\epsilon = \epsilon_i + \epsilon_j - \epsilon_k - \epsilon_l$, $\bar{n}_k = 1 - n_k$ and η is a small positive number. In this expression, we neglect a small shift of the frequency ω_λ arising from the principle value part of the damping term.

Solving eq.(9) for the amplitudes by Fourier transform, the response of the system to the external perturbation F can be expressed as

$$\delta\rho(\omega) = R(\omega, T)\cdot F \qquad (5.11)$$

where $R(\omega, T)$ denotes the finite temperature extended RPA response function including damping. The strength distribution of the RPA response is obtained by the imaginary part of the response function,

$$\begin{aligned} S(\omega, T) &= -\tfrac{1}{\pi}Tr\{\ F^\dagger ImR(\omega, T)F\} \qquad (5.12)\\ &= \tfrac{1}{\pi}\sum_{\lambda>0}\ \{|< [O_\lambda, F] >_0|^2 D(\omega - \omega_\lambda) - \\ &\qquad |< [O_\lambda^\dagger, F] >_0|^2 D(\omega + \omega_\lambda)\Big\} \end{aligned}$$

where the sum goes over the positive frequency modes and

$$D(\omega - \omega_\lambda) = \frac{\Gamma_\lambda/2}{(\hbar\omega - \hbar\omega_\lambda)^2 + (\Gamma_\lambda/2)^2}. \qquad (5.13)$$

In reference [13], neglecting the depletion of collective amplitudes in the collision term, the damping width of RPA modes are calculated in

Thomas-Fermi approximation by substituting $\omega = \omega_\lambda$ in eq.(10). Here, we take into account for depletion of the collective amplitude in the collision term, and calculate the damping width by substituting $\omega = \omega_\lambda - \frac{i}{2}\Gamma_\lambda$. Then, the expression becomes a secular equation for the damping width.

3. RESULTS AND CONCLUSIONS

We calculate the isoscalar monopole, isoscalar quadrupole and isovector dipole excitations in ^{40}Ca at several temperatures. We use the Skyrme interaction SGII for the Hartree-Fock and RPA calculations [17] and we neglect the temperature dependence of single particle energies and wave functions. We determine the hole states by solving the Hartree-Fock problem in coordinate representation. Then, the particle states are generated by diagonalizing the Hartree-Fock Hamiltonian in a large harmonic oscillator representation by including 12 major shells. The RPA strength distributions of the monopole $F_0(r) = r^2$, dipole $F_1(\mathbf{r}) = \tau_z r Y_{10}(\hat{\mathbf{r}})$ (in isospin symmetric systems $N = Z$), and quadrupole $F_2(\mathbf{r}) = r^2 Y_{20}(\hat{\mathbf{r}})$ excitation operators at temperatures $T = 0, 2, 4$ MeV are shown in figure 5.1. As seen from the top panel, the monopole strength at $T = 0$ MeV exhibit a large Landau spreading over a broad energy region $E = 16-28$ MeV with an average energy $E = 21.5$ MeV. The recent experimental data also show a broad resonance around a peak value of 17.5 MeV[18]. For increasing temperature, transition strength spread a broader range towards lower energies. As shown in the middle panel, the strength distribution of isovector dipole shows a weaker temperature dependence than monopole. At $T = 0$, the dipole strength is concentrated at range $E = 16-23$ MeV. The Landau width is large and is spreading for increasing temperature. However, the average energy of the main peak remains nearly constant around $E = 16.5$ MeV. The experimental data shows a broad resonance at around 20 MeV [19] with a width close to 6 MeV. As illustrated at the bottom panel of figure 5.1, the RPA result at $T = 0$ MeV gives a very collective quadrupole mode peaked at $E = 17.5$ MeV, which agrees well the experimental finding of an average energy 17 MeV and the calculations of Sagawa and Bertsch [20]. At higher temperatures in addition to p-h excitations, p-p and h-h excitations become possible. The p-p and h-h configurations mainly change the strength distribution at low energy side at $E = 4$ MeV. As a result, the giant resonance has less transition strength.

We obtain the collisional damping widths of the collective states by calculating the expression (10) and solving the associated secular equation by graphical method. We perform the sums over single particle

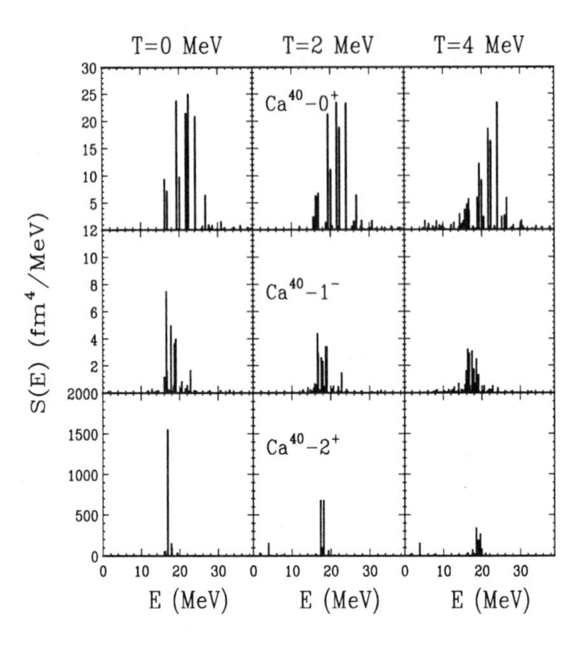

Figure 5.1 RPA strength distributions in Ca^{40} as a function of the energy at temperatures $T = 0, 2, 4$ MeV for isoscalar monopole (top), isovector dipole (middle) and isoscalar quadrupole (bottom) excitations.

states explicitly using the projection of the total spin, m, as one of the explicit quantum numbers as done in [21]. Figure 5.2 shows the damping widths as a function of temperature, that are averaged over several nearby states with strengths more than 10% of the EWSR. The results for monopole, dipole and quadrupole are indicated in the top, middle and bottom panels, respectively. The incoherent damping widths at low temperatures are, in general, small and thus leaving room for a possible coherence effect of doorway states in the description of the damping properties. In particular in the case of dipole mode, since there is no odd parity 2p-2h states available in the vicinity of the collective energy, the collisional damping width vanishes at $T = 0$ MeV. This behavior is a particular quantum feature due to shell effects in the extended RPA calculation of double magic light nuclei, and it can not be described in the framework of semi-classical approaches. For increasing temperature the collisional damping becomes large and may even dominate the spreading width since the coherence effect is expected to diminish rather rapidly. Temperature dependence of the damping width in quantal calculations is a more complex than the simple quadratic increase predicted by the semi-classical calculations. An interesting property of the collisional damping is that it may saturate for increasing temperature. In fact, our calculations indicate that the damping width of giant quadrupole saturates around $T = 3-4$ MeV, however a saturation of the giant monopole and dipole modes is not visible at these temperatures.

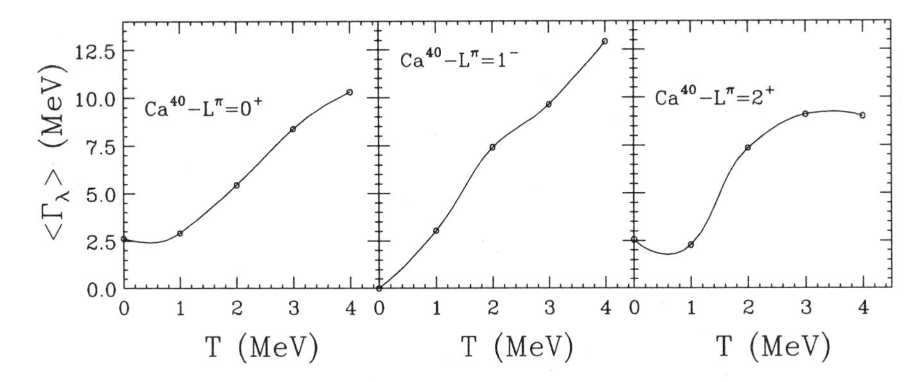

Figure 5.2 Collisional damping widths that are averaged over nearby states with more than 10% of the EWSR, for monopole, dipole and quadrupole modes as a function of temperature.

Figure 5.3 shows the strength distributions including the collisional damping. The giant dipole strength at $T = 0$ MeV is smoothed by performing an averaging with a Lorentzian weight with a width of 0.5 MeV. The excitation strengths become broader for increasing temperature. The peak position of the monopole resonance does not change much, but the peak position of dipole slightly shifts down and quadrupole slightly shifts up in energy. This is a signature of the reduction of the collectivity of those states with temperature because the peak energy moves back towards the single particle expectations.

There are important quantal effects in the collective behavior of a hot nuclear system as illustrated in [22]. Investigations presented here, also indicates that, the quantal effects has a large influence on the damping properties of collective excitations at low temperatures, which may even persist at relatively high excitations. The magnitude of the collisional damping is rather sensitive to the effective residual interactions, for which an accurate information is not available. The effective Skyrme force is well fitted to describe the nuclear mean-field properties, but not the in-medium cross-sections and damping properties. Therefore, a systematic study of the effective interactions in this context is clearly called for. However, our investigation, while remains semi-quantitative, gives a valuable insight on the quantal properties of collective excitations at finite temperature.

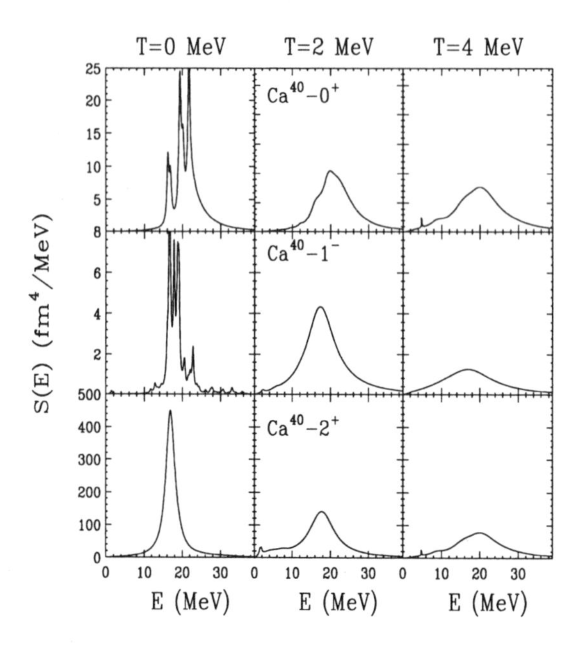

Figure 5.3 Extended RPA strength distributions in Ca^{40} as a function of the energy at temperatures $T = 0, 2, 4$ MeV for isoscalar monopole (top), isovector dipole (middle) and isoscalar quadrupole (bottom) excitations.

Acknowledgments

One of us (S. A.) gratefully acknowledges GANIL Laboratory for a partial support and warm hospitality extended to him during his visits to Caen. This work is supported in part by the US DOE grant No. DE-FG05-89ER40530.

References

[1] Ramakrishan, E., et al. (1996) Phys. Rev. Lett. **76**, 2025.

[2] Ramakrishan, E., et al. (1996) Phys. Lett. **B383**, 252.

[3] Hofmann, H.J., et al. (1994) Nucl. Phys. **A571**, 301.

[4] Bertsch, G.F. and Broglia, R.A. (1994).*Oscillation in finite quantum systems*, Cambridge.

[5] Chomaz, Ph. (1996) Ann. Phys. Fr. **21**, 669.

[6] Ormand, W.E., Bortignon, P.F. and Broglia, R.A. (1996) Phys. Rev. Lett. **77**, 607.

[7] Dang, N.D. and Arima, A. (1998) Phys. Rev. Lett.**80**, 4145.

[8] Smerzi, A., Bonasera, A. and DiToro, M. (1991) Phys. Rev. **C44**, 1713.

[9] Ando, K., Ikeda, A and Holzwarth, G. (1983) Z. Phys. **A310**, 223.

[10] Ayik, S. and Dworzecka, M. (1985) Phys. Rev. Lett. **54**, 534; (1985) Nucl. Phys. **A440**, 424.

[11] Reinhard, P.G., Yadav, H.L. and Toepler, C. (1986) Nucl. Phys. **A458**, 301.

[12] Belkacem, M, Ayik, S. and Bonasera, A. (1995) Phys. Rev. **C52**, 2499.

[13] Ayik, S., Yilmaz, O., Gokalp A. and Schuck, P. (1998) Phys. Rev. **C 58**, 1594 .

[14] Lacroix, D., Chomaz, Ph. and Ayik, S. (1998) Phys. Rev. **C 58**, 2154.

[15] Goeke, K. and Reinhard, P.-G. (1982). *Time-Dependent Hartree-Fock and Beyond.* Proceedings, Bad Honnef, Germany.

[16] Vautherin, D and Vinh Mau, N. (1983) Phys. Lett. **B120**, 261; (1984) Nucl. Phys. **A422**, 140.

[17] Van Giai, N. and Sagawa, H. (1981) Nucl. Phys. **A371**, 1.

[18] Youngblood, D.H., Lui, Y.-W. and Clark, H.L. (1997) Phys. Rev. **C55**, 2811.

[19] Berman, B.L. and Fultz, F.C. (1975) Rev. Mod. Phys. **47**, 713.

[20] Sagawa, H. and Bertsch, G.F. (1984) Phys. Lett. **B146**, 138.

[21] Catara, F., Chomaz, Ph. and Van Giai, N. (1989) Phys. Lett. **233B**, 6.

[22] Lacroix, D. and Chomaz, Ph. (1998) Nucl. Phys. **A639**, 85.

Chapter 6

IN-MEDIUM KAON PRODUCTION IN HEAVY ION COLLISIONS AT SIS ENERGIES

C. Fuchs
Institut für Theoretische Physik
Universität Tübingen,
D-72076 Tübingen, Germany
christian.fuchs@uni-tuebingen.de

Abstract The influence of the chiral mean field on the collective motion of kaons in relativistic heavy ion reactions, such as transverse flow, out-of-plane flow (squeeze-out) and radial flow, is investigated. The considerd energy range (SIS energies) is 1-2 GeV/nucleon. The kaon dynamics is described with relativistic mean fields originating form chiral lagrangiens. We adopt a covariant quasi-particle picture including scalar and vector fields and compare this to a treatment with a static potential like force. The comparison to the available data (K^+) measured by FOPI and KaoS strongly favors the existence of an in-medium potential. However, using full covariant dynamics makes it more difficult to describe the data which indicates that the mean field level is not sufficient for the description of the in-medium kaon dynamics.

Keywords: Relativistic heavy ion reactions, kaon production, chiral mean field

1. INTRODUCTION

In recent years strong efforts have been made towards a better understanding of the medium properties of kaons in dense hadronic matter. This feature is of particular relevance since the kaon mean field is related to chiral symmetry breaking [1]. The in-medium effects give rise to an attractive scalar potential which is in first order proportional to the kaon-nucleon Sigma term Σ_{KN}. A second part, stemming form the interaction with vector mesons [1, 2, 3], is repulsive for K^+ and, due to G-parity conservation, attractive for antikaons K^-. Experimental groups have extensively searched for signatures of these kaon-nucleus potentials in heavy ion reactions at intermediate energies [5, 6, 7]. However, the theoretical uncertainties of the knowledge of the elementary production cross sections are still too large to draw definite conclusions from experimental yields. The consideration of K^+/K^- ratios [4, 9] is one way to circumvent this problem since such uncertainties - at least partially - cancel out. Another alternative reported in the present work is the study of the in-medium kaon dynamics. It has been found that in particular the collective motion of kaons in the dense hadronic environment is sensitive to medium effects [4, 8, 9, 10, 11, 12]. Three types of collective motion in heavy ion reactions, namely the radial flow [12], the emission out of the reaction plane (squeeze-out) [13] and the in-plane flow (transverse flow) [8, 10, 14] are discussed. The interaction of the kaons with the dense hadronic medium is described on the mean field level based on chiral models [1, 8] including also effects which originate from a fully covariant treatment and are not included in standard approaches with static potentials.

2. COVARIANT KAON DYNAMICS

From the chiral Lagrangian [1] the field equations for the $K^{\pm}$–mesons are derived from the Euler-Lagrange equations [8]

$$\left[\partial_\mu \partial^\mu \pm \frac{3i}{4f_\pi^2} j_\mu \partial^\mu + \left(m_{\rm K}^2 - \frac{\Sigma_{\rm KN}}{f_\pi^2}\rho_{\rm s}\right)\right] \phi_{{\rm K}^\pm}(x) = 0 \quad , \qquad (6.1)$$

where $\rho_{\rm s}$ is the baryon scalar density and j_μ the baryon four-vector current. Introducing the kaonic vector potential $V_\mu = 3/(8f_\pi^2)j_\mu$ Eq. (6.1) can be rewritten in the form [14]

$$\left[(\partial_\mu \pm iV_\mu)^2 + m_{\rm K}^{*2}\right] \phi_{{\rm K}^\pm}(x) = 0 \quad . \qquad (6.2)$$

Thus, the vector field is introduced by minimal coupling into the Klein-Gordon equation. The effective mass $m_{\rm K}^*$ of the kaon is then given by [3, 14]

$$m_{\rm K}^* = \sqrt{m_{\rm K}^2 - \frac{\Sigma_{\rm KN}}{f_\pi^2}\rho_{\rm s} + V_\mu V^\mu} \quad . \qquad (6.3)$$

This effective quasi-particle mass (6.3) is a Lorentz scalar and is equal for K^+ and K^-. It should not be mixed up with the quantity, i.e. kaon energy at zero momentum $\omega(\mathbf{k}=0) = m_{\rm K}^* \pm V_0$ for $K^{\pm}$, which is sometimes denoted as in-medium mass [2, 8, 9, 15] and which determines the shift of the corresponding production thresholds (see discussion after Fig. 2). These two quantities, namely the energy at zero momentum and the in-medium quasiparticle mass $m_{\rm K}^*$ are compared in Fig.6.1. For the

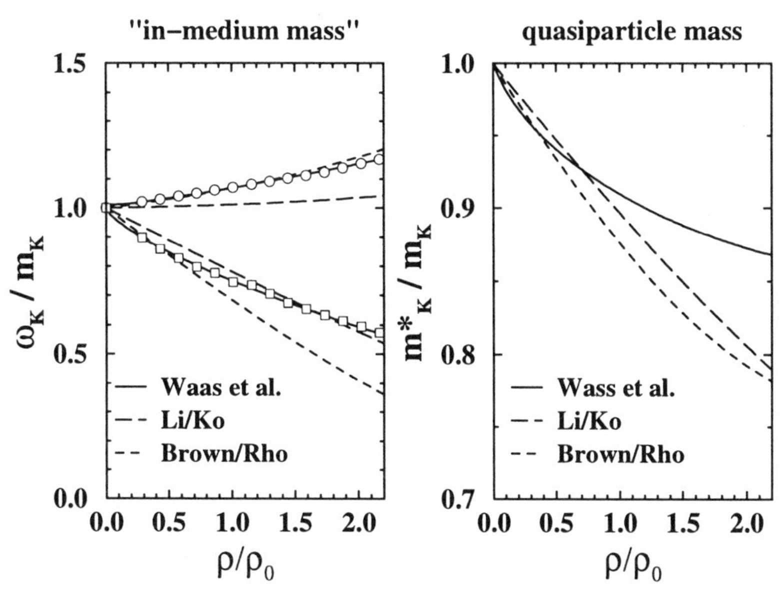

Figure 6.1 In-medium kaon mass in nuclear matter for various mean fields derived from chiral lagrangiens. The right panel shows the kaon energy at zero momentum which is often denoted as in-medium mass in the literature (the upper lines refer to K^+, the lower ones to K^-). The right panel shows the quasi-particle mass given by Eq. (6.3) which is equal for K^+ and K^-.

modification of the kaon in-medium properties we parameterize results obtained form coupled channel calculation in chiral perturbation theory [15] (ChPT) and to two more simple mean field models suggested by Li and Ko [8] (MF) as well as by Brown and Rho [2] (MF2). The quasi-particle mass $m^*_{\rm K}$ is equal for K^+ and K^- and generally reduced inside the nuclear medium. However, in the approach of Ref. [15] (Waas et al.) this reduction is much weaker than in the simple MF parameterization where the scalar field is in first order proportional to the scalar nucleon density.

Introducing an effective momentum $k^*_\mu = k_\mu \mp V_\mu$ for $K^+(K^-)$, the Klein-Gordon equation (6.2) reads in momentum space

$$\left[k^{*2} - m^{*2}_{\rm K}\right] \phi_{{\rm K}^\pm}(k) = 0 \tag{6.4}$$

which is just the mass-shell constraint for the quasi-particles inside the nuclear medium. The quasi-particles can now be treated like free particles. In nuclear matter at rest the spatial components of the vector potential vanish, i.e. $\mathbf{V} = 0$, and Eq. (6.2) reduces to the expression already given in Ref. [8]. The covariant equations of motion for the kaons obtained in the classical (testparticle) limit are analogous to the corresponding relativistic equations for baryons and read

$$\frac{dq^\mu}{d\tau} = \frac{k^{*\mu}}{m_K^*} \quad , \quad \frac{dk^{*\mu}}{d\tau} = \frac{k_\nu^*}{m_K^*} F^{\mu\nu} + \partial^\mu m_K^* \quad . \tag{6.5}$$

Here $q^\mu = (t, \mathbf{q})$ are the coordinates in Minkowski space and $F^{\mu\nu} = \partial^\mu V^\nu - \partial^\nu V^\mu$ is the field strength tensor for K^+. For K^- where the vector field changes sign the equation of motion are identical, however, $F^{\mu\nu}$ has to be replaced by $-F^{\mu\nu}$. The structure of Eqs. (6.5) becomes more transparent considering only the spatial components

$$\frac{d\mathbf{k}^*}{dt} = -\frac{m_K^*}{E^*}\frac{\partial m_K^*}{\partial \mathbf{q}} \mp \frac{\partial V^0}{\partial \mathbf{q}} \mp \partial_t \mathbf{V} \pm \frac{\mathbf{k}^*}{E^*} \times \left(\frac{\partial}{\partial \mathbf{q}} \times \mathbf{V}\right) \tag{6.6}$$

where the upper (lower) signs refer to K^+ (K^-). The last term in Eq. (6.6) gives rise to an additional momentum dependence. Such a velocity dependent ($\mathbf{v} = \mathbf{k}^*/E^*$) Lorentz force is a genuine feature of relativistic dynamics as soon as a vector field is involved. If the equations of motion are, however, derived from a static potential [8, 4, 9]

$$U(\mathbf{k}, \rho) = \omega(\mathbf{k}, \rho) - \omega_0(\mathbf{k}) = \sqrt{\mathbf{k}^2 + m_K^{*2}} \pm V_0 - \sqrt{\mathbf{k}^2 + m_K^2} \tag{6.7}$$

the Lorentz-force contribution is missing. The same holds for non-relativistic approaches [10, 12]. However, the kaon production is not affected by the spatial components since the corresponding thresholds are only shifted by the static part U.

The influence of the K^+ in-medium potential on the yields is demonstrated in Fig.6.1 for the reaction $C+C$ at 2.0 A.GeV. To compare to the KaoS data [16] a $\Theta_{\text{Lab}} = 40^o \pm 4^o$ polar angular cut was applied. We distinguish three different cases: No medium effects at all, the potential is only included in the propagation, and finally a full calculation where the potential is included in the threshold as well as in the propagation. Without medium effects the low p_t-region of the spectrum is well reproduced but the high p_t part is underpredicted. On the other hand, the potential entering into the threshold strongly suppresses the low energetic kaons. If the potential also acts on the propagation it makes the K^+ spectrum harder which is due to the repulsive forces. However, the uncertainty in the description of the cross section is still too large in order to draw definite conclusions on the importance of medium effects. Although our calculation is in reasonable agreement with the results of other groups [4, 9], there remain still discrepancies which reflect the uncertainties in the theoretical knowledge of the elementary production cross sections.

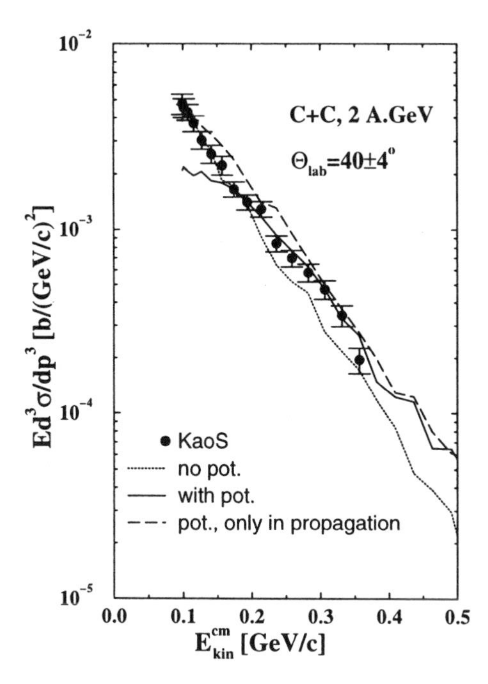

Figure 6.2 Invariant cross section of the K^+ production in C+C collisions at 2 A.GeV. The calculations are performed without any medium effects (dotted), with full medium effects (solid), and including the in-medium potential only for the propagation of the kaons but neglecting it in the production thresholds (long-dashed). A $\Theta_{\rm lab} = 40 \pm 4^o$ polar angular cut has been applied in order to compare to the corresponding KaoS data taken from Ref. [16].

3. COLLECTIVE FLOW OF KAONS

3.1 RADIAL FLOW

One way to obtain information on the collective motion is to investigate particle multiplicities as a function of the transverse mass m_t [12]. In the left panel of Fig.6.3 the transverse mass spectrum of K^+ mesons emitted at midrapidity (-0.4<$y_{c.m.}/y_{proj}$<0.4) is shown for a central (b=3 fm) Au+Au reaction at 1 A.GeV. In the presence of a kaon potential (MF2, taken from [2]) the kaon m_t spectrum clearly exhibits a "shoulder-arm" shape which deviates from a pure thermal picture. On the other hand the calculation without any potential can be well described by a Boltzmann distribution, which is more or less a straight line if plotted logarithmically. Thus, the "shoulder-arm" structure is caused by the mean field rather than by a collective expansion of the kaon sources. The K^+ mesons experience an acceleration due to the repulsive potential as they propagate outwards from the participant region. To extract the collective component from the kaon spectrum, the QMD results can by fitted by a Boltzmann distribution including a common radial expansion [12]. The fit parameters are the temperature T and the common radial velocity β. With $E = m_t \cosh y$, $p = \sqrt{{p_t}^2 + {m_t}^2 sinh^2 y}$,

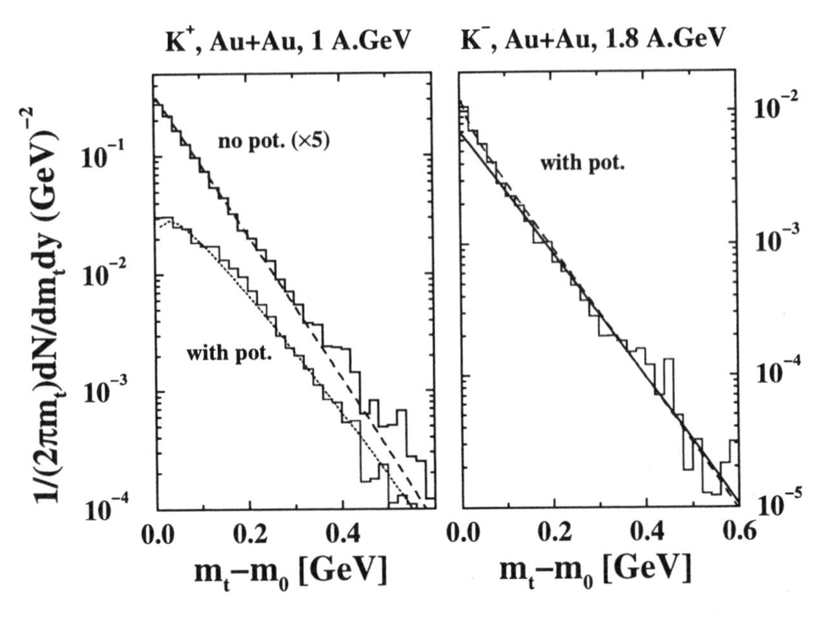

Figure 6.3 Influence of the in-medium potential on the kaon m_t spectrum. The left panel shows two calculations for K^+ with and without repulsive potential for a central (b=3 fm) Au+Au reaction at 1 A.GeV. For a better representation the results without potential are multiplied by a factor of 5. The lines are Boltzmann fits to the calculations including (lower curve) and without (upper curve) a radial flow. The right panel shows the results for K^- including the attractive potential in a semi-central (b=5 fm) Au+Au reaction at 1.8 A.GeV. The straight line is a Boltzmann fit without radial flow, the dashed line includes a virtual, inverse radial flow fit.

$\alpha = \gamma\beta p/T$, $\gamma = (1-\beta^2)^{-1/2}$ the distribution reads

$$\frac{d^3N}{d\phi dy m_t dm_t} \sim \tag{6.8}$$
$$e^{-(\frac{\gamma E}{T}-\alpha)}\{\gamma^2 E - \gamma\alpha T(\frac{E^2}{p^2}+1) + (\alpha T)^2\frac{E}{p^2}\}\frac{\sqrt{(\gamma E-\alpha T)^2 - m^2}}{p}$$

The fit yields $\beta = 0.11$ and $T = 62 MeV$. The attractive potential for K^- leads on the other hand to an enhancement of low energetic particles and a concave spectrum seen on the right panel of Fig.6.3. Turning the sign of the flow velocity β in Eq. (6.8) the calculation can also be described by a radially imploding source, i.e. by a virtual radial flow [12]. Thus the effect of the kaon potential on the collective motion can clearly be distinguished from the thermal contribution.

3.2 SQUEEZE-OUT

At SIS energies shadowing dominates the reaction of nucleons and pions and leads to an enhanced particle emission out of the reaction plane (squeeze-out). For K^+ mesons the shadowing effect is small due to the moderate absorption cross section $\sigma_{K^+N} \approx 10$ mb. The absorption cross for the K^- is much larger, $\sigma_{K^-N} \approx 50$ mb, and thus the K^- dynamics will be dominated by shadowing effects [13]. In Fig.6.4 the squeeze-out

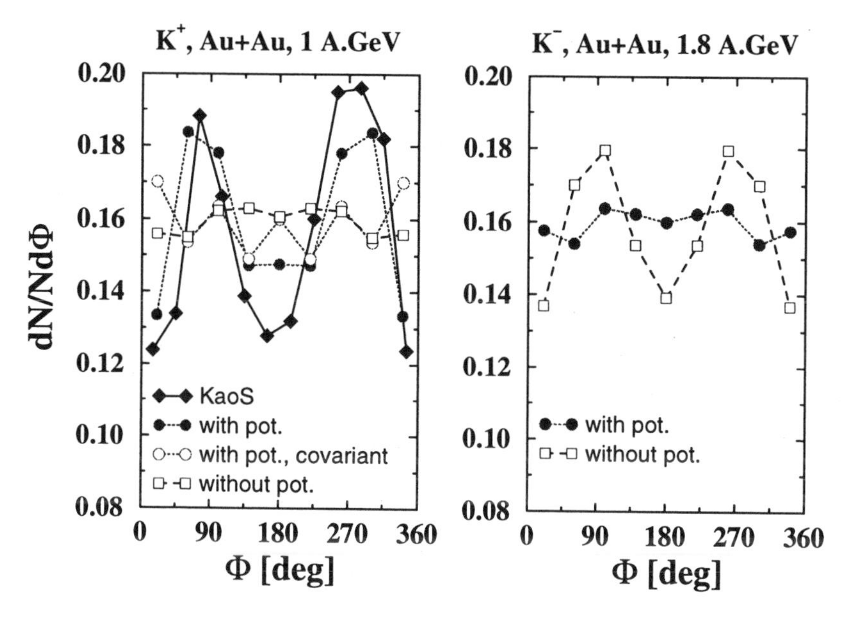

Figure 6.4 Influence of the in-medium potential on the kaon squeeze-out. The left panel compares calculations for K^+ for a semi-central (b=6 fm) Au+Au reaction at 1 A.GeV to the KaoS data (diamonds) from [7]. The calculations are performed including (full circles) and without (squares) the static kaon potential and using full covariant kaon dynamics, i.e. the kaon potential with Lorentz forces (open circles). The right panel shows calculations for K^- for a semi-central (b=8 fm) Au+Au reaction at 1.8 A.GeV without (squares) and including the kaon potential (full circles).

for mid-rapidity ($-0.2 < (Y/Y_{proj})^{cm} < 0.2$) K^+ mesons in a semi-central (b=6 fm) Au+Au reaction at 1 A.GeV incident energy is shown. In addition a transverse momentum cut of $P_T > 0.2$GeV/c has been applied in order to compare with the KaoS data [7]. The QMD calculations are performed for three different cases: (1) without any in-medium effects, (2) including the K^+ potential $U(\rho, \mathbf{k})$, Eq. (6.7), but neglecting the space-like components of the repulsive vector potential $\mathbf{V}$ and (3) with covariant in-medium dynamics, i.e. retaining also the space-like vector contribution in Eq. (6.6). The potential is taken from Ref. [2] (MF2).

The enhanced out-of-plane emission of K^+ mesons is mainly a result of the kaon potential. Without any medium effects the K^+ emission is nearly azimuthally isotropic. Similar effects were also found be the Stony Brook group [7]. However, the influence of the space-like components of the repulsive vector potential destroys the preferential emission of K^+ mesons out of the reaction plane and thus also the agreement with the data. Concerning K^- mesons the situation is just opposite. In the left panel of Fig.6.4 the azimuthal distributions of K^-s emitted at midrapidity (-0.2 $< (Y/Y_{proj})^{cm} <$ 0.2) in a semi-peripheral (b=8 fm) Au+Au reaction at 1.8 A.GeV are shown using a $P_T >$ 0.5 GeV cut. The K^-s are strongly scattered or absorbed in the nuclear medium. If there is no in-medium potential the emission at midrapidity exhibits an out-of-plane preference much like pions [17]. Now the in-medium potential (the space-like components of the attractive vector part are again neglected) reduces dramatically the out-of-plane K^- abundance ($\phi = 90^0$ and 270^0), and leads thereby to a nearly isotropic emission. Hence, the medium effects act opposite on K^+ and K^- – at least in the non-covariant description – and comparing the out-of-plane emission for both types of mesons should yield rather conclusive information form the experimental side.

3.3 TRANSVERSE FLOW

In Fig.6.5 the transverse flow of K^+ mesons in Ni+Ni collisions at 1.93 A.GeV is compared to the FOPI data [5]. The results are obtained for impact parameters $b \leq 4$ fm and with a transverse momentum cut $P_T/m_K > 0.5$. Without medium effects a clear flow signal is observed which reflects the transverse flow of the primary sources of the K^+ production. The dominantly repulsive character of the in-medium potential (6.7) tends to push the kaons away from the spectator matter which leads to a zero flow around midrapidity. The situation changes, however, dramatically when the full Lorentz structure of the mean field is taken into account according to Eq. (6.6). The influence of the repulsive part of the potential, i.e. the time-like component, on the in-plane flow is almost completely counterbalanced by the velocity dependent part of the interaction. Hence, no net effect of the potential is any more visuable. This feature is rather independent on the actual strength of the potential as can be seen from the right panel of Fig.6.5 where the same calculation is performed for Ni+Ni at 1.93 A.GeV and an fixed impact parameter b=3 fm. Although the two types of potentials vary considerably in the strength which leads to significantly different results in the case of the static-potential approach (top), this effect is completely

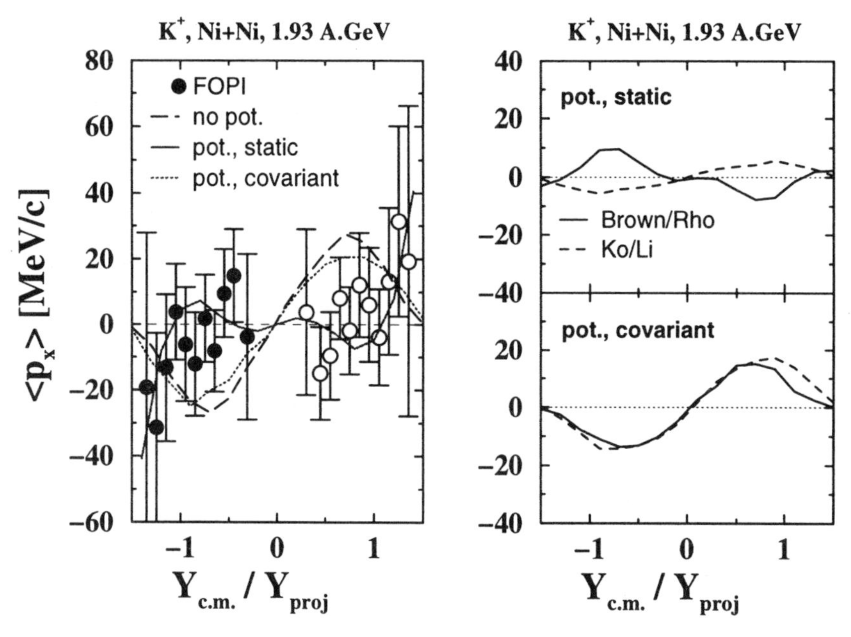

Figure 6.5 Influence of the in-medium potential on the K^+ transverse flow. The left panel compares calculations for K^+ for a (b¡4 fm) Ni+Ni reaction at 1.93 A.GeV to the FOPI data [5]. The calculations are performed including (solid) and without (dashed) the kaon potential and using full covariant kaon dynamics, i.e. the kaon potential with Lorentz forces (dotted). The right panel shows the flow for the same reaction at b=3 fm for two types of kaon mean fields, again with full covariant kaon dynamics (bottom) and using only the static potential part (top).

counterbalanced by the Lorentz-force contribution included in the lower figure.

4. SUMMARY

Mean fields based on chiral perturbation theory were applied to study the kaon dynamics in heavy ion collisions. A covariant quasi-particle formulation was used. As a consequence of the full relativistic dynamics a Lorentz-force term appears which is missing in a description with a potential-like force only. In agreement with other works we find that the collective motion of kaons is strongly influenced by the mean field in dense matter. The comparison with data for the K^+ squeeze-out and the K^+ transverse flow strongly favors the existence of such in-medium potentials as predicted form the chiral models. However, applying the full relativistic dynamics the description of the data is more difficult since we observe strong cancellation effects connected to the Lorentz-force contribution. Thus we conclude that the present mean field description might be too simple and higher order terms from the chiral expansion should be taken into account as well.

References

[1] Kaplan, B.D., and Nelson, A.E. (1986) Phys. Lett. **B 175** 57; (1987) Phys. Lett. **B 192** 193.

[2] Brown, G.E., and Rho, M. (1996) Nucl. Phys. **A 596** 503.

[3] Schaffner, J. , Bondorf, J., Mishustin, I,N. (1997) Nucl. Phys. **A 625** 325.

[4] Li, G.Q., Lee, C.-H., Brown, G.E. (1997) Nucl. Phys. **A 625** 372; (1997) Phys. Rev. Lett. **79** 5214.

[5] Ritman, J.L., and the FOPI Collaboration (1995) Z. Phys. **A 352** 355.

[6] Barth, R., and the KaoS Collaboration (1997) Phys. Rev. Lett. **78** 4007.

[7] Shin, Y., and the KaoS Collaboration (1998) Phys. Rev. Lett. **81** 1576.

[8] Li, G.Q., Ko, C.M., and Li, Bao-An (1995) Phys. Rev. Lett. **74** 235; Li, G.Q., Ko, C.M. (1995) Nucl. Phys. **A 594** 460.

[9] Cassing, W., and Bratkovskaya, E.L. (1999) Phys. Reports **308** 65.

[10] Wang, Z.S., Faessler, Amand, Fuchs, C., Uma Maheswari, V.S., Kosov, D.S. (1997) Nucl. Phys. **A 628** 151.

[11] Ko, C.M., Li, G.Q. (1996) J. Phys. **G 22** 1673.

[12] Wang, Z.S., Faessler, Amand, Fuchs, C., Uma Maheswari, V.S., and Waindzoch, T. (1998) Phys. Rev. **C 57** 3284.

[13] Wang, Z.S., Fuchs, C., Faessler, Amand, Gross-Boelting, T. (1998) nucl-th/9809043

[14] Fuchs, C., Kosov, D.S., Faessler, Amand, Wang, Z.S. and Waindzoch, T. (1998) Phys. Lett. **B 434** 245.

[15] Waas, T., Kaiser, N., and Weise, W. (1996) Phys. Lett. **B 379** 34.

[16] Laue, F., and the KaoS Collaboration (1999) Phys. Rev. Lett. in press.

[17] Brill, D., and the KaoS Collaboration (1993) Phys. Rev. Lett. **71** 336.

Chapter 7

LEADING SOFT GLUON PRODUCTION IN HIGH ENERGY NUCLEAR COLLISIONS

Xiaofeng Guo
Department of Physics and Astronomy
University of Kentucky
*Lexington, KY 40506, USA**

*This work is supported in part by the U.S. Department of Energy under Grant Nos. DE-FG02-93ER40764 and DE-FG02-96ER40989.

Abstract The leading soft gluon p_T distribution in heavy ion collisions was obtained by Kovner, McLerran, and Weigert after solving classical Yang-Mills equations. I show explicitly this result can be understood in terms of conventional QCD perturbation theory. I also demonstrate that the key logarithm in their result represents the logarithm in DGLAP evolution equations.

Keywords: RHIC, quarks, gluons, minijet, nuclear collisions

1. INTRODUCTION

In ultra-relativistic heavy ion collisions, after initial collisions, the parton system will be dominated by gluons [1]. Understanding the distribution of soft gluons formed in the initial stage of the collision is particularly interesting and important for studying the formation of quark-gluon plasma. In terms of conventional QCD perturbation theory, a calculable cross section in high energy hadronic collisions is factorized into a single collision between two partons multiplied by a probability to find these two partons of momentum fractions x_1 and x_2, respectively, from two incoming hadrons. The probability is then factorized into a product of two parton distributions $\phi(x_1)$ and $\phi(x_2)$, which are probabilities to find these two partons from the respective hadrons [2]. Because of extremely large number of soft gluons in heavy ion beams, it is natural to go beyond the factorized single-scattering formalism to include any possible multiple scattering, and long range correlations between soft gluons from two incoming ions.

McLerran and Venugopalan (MV) developed a new formalism for calculation of the soft gluon distribution for very large nuclei [3, 4]. In this approach, the valence quarks in the nuclei are treated as the classical source of the color charges. They argued that the valence quark recoil can be ignored in the limit when the gluons emitted are soft. In addition, because of the Lorentz contraction, the color charge of the valence quarks is treated approximately as an infinitely thin sheet of color charge along the light cone. With these assumptions, the gluon distribution function for very large nuclei may be obtained by solving the classical Yang-Mills Equation [4, 5]. Using the classical glue field generated by a single nucleus obtained in the MV formalism as the basic input, Kovner, McLerran, and Weigert (KMW) computed the soft gluon production in a collision of two ultra-relativistic heavy nuclei by solving the classical Yang-Mills equations with the iteration to the second order [6]. The two nuclei are treated as the infinitely thin sheets of the classical color charges moving at the speed of light in the positive and the negative z directions, respectively. Following this approach, the distribution of soft gluons at the rapidity y and the transverse momentum p_T in nuclear collisions can be expressed as [6, 7]

$$\frac{d\sigma}{dyd^2p_T} = \frac{2g^6}{(2\pi)^4}\left(\frac{N_q}{2N_c}\right)^2 N_c(N_c^2-1)\,\frac{1}{p_T^4}\,\ell n\left(\frac{p_T^2}{\Lambda_{cutoff}^2}\right) . \tag{7.1}$$

where g is the strong coupling constant, $N_c = 3$ is the number of the color, and Λ_{cutoff} is a cutoff mass scale [6]. In Eq. (7.1), $N_q/(2N_c) = S_T\mu^2$, where μ^2 is the averaged color charge squared per unit area of the valence quark, and S_T is the transverse area of the nuclei. N_q is the total number of the quarks in the color charge source. Eq. (7.1)) is potentially very useful in estimating the production of mini-jet rates, and the formation of the possible quark-gluon plasma at RHIC [8].

In the following, I show that this result can be understood in terms of QCD perturbation theory. I will also explore under what kind of approximation this result matches the conventional perturbative calculation [9].

2. PARTONIC PROCESS $QQ \to QQG$

KMW's derivation for Eq. (7.1) is based on the following physical picture: in ultra-relativistic heavy ion collisions, gluons are produced

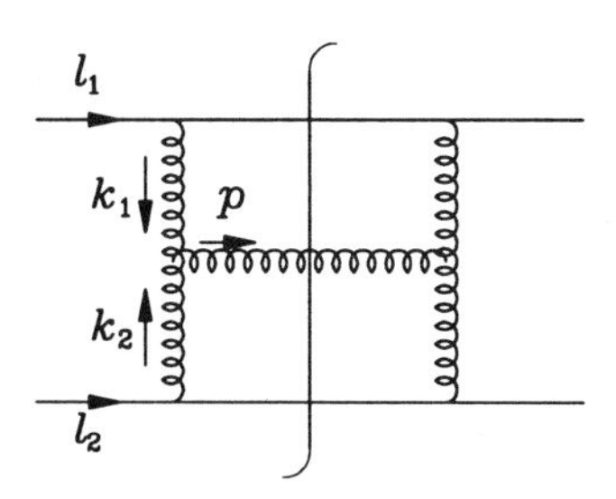

Figure 7.1 Square of the leading Feynman diagram to the process: $qq \to qqg$.

by the fields of two strongly Lorentz contracted color charge sources, which are effectively equal to the valence quarks of two incoming ions. In order to understand KMW's result in the language of perturbative QCD, let us consider a specific partonic process: $qq \to qqg$, as sketched in Fig. 7.1. If we assume that the incoming quarks qq are the valence quarks in the initial color charge sources, the partonic subprocess in Fig. 7.1 mimics the physical picture adopted in KMW's derivation. In this section, I show how to extract the leading contribution to the soft gluon production from this diagram. However, as a Feynman diagram in QCD perturbation theory, the single diagram shown in Fig. 7.1 is not gauge invariant. In next section, I show that with a proper choice of gauge and certain approximation, other diagrams are suppressed in soft gluon limit.

In general, the invariant cross section of the gluon production, shown in Fig. 7.1, can be written as

$$d\sigma_{qq\to g}/dyd^2p_T = \frac{1}{2s}\,|\overline{M}|^2\,dps\;, \tag{7.2}$$

where $s = (l_1 + l_2)^2$, and $|\overline{M}|^2$ is matrix element square with the initial-spin averaged and the final-spin summed. l_1 and l_2 as the momenta of the two incoming quarks respectively. In Eq. (7.2), the phase space

$$\begin{aligned} dps \;\;\propto\;\; & \frac{d^4k_1}{(2\pi)^4}(2\pi)\delta((l_1-k_1)^2) \times \frac{d^4k_2}{(2\pi)^4}(2\pi)\delta((l_2-k_2)^2) \\ & \times(2\pi)^4\delta^4(k_1+k_2-p)\;, \end{aligned} \tag{7.3}$$

where k_1 and k_2 are the momenta for the two gluons emitted from the initial quarks. Because of the gluon propagators, as shown in Fig. 7.1, the matrix element square $|\overline{M}|^2$ has the following pole structure:

$$poles = \frac{1}{k_1^2+i\epsilon}\,\frac{1}{k_1^2-i\epsilon}\,\frac{1}{k_2^2+i\epsilon}\,\frac{1}{k_2^2-i\epsilon}\,. \tag{7.4}$$

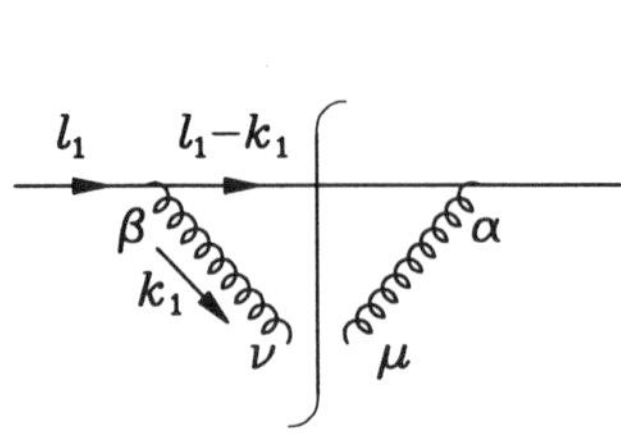

Figure 7.2 Feynman diagram for quark splitting: $P_{l_1 \to k_1}(x, k_{1T} < p_T)$.

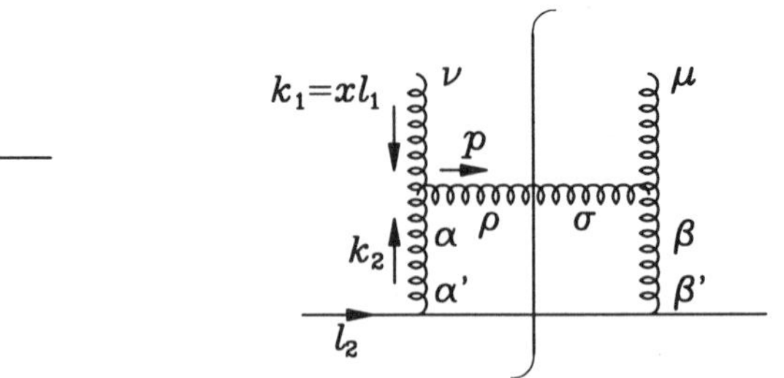

Figure 7.3 Leading Feynman diagram contributing to the hard scattering part $H(xl_1, l_2, p)$.

When integrating over the phase space, we see that the leading contribution comes from the terms with $k_1^2 \to 0$ or $k_2^2 \to 0$ limit. Note that k_1^2 and k_2^2 can not be zero at the same time, because k_1 and k_2 come from different directions, and we have the on-shell condition of $p^2 = (k_1 + k_2)^2 = 0$. Therefore, to calculate the leading contribution, we can first calculate the diagram in $k_1^2 \to 0$ limit. The total leading contribution is just twice of it, because the diagram is symmetric for k_1 and k_2.

To derive the leading contribution at $k_1^2 \to 0$ limit, we perform the collinear approximation $k_1 \approx xl_1 + O(k_{1T})$, with $k_{1T} \sim \Lambda_{cutoff} << p_T$, where Λ_{cutoff} is a collinear cutoff scale [10]. This approximation means that the leading contribution is from the phase space where almost all transverse momentum of the final-state gluon comes from the gluon of k_2, and k_1 is almost collinear to l_1. After such collinear approximation, the cross section in Eq. (7.2) can be approximately written in a factorized form [2]:

$$\frac{d\sigma_{qq\to g}}{dy dp_T^2} \approx 2\left(\frac{1}{2(2\pi)^3}\frac{1}{2s}\right)\int \frac{dx}{x} P_{l_1\to k_1}(x, k_{1T} < p_T)\, H(xl_1, l_2, p) + O\left(\frac{\Lambda_{cutoff}^2}{p_T^2}\right), \tag{7.5}$$

where the overall factor of 2 is due to the fact that the leading contribution come from two regions corresponding to $k_1^2 \to 0$ and $k_2^2 \to 0$, respectively. In Eq. (7.5), $P_{l_1\to k_1}(x, k_{1T} < p_T)$ represents the probability of finding an almost collinear gluon with the momentum fraction x from an incoming quark of momentum l_1, and is represented by Fig. 7.2. $H(xl_1, l_2, p)$ in Eq. (7.5) is effectively the hard scattering between the gluon of $k_1 = xl_1$ and the incoming quark of l_2, and it is represented by the diagram shown in Fig. 7.3.

Under the collinear expansion $k_1 \approx xl_1$, the gluon line which connects the partonic parts $P_{l_1 \to k_1}$ and $H(xl_1, l_2, p)$ is effectively on the mass-shell, and therefore, the partonic parts, P and H in Eq. (7.5), are separately gauge invariant. The quark splitting function $P_{l_1 \to k_1}(x, k_{1T} < p_T)$ can be calculated in $n \cdot A = 0$ gauge. We have [9]

$$\begin{aligned} & P_{l_1 \to k_1}(x, k_{1T} < p_T) \\ = & \frac{N_c^2 - 1}{2N_c} \left(\frac{g^2}{8\pi^2} \frac{1 + (1-x)^2}{x} \right) \ell n \left(\frac{p_T^2}{\Lambda^2_{cutoff}} \right) . \end{aligned} \tag{7.6}$$

In $\bar{n} \cdot A = 0$ gauge, we have the hard scattering function [9]

$$\begin{aligned} H(xl_1, l_2, p) = & (2\pi) 4 g^4 \left(\frac{1}{2} \right) \left(\frac{p_+}{xl_+} \right)^2 \frac{1}{p_T^4} \frac{1}{s - 2l_+ p_-} \delta(x - \frac{2p_+ l_-}{s - 2l_+ p_-}) \\ & \times \left[(xs - 2xl_+ p_-)^2 + (2xl_+ p_-)(2p_+ p_-) \right] . \end{aligned} \tag{7.7}$$

For soft gluon production, we define the soft gluon limit as

$$\frac{p_-}{l_-} \ll 1 \quad \text{and} \quad \frac{p_+}{l_+} \ll 1 . \tag{7.8}$$

Combining Eq. (7.6), Eq. (7.7) and Eq. (7.5), and taking the soft gluon limit, we have [9]

$$\frac{d\sigma_{qq \to g}}{dy d^2 p_T} = \frac{2g^6}{(2\pi)^4} \left(\frac{1}{2N_c} \right)^2 N_c (N_c^2 - 1) \left(\frac{1}{p_T^4} \right) \ell n \left(\frac{p_T^2}{\Lambda^2_{cutoff}} \right) . \tag{7.9}$$

Eq. (7.9) reproduced Eq. (7.1), which is obtained by solving classical Yang-Mills equations, at $N_q = 1$. If we consider the total number of the quarks in the charge sources of both sides, we need to multiply N_q^2 to Eq. (7.9).

3. GAUGE INVARIANCE

For the gluon production in the process $qq \to qqg$, the single diagram shown in Fig. 7.1 is not gauge invariant. In general, we also need to consider the radiation diagrams shown in Fig. 7.4. Similar to the diagram in Fig. 7.1, the contribution of Fig. 7.4a and Fig. 7.4b can also be written in the same factorized form:

$$\begin{aligned} E \frac{d\sigma^{\text{rad}}_{qq \to g}}{d^3 p} \approx & \frac{1}{2(2\pi)^3} \frac{1}{2s} \int \frac{dx}{x} P_{l_1 \to k_1}(x, k_{1T} < p_T) \, H_i(xl_1, l_2, p) \\ & + O(\frac{\Lambda^2_{cutoff}}{p_T^2}) , \end{aligned} \tag{7.10}$$

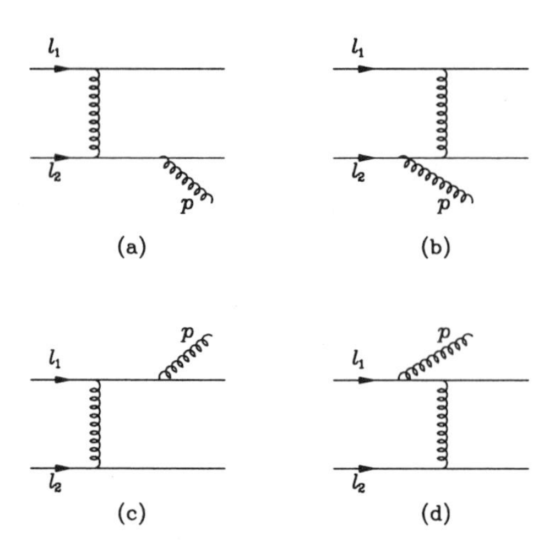

Figure 7.4 The rest of Feynman diagrams contributing to the process: $qq \to qqg$, in addition to the diagram in Fig. 1.

with $i = a, b$. $P_{l_1 \to k_1}(x, k_{1T} < p_T)$ is the quark splitting function given by Eq. (7.6). $H_a(xl_1, l_2, p)$ and $H_b(xl_1, l_2, p)$ are the hard scattering parts from the diagrams in Fig. 7.4a and Fig. 7.4b, and they are represented by Fig. 7.5a and Fig. 7.5b, respectively. With the contribution from diagrams in Fig. 7.4, Eq. (7.5) changes to

$$
\begin{aligned}
E\frac{d\sigma_{qq\to g}}{d^3p} \approx\; & 2\left(\frac{1}{2(2\pi)^3}\,\frac{1}{2s}\right)\int \frac{dx}{x}\, P_{l_1\to k_1}(x, k_{1T} < p_T) \\
& \times [H(xl_1, l_2, p) + H_a(xl_1, l_2, p) + H_b(xl_1, l_2, p) \\
& + \text{interference terms}] + O(\frac{\Lambda^2_{cutoff}}{p_T^2})\,, \qquad (7.11)
\end{aligned}
$$

with the approximation $k_1 = xl_1 + O(k_{1T})$. Feynman diagrams shown in Fig. 7.3 and Fig. 7.5 form a gauge invariant subset for calculating the hard scattering parts, $H(xl_1, l_2, p)$'s in Eq. (7.11). With our choice of gauge, $H_i/H \sim p_-/l_- << 1$ in the soft gluon limit. Therefore, the contribution from diagrams in Fig. 7.4a and Fig. 7.4b can be neglected in comparison with the contribution from the diagram in Fig. 7.1 at $k_1^2 \sim 0$.

Similarly, the contributions from diagrams shown in Fig. 7.4c and Fig. 7.4d can be neglected in the soft gluon limit, when compared with the contribution from the diagram in Fig. 7.1 at $k_2^2 \sim 0$. Therefore, with the approximation of $k_1^2 \sim 0$ (or $k_2^2 \sim 0$) and the soft gluon limit, and a proper choice of the gauge, the contributions *extracted* from the

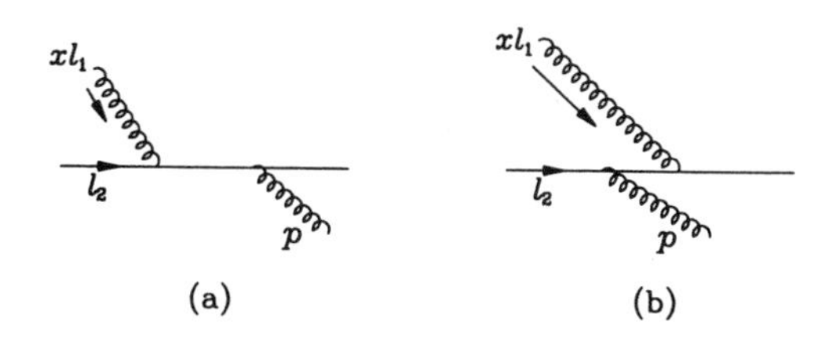

Figure 7.5 Feynman diagrams contributing to $H_a(xl_1, l_2, p)$ (a), and $H_b(xl_1, l_2, p)$ (b).

diagram in Fig. 7.1 to the leading soft gluon production in Eq. (7.1) is gauge invariant.

4. DISCUSSION

When we consider the collision between two nuclei, we can treat the two incoming quarks in Fig. 7.1 as coming from two nuclei respectively. In this picture, the number of the valence quark N_q is replaced by the quark distribution in the nuclei. In terms of the parton model, the cross section between the two nuclei A and B can be expressed in the following form:

$$\frac{d\sigma_{AB\to g}}{dydp_T^2} = \int dz_1\, dz_2\, f_{q/A}(z_1)\, f_{q/B}(z_2)\, E\frac{d\sigma_{qq\to g}}{d^3p} . \qquad (7.12)$$

Here z_1 and z_2 are the momentum fractions of the quarks, and $f_{q/A}(z_1)$ and $f_{q/B}(z_2)$ are the quark distributions (or quark number densities) of the two nuclei. If we denote p_A and p_B as the momenta for the two nuclei respectively, then $z_1 = l_1/p_A$ and $z_2 = l_2/p_B$. Substituting Eq. (7.5) into Eq. (7.12), we have

$$\begin{aligned}\frac{d\sigma_{AB\to g}}{dydp_T^2} \approx\ & \frac{1}{2(2\pi)^3}\frac{1}{2S}\int\frac{dz_1}{z_1}\frac{dz_2}{z_2} \\ & \times\left[\int\frac{dx_1}{x_1}\, f_{q/A}(z_1)f_{q/B}(z_2)\right. \\ & \qquad \times P_{l_1\to k_1}(x_1, k_{1T} < p_T)\, H(x_1l_1, l_2, p) \\ & + \int\frac{dx_2}{x_2}\, f_{q/A}(z_1)f_{q/B}(z_2) \\ & \qquad \left.\times P_{l_2\to k_2}(x_2, k_{2T} < p_T)\, H(l_1, x_2l_2, p)\right] \qquad (7.13)\end{aligned}$$

where the overall factor 2 in Eq. (7.5) is now represented by the two terms, and $S = (p_A + p_B)^2 \approx 2p_A \cdot p_B$. In Eq. (7.13), $x_i = k_i/l_i$ with

$i = 1, 2$, and $l_1 = z_1 p_A$ and $l_2 = z_2 p_B$. If we denote the momentum fraction of gluon k_1 with respect to p_A as $z'_1 = k_1/p_A$, and k_2 with respect to p_B as $z'_2 = k_2/p_B$, we can rewrite Eq. (7.13) in terms of z'_1 and z'_2:

$$\begin{aligned}\frac{d\sigma_{AB\to g}}{dy dp_T^2} \approx & \frac{1}{2(2\pi)^3}\frac{1}{2S} \\ & \times\left\{\int \frac{dz'_1}{z'_1}\frac{dz_2}{z_2}\left[\int \frac{dz_1}{z_1} f_{q/A}(z_1)\, P_{l_1\to k_1}(z'/z_1, k_{1T} < p_T)\right]\right. \\ & \times f_{q/B}(z)\, H(z'_1 p_A, z_2 p_B, p) \\ & + \int \frac{dz'_2}{z'_2}\frac{dz_1}{z_1}\left[\int \frac{dz_2}{z_2} f_{q/B}(z_2)\, P_{l_1\to k_1}(z'_2/z_2, k_{2T} < p_T)\right] \\ & \left.\times f_{q/A}(z_1)\, H(z_1 p_A, z'_2 p_B, p)\right\} \end{aligned} \quad (7.14)$$

According to the QCD factorization theorem [2], we see that the part inside the square brackets is actually the gluon distribution from nuclei A (or B) at the factorization scale $\mu_F^2 = p_T^2$, with only the quark splitting function [11],

$$\begin{aligned} f_{g/A}(z'_1, \mu_F^2 = p_T^2) = & \int \frac{dz_1}{z_1} f_{q/A}(z_1) P_{l_1\to k_1}(z'_1/z_1, k_{1T} < p_T) \\ & + \text{term from gluon splitting.} \end{aligned} \quad (7.15)$$

Using Eq. (7.15), we can then reexpress Eq. (7.14) as:

$$\begin{aligned}\frac{d\sigma_{AB\to g}}{dy dp_T^2} \approx & \frac{1}{2(2\pi)^3}\frac{1}{2S} \\ & \times \int \frac{dz'}{z'}\frac{dz}{z}\left[f_{g/A}(z', \mu_F^2 = p_T^2) f_{q/B}(z) H(z' p_A, z p_B, p)\right. \\ & \left. + f_{q/A}(z) f_{g/B}(z', \mu_F^2 = p_T^2) H(z p_A, z' p_B, p)\right], \end{aligned} \quad (7.16)$$

which is the factorized formula for two-to-two subprocesses in the conventional perturbative QCD for the nucleus-nucleus collisions. In KMW formalism, only the valence quark color charge was used as the source of the classical charge of colors. As a result, the gluon splitting term in Eq. (7.15) is neglected for the distribution $f_{g/A}$.

From the above comparison, we conclude that by solving the classical Yang-Mills Equation to the second order in iteration, KMW's result is consistent with conventional perturbative QCD at the leading logarithmic approximation. The logarithmic dependence shown in KMW's result basically describes the logarithmic DGLAP evolution of the quark distributions [12]. However, the iteration in this approach is different from the expansion series in conventional perturbative QCD. The McLerran-Venugopalan formalism was later further developed to include the harder gluons into the charge density μ^2 and treat the charge source as an extended distribution which depends on the rapidity [13]. It will be potentially very useful if this new approach, after including higher orders of iteration, can include the parton recombination [14] and other non-perturbative effects which are not apparent in the normal perturbative calculation.

References

[1] K.J. Eskola (1997), nucl-th/9705027; and the references therein.

[2] J. C. Collins, D. E. Soper and G. Sterman (1989), in *Perturbative Quantum Chromodynamics*, ed. A. H. Mueller. Singapore: World Scientific.

[3] L. McLerran and R. Venugopalan (1994), Phys. Rev. D **49**, 2233.

[4] L. McLerran and R. Venugopalan (1994), Phys. Rev. D **49**, 3352.

[5] L. McLerran and R. Venugopalan (1994), Phys. Rev. D **50**, 2225.

[6] A. Kovner, L. McLerran, and H. Weigert (1995), Phys. Rev. D**52**, 3809.

[7] M. Gyulassy and L. McLerran (1997), Phys. Rev. C **56**, 2219.

[8] X.-N. Wang (1997), Phys. Rep. **280**, 287.

[9] X.-F. Guo (1998), hep-ph/9812257 (to appear in Pgys. Rev. D).

[10] G. Sterman (1996), in *QCD and Beyond (TASI '95)*, ed. D.E. Soper. Singapore: World Scientific.

[11] For example, see R. Field (1989), *Application of Perturbative QCD.* New York: Addison-Wesley.

[12] G. Altarelli and G. Parisi (1977), Nucl. Phys. B **126**, 298; Yu.L. Dokshitser (1977), Sov. Phys. JETP **46**, 641; V.N. Gribov and L.N. Lipatov (1972), Sov. J. Nucl. Phys. **15**, 438, 675.

[13] J. Jalilian-Marian, A. Kovner, L. McLerran, and H. Weigert (1997), Phys. Rev. D **55**, 5414; A. Ayala, J. Jalilian-Marian, L. McLerran, and R. Venugopalan (1996), Phys. Rev. D **53**, 458.

[14] A.H. Mueller and J.-W. Qiu (1986), Nucl. Phys. **B268**, 427; J.-W. Qiu (1987), Nucl. Phys. **B291**, 746.

Chapter 8

UNIVERSALITY IN FRAGMENT INCLUSIVE YIELDS FROM AU + AU COLLISIONS

A. Insolia[1], C. Tuvè[1], S. Albergo[1], F. Bieser[2], F. P. Brady[5], Z. Caccia[1] , D. Cebra[2,4], A. D. Chacon[6], J. L. Chance[5], Y. Choi[4] * , S. Costa[1], J. B. Elliott[4], M. Gilkes[4], J. A. Hauger[4], A. S. Hirsch[4], E. L. Hjort[4], M. Justice[3], D. Keane[3], J. Kintner[5], M. Lisa[2], H. S. Matis[2], M. McMahan[2], C. McParland[2], D. L. Olson[2], M. D. Partlan[5], N. T. Porile[4], R. Potenza[1], G. Rai[2], J. Rasmussen[2], H. G. Ritter[2], J. L. Romero[5], G. V. Russo[1], R. Scharenberg[4], A. Scott[3], Y. Shao[3], B. K. Srivastava[4], T. J. M. Symons[2], M. L. Tincknell[4] †, S. Wang[3] ‡, P. G. Warren[4], H. H. Wieman[2], K. L. Wolf[6]

(EOS Collaboration)

[1] *Università di Catania & INFN, 95129 Catania, Italy*
[2] *Nuclear Science Division, LBNL, Berkeley, California, 94720*
[3] *Kent State University, Kent, Ohio, 44242*
[4] *Purdue University, West Lafayette, Indiana, 47907-1396*
[5] *University of California, Davis, California, 95616*
[6] *Texas A&M University, College Station, Texas, 77843*

*† Current address: Sung Kwun Kwan University, Suwon 440-746, Republic of Korea
†⋆ Current address: Lincoln Laboratory, S1-257, 244 Wood Street, Lexington, MA 02173-9108
‡‡ Current address: Harbin Institute of Technology, Harbin 150001, People's Republic of China

Abstract The inclusive light fragment ($Z \leq 7$) yield data in Au+Au reactions, measured by the EOS Collaboration at the LBNL Bevalac, are presented and discussed. For peripheral collisions the measured charge distributions develop progressively according to a power law which can be fitted by a single τ exponent independently of the bombarding energy in the range 250 - 1200 A MeV. In addition to this universal feature, we observe that the location of the maximum in the individual yields of different charged fragments shift towards lower multiplicity as the fragment charge increases from $Z = 3$ to $Z = 7$. This trend is common to all six measured beam energies. Finally, a monotonic increase is observed in the $T_{He,DT}$ thermometer, built from double yield ratios. We find that our data bring additional experimental evidence of a liquid - gas phase transitions in nuclei, already reported by the EOS Collaboration for the Au + C reaction.

Keywords: BEVALAC, fragmentation, yields, IMF production, nuclear temperature

1. INTRODUCTION

The pioneering work of the Purdue group [1], with the recognition of a power law dependence in the fragment yield in $p + Xe$ and $p + Kr$ reactions, was recognized early on as one of the possible signatures that, under proper conditions, nuclear matted could undergo a phase transitions[2]. Since then a lot of scientific effort has been devoted to the subject [3], along with the experimental discovery of universal features in nucleus - nucleus collisions in the energy range 100 A - 1000 A MeV. In particular, the EOS Collaboration, in an event-by-event analysis of the Au + C reaction at 1 A GeV, has been able to characterize the experimental data in terms of critical exponents [4] thus bringing the first clear evidence for a second order liquid - gas phase transition in multifragmentation reaction between heavy ions. The purpose of the present analysis of Au + Au collisions, in the energy range $250A - 1200AMeV$ is to show that the inclusive yields produced by the fragmentation of the projectile spectator carry most of the universal features, already recognized by different groups as signatures for a critical behaviour of nuclear matter [3, 4, 5]. The fingerprints of the *liquid - gas* phase transition will be therefore found in the power law trend observed in the inclusive yields for peripheral collisions, in the moments of charge distributions, as well as, with a word of caution, in the observed trend of the nuclear temperature (the double yield ratios [6]) versus multiplicity.

2. INCLUSIVE YIELDS AND CONDITIONAL MOMENTS

The EOS Collaboration has recently measured fragment production in Au + Au collisions at the LBNL Bevalac beam energies of 0.25, 0.4, 0.6, 0.8, 1.0 and 1.15 A GeV. The experiment was conducted with the EOS

Time Projection Chamber (TPC), a multiple sampling ionization chamber (MUSIC) as well as a TOF wall and a neutron detector (MUFFINS). A brief description of the TPC and MUSIC detectors can be found in ref. [7]. Details about the MUFFINS spectrometer are to be found in ref. [8, 9]. Only data from the TPC detector will be used in the present analysis. The TPC [7, 10] afforded almost complete coverage in the forward hemisphere. Particle identification is achieved via specific energy loss along particle tracks. The EOS TPC resolution allows one to identify particles up to $Z = 7$, measuring their momentum with virtually no p_t cut. According to the Plastic Ball analysis prescription [11], in order to define the centrality of the collision, the multiplicity spectrum is divided in 8 bins from 0 to the maximum multiplicity. The latter is defined as the multiplicity value, near the end of the multiplicity distribution, where the distribution falls down to 1/2 of its plateau value [11]. Multiplicities greater than the so defined M_{max} contribute to bin 9. This multiplicity scale is therefore a sort of scale on excitation energy even if we cannot define an exact correspondence among multiplicity bins and excitation energy.

The measured inclusive rapidity distribution, see fig. 1, or the scatter plot $p_t - y'$ in the transverse momentum - normalized rapidity plane, allow to define proper cuts to select the fragmentation of the projectile spectators. We have used a sharp cut near beam rapidity, $y' \geq 0.60$. It is seen, in fig. 1, that rapidity spectra for $Z = 1, 2$ exhibit a large prompt component at lower rapidity. For larger Z values the fragments are produced with a rapidity close to the beam rapidity with an almost simmetric distribution. The problem posed by the presence of prompt particles has been discussed widely in the literature [7, 12, 13]. In the present analysis the prompt component is strongly suppressed by the sharp cut close to beam rapidity that we use to select projectile spectator fragmentation. In addition, one should recall the result of ref. [7], where it has been shown, for the Au + C reaction at $1AGeV$ that the total multiplicity is proportional to the second stage multiplicity. This fact makes less crucial the problem of a complete suppression of the prompt component. The measured inclusive yields for the six beam energies are reported in fig. 2 versus the fragment charge Z.

We see that for central collisions (multiplicity bins 8 or 9, see fig.2) an almost a linear dependence (in semilog scale) is observed, while approaching the semi-peripheral domain a power law starts to develop. This is nicely seen, in the figure, for the multiplicity bins $M_{bins} = 3$, 4 and 5, where the yields fall one on top of the other. For these bins, the experimental dependence on Z can well be fitted with a power law of the type $Y = q_0 Z^{-2.1}$. Moreover, the observed power law is independent

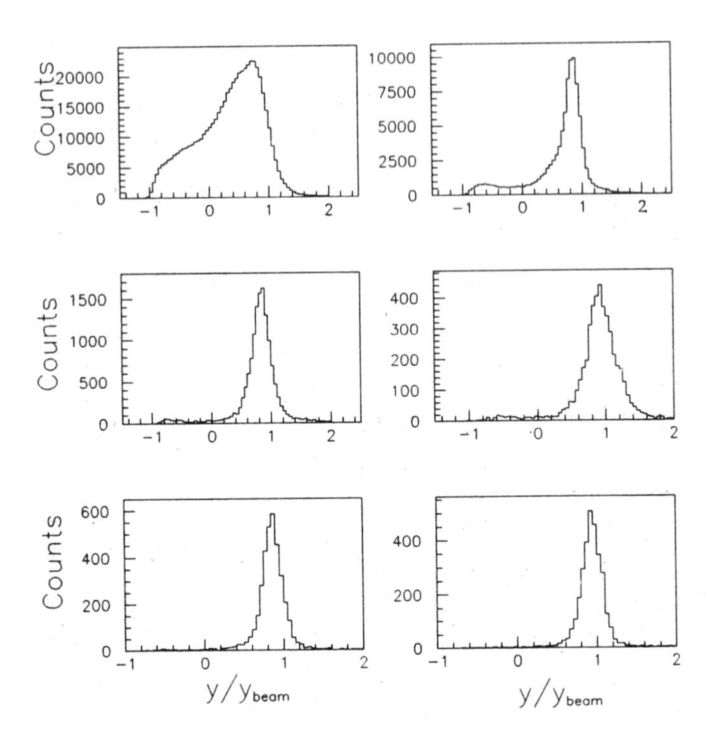

Figure 8.1 Typical inclusive particle rapidity distributions for Au + Au, where $y' = y/y_{beam}$, at $E_{lab} = 1000\ A\ MeV$, with the proper selection for peripheral collisions ($M_{bin} = 3, 4, 5$).

of the beam energy in the range of the experiment. This shows that the main mechanism for fragment production from the projectile remnant is very much the same in this energy range. This power-law-like distribution has been already reported for Au + Au reactions by the ALADiN Collaboration [14]. Indeed, they found that the fragment yields at 1000 MeV/amu gated with a constraint on large impact parameter (peripheral collisions) show a power law dependence on the fragment charge. This is at variance with the trend observed at 100 MeV/amu in which, selecting central collisions, an exponential behaviour was observed [14]. This feature of the multifragmentation phenomenon has been, on the other hand, interpreted as a possible signature for a continuous liquid-gas phase transition [1, 3]. Actually, if a liquid - gas phase transition takes place for a given value of excitation energy transferred to the system, one has to expect a typical power law dependence of the relevant physical quantity which, in our case, is just the fragment distribution.

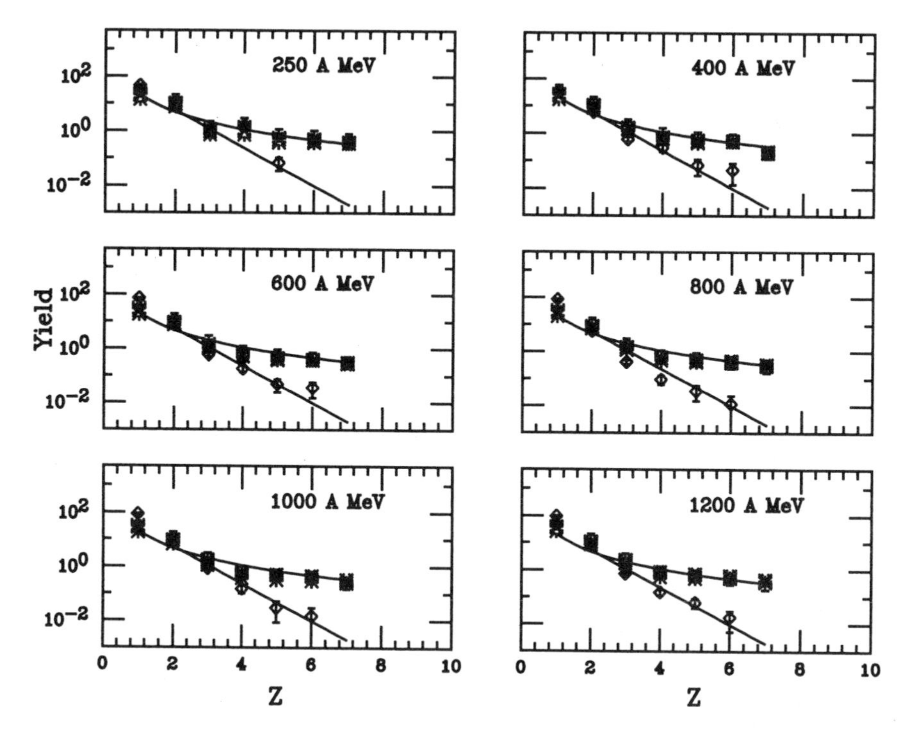

Figure 8.2 Charge Yields for different multiplicity bins. Data points are represented by different symbols for different multiplicity bins. However, data corresponding to peripheral collisions (multiplicity bins 3,4 and 5) are almost undistinguishable for they fall one on top of the other. Only one multiplicity bin corresponding to central collisions ($M_{bin} = 9$ at 250 and 400 MeV, or $M_{bin} = 8$ at the remaining energies) is reported. The latter shows an almost exponential decrease of the yield (linear behaviour in semilog scale). All curves at all beam energies correspond to a power law $Y = q_0 Z^{-2.1}$ and to an exponential $Y = Y_0 e^{-1.58*Z}$, for the peripheral and central collisions, respectively. The numerical values of q_0 and Y_0 are the same for the six beam energies.

Different model calculations have reported very clearly this behaviour. M. Baldo et al. [15] find that deterministic chaos inside the spinodal zone is associated with the multifragmentation and the predicted fragment distribution shows a power law trend with an exponent τ very close to what was previously reported by the analysis of Au + C EOS data at 1.0 A GeV [3, 4]. Furthermore, A. Bonasera et al. [16], in the frame of a simple purely classical molecular dynamic model, have found that a system of 100 particles exhibits an expanding scenario that for a given initial temperature produces multifragmentation. The mass distribution displays a power law behaviour $Y(A) \simeq A^{-\tau}$ with $\tau = 2.23$ [16]. This is close to the value predicted by the Fisher droplet model[17, 18] and, again, it is very close to the critical exponents previously reported by the EOS Collaboration. Our observation of a power law for semi-peripheral collisions in Au + Au at different energies is therefore consistent with a liquid - gas phase - transition interpretation of the multifragmentation.

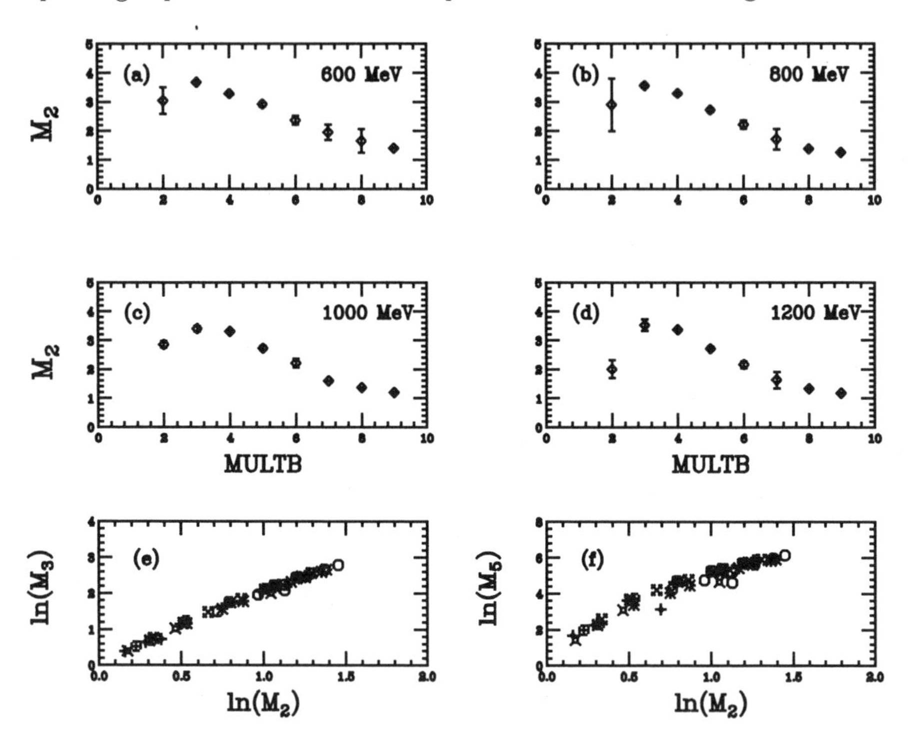

Figure 8.3 (a,b,c,d) Second moment of the charge distribution for the indicated beam energies.. The maximum is observed at $M_{bin} = 3$. (e) Correlation between the third and the second moment. (f) Correlation between the fifth and the second moment.

The second moment of the charge distribution (for a beam energies) are given in figs. 3(a,b,c,d). The correlations between different moments are shown in fig. 3(e) and 3(f). One can easily recognize that a maximum appears at $M_{bin} = 3$. This is the first multiplicity bin for which we observe the universal trend in the yield versus Z. In particular, as expected, the second moment is characterized by a maximum which tends to soften when a mass cutoff is applied [20]. In our case this comes in a natural way due to the detector cut at charge $Z \geq 8$. A correlation among different moments, third versus second one, fig. 3(e), and fifth versus second one, in fig. 3(f), is nicely seen, in agreement with the percolation model calculations [19, 18, 20]. The slopes are consistent with the τ exponent which best fits the yields for $M_{bin} = 3, 4, 5$ in fig. 2.

3. INTERMEDIATE MASS FRAGMENTS

The integrated intermediate mass ($Z = 3-7$) fragment yield is almost energy independent for peripheral collisions, while it is suppressed as the beam energy increases for central collisions, as seen in fig. 4.

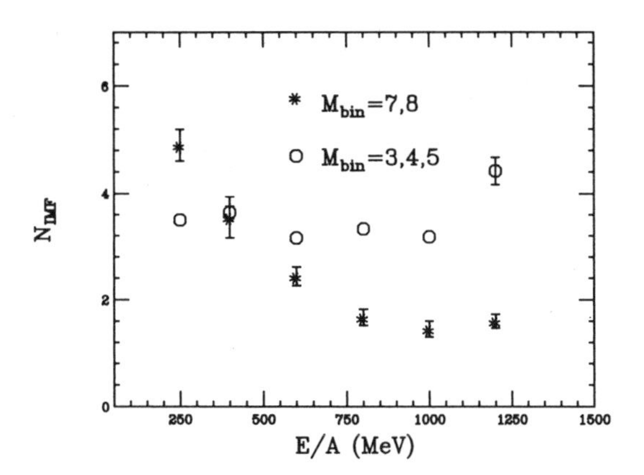

Figure 8.4 Energy depedence of the IMF (Z=3-7) production. The peripheral integrated yields are averaged over the multiplicity bins 3,4 and 5. For central collisions a mean between the bins 7 and 8 has been reported.

It is quite interesting to look at the individual fragment yields versus multiplicity. The data are reported in fig. 5, where only two typical values ($Z = 3$ and $Z = 6$) are considered for simplicity. All six energies show similar features. The same type of comments as for the results of fig. 1 of ref. [21], in which multifragmentation in the reaction Au + C at 1 $A\ MeV$ was studied, applies. Indeed, increasing multiplicity corresponds to events with greater excitation energy of the projectile remnant. This produces the rise of the number of each species of IMF as the multiplicity increases from the lower M_{bin} values. We do observe a strict ordering: lighter species is always more abundant than the heavier. As noticed in [21], all IMF yields decrease at the highest multiplicity. For those M_{bin} values, the excitation energy is so large that only the lightest fragments can survive. Finally we observe that, in spite of the unavoidable smearing produced by the fact that we average the individual yields over many events within the multiplicity bins, the location of the maximum still shifts towards lower multiplicity for the larger Z values. This is a very important feature for it allowed to extract for the first time the critical exponent σ in the Au + C analysis at 1 $A\ GeV$ [21]. In Au + Au inclusive individual yields, with which this paper is concerned, the shift of the maximum location appears to be, again, independent of the beam energy along with the expected and well known *rise and fall* of the fragment production versus the centrality of the collision.

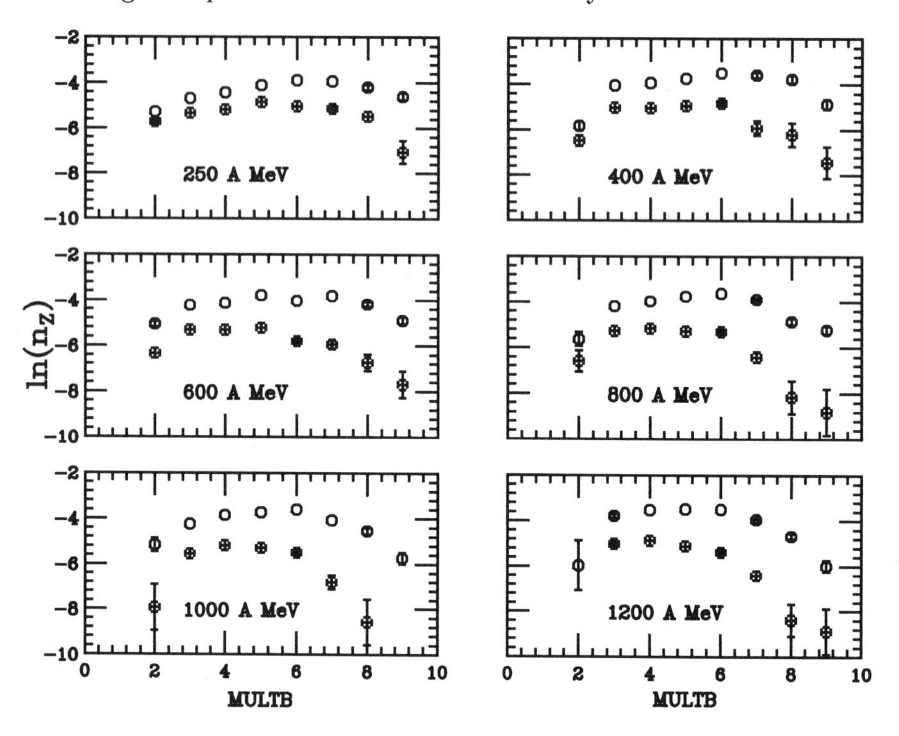

Figure 8.5 Individual yields $ln(n_Z)$ versus multiplicity. The number of fragment for each species has been divided by $Z = 79$ (the nuclear charge of gold).

4. NUCLEAR TEMPERATURE

A lot of interest has been raised by the possibility to characterize the remnant by determining its nuclear temperature. Different hadronic thermometers have been used so far [5] (slope parameter, isotopic composition, relative population of states). Different thermometers usually produce different temperatures. This is especially true for the inverse slope parameter determination. We will present our own results with the widely used double yield ratio [6] thermometer. In this case one has to assume that the nuclear system finds itself at low density and in chemical and thermal equilibrium. Under those conditions it has been shown [6] that the temperature can be obtained through the double yield ratio of two isotope pairs, (Y_1/Y_2) and (Y_3/Y_4), differing by the same number of neutrons and/or protons. We have chosen both the $T_{He,DT}$ and the $T_{He,Li}$ thermometers. Different problems arise with these two different thermometers. The $T_{He,DT}$ could be partially affected by uncertainty in the prompt particles separation (mainly for Z=1 and Z=2) from the corresponding $Z = 1, 2$ fragments emitted by the remnant decay in the second stage [7]. On the other end, the $T_{He,Li}$ could be severely biased by side feeding effects which could modify the yield ratio [5, 7, 22]. We have found that the $T_{He,DT}$ is actually more reliable due to its insensitivity to varying the cut used to suppresse the prompt particles (either 60% or 75% of the beam rapidity) and to side feeding effects. Our results are reported in fig. 6. The left panel shows $T_{He,DT}$ vs. multiplicity bin for all six beam energies. In spite of differences between the nuclear temperature for the various energies, one should notice the common general trend. It resembles the one measured in the Au + C reaction at 1 A GeV [7], even if with larger absolute values and it clearly shows a monotonic increase of the temperature versus multiplicity. The energy dependence is better seen in the right panel of fig. 6, for two given multiplicity bins corresponding to peripheral ($M_{bin} = 3$) and central collisions ($M_{bin} = 7$), respectively. One can notice an overall trend of increasing temperatures with increasing beam energy. The behaviour observed at $1200 AMeV$ could be due to the larger production of Deuteron with respect to Triton not counterbalanced by a corresponding enhancement of the ^3He yield. This shows the limits of using the double yield ratio thermometer over such a large range of energies. The $T_{He,Li}$ thermometer shows a stronger multiplicity dependence especially at the intermediate energies of 600 A MeV and 800 A MeV. This is not clearly understood at the moment. We have analyzed the sensitivity to different prompt particle removal cuts and found appreciable differences for the two cases $y' \geq 0.60$ or $y' \geq 0.75$. In addition, it is well known that a correction for side feeding, like the one proposed by Tsang et al. [22], could produce major effects for higher multiplicity bins, affecting more the higher temperature values than the lower ones. Work is in progress on this point.

5. SUMMARY

In conclusion, we have shown that EOS data for the inclusive yield of fragments ($Z = 1 - 7$) in Au + Au, from $E_{beam} = 250$ A MeV up to $E_{beam} = 1200$ A MeV, show a typical power law distribution when triggering on peripheral collisions ($M_{bin} = 3, 4, 5$) in agreement with previous model calculations as well as with other experimental data. The second moment of the charge distribution shows a maximum at the $M_{bin} = 3$ for all considered energies. In addition to this universal feature, we have found that the location of the maximum in the individual yields of different charged fragments shift towards lower multiplicity as the

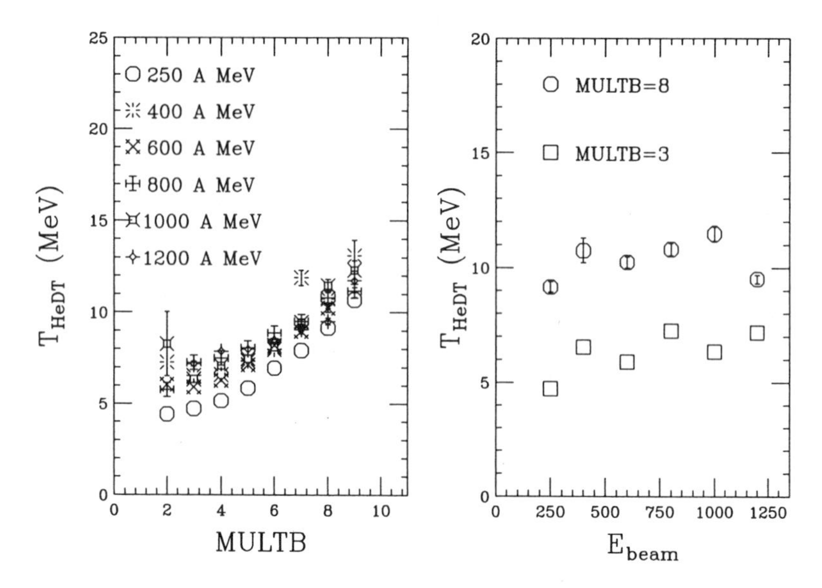

Figure 8.6 left panel: $T_{He,DT}$ thermometer built from the double yield ratios versus multiplicity bins; right panel: Energy dependence of the measured temperature for two selected muliplicity bins, corresponding to peripheral ($M_{bin} = 3$) and central collisions ($M_{bin} = 7$), respectively.

fragment charge increases from $Z = 3$ to $Z = 7$. This trend is common to all six energies. Finally, a monotonic increase vs. multiplicity bins was observed in the $T_{He,DT}$ thermometer built from double yield ratios. We conclude that our data bring additional experimental evidence of a liquid - gas phase transitions in nuclei, as previously reported by the EOS Collaboration for the Au + C reaction.

Acknowledgments

Two of the authors (A.I. and C.T.) thank Andrea Rapisarda for many useful discussions.

References

[1] J. E. Finn et al., Phys. Rev. Lett. **49** (1982) 1321

[2] The Purdue group concluded writing ... *whether or not we have observed such a phase transition in analogy with classical gases is still an open question* ...[1].

[3] *Critical Phenomena and Collective Observables*, Proc. of the Int. Conf. CRIS'96, Acicastello (Italy, 1996); Ed.s S. Costa, S. Albergo, A. Insolia and C. Tuvé, World Scientific (1996)

[4] M. Gilkes et al. (EOS Collaboration), Phys. Rev. Lett. **73** (1994) 1590

[5] J. Pochodzalla , Prog. Part. Nucl. Phys. **39** (1997) 443

[6] S. Albergo, S. Costa, E. Costanzo and A. Rubbino, Il Nuovo Cim. A**89** (1985) 1

[7] J. A. Hauger, P. Warren et al. (EOS Collaboration), Phys. Rev. C**57** (1998) 764

[8] C. Tuvè et al. (Transport Collaboration), Phys. Rev. C**56** (1997) 1057; Phys. Rev. C**59** (1999) 233

[9] S. Albergo et al., Nucl. Instr. Meth. A**311** (1992) 280;
J. Engelage et al. (Transport Collaboration), Radiat. Measurements **27** (1997) 549

[10] G. Rai et al., IEEE Transactions on Nuclear Science **37** (1990) 56

[11] H. H. Gutbrod, A. M. Poskanzer and H. G. Ritter, Rep. Prog. Phys. **52** (1989) 1267

[12] W. Bauer and W. A. Friedman, Phys. Rev. Lett. **75** (1995) 767; W. Bauer and A. Botvina, Phys. Rev. C**75** (1995) R1760

[13] M.L. Gilkes et al., Phys. Rev. Lett. **75** (1995) 768

[14] G.J. Kunde et al., Phys. Rev. Lett. **74** (1995) 38

[15] M. Baldo, G.F. Burgio and A. Rapisarda, Phys. Rev. C**51** (1995) 198;

[16] V. Latora, M. Belkacem and A. Bonasera, Phys. Rev. Lett. **73** (1994) 1765; M. Belkacem, V. Latora and A. Bonasera, Phys. Rev. C**52** (1995) 271;

[17] M. E. Fisher, Rep. Prog. Phys. **30** (1987) 615

[18] D. Stauffer and A. Aharony, *Introduction to Percolation Theory* (Taylor & Francis, London, 1994).

[19] X. Campi, J. Phys. A **19** (1986) L917; Phys. Lett. B**208** (1988) 351

[20] W. Bauer, Phys. Rev. C**38** (1988) 1297

[21] J.B. Elliott et al. (EOS Collaboration), Phys. Lett. **B381** (1996) 35

[22] M. B. Tsang et al., in *Critical Phenomena and Collective Observables*, Proc. of the Int. Conf. CRIS'96, Acicastello (Italy, 1996); Ed.s S. Costa, S. Albergo, A. Insolia and C. Tuvé, World Scientific (1996).

Chapter 9

THE PHENIX EXPERIMENT

K.F. Read[1,2] for the PHENIX Collaboration*
[1] *Physics Department, Oak Ridge National Laboratory, Oak Ridge, TN 37831, USA*
[2] *Department of Physics, University of Tennessee, Knoxville, TN 37996, USA*
readkfjr@ornl.gov

*For the complete PHENIX Collaboration author list, please refer to Ref. 1. See *http://www.phenix.bnl.gov* for extensive current information about PHENIX.

Abstract The PHENIX experiment at RHIC is currently under construction with data collection planned to start in 1999. The heavy ion and spin physics goals of PHENIX are described. We discuss the experiment's capabilities to address these physics goals. Highlights of the present status of construction and installation are presented.

Keywords: PHENIX, RHIC, quark gluon plasma, heavy-ion collisions, muons, spin

1. INTRODUCTION

The PHENIX experiment [1, 2] is currently under construction at the BNL Relativistic Heavy Ion Collider (RHIC). Data collection is planned to start in 1999. We describe the physics goals of the experiment and how they lead to its design philosophy. These physics goals consist of two broad complementary programs of experimental investigation, namely the PHENIX heavy ion and spin physics programs. We separately discuss these experimental programs and emphasize the experiment's capabilities to address them. In particular, the experiment is able to identify and trigger on exclusive leptons, photons, and high p_T hadrons with excellent momentum and energy resolution. Highlights of the present status of construction and installation are presented.

2. HEAVY ION PHYSICS PROGRAM

The primary purpose of the PHENIX experiment is to detect and characterize a predicted new state of matter, the Quark Gluon Plasma [3] (QGP), via simultaneous measurement of various signatures as a function of energy density in $A + A$, $p + A$, and pp collisions.

These signatures include deconfinement as manifested in differential suppression of J/ψ and ψ' production; chiral symmetry restoration as manifested in a modification of the effective ϕ meson mass, width, and branching ratios; thermal radiation of real and virtual photons from a hot initial state; enhanced production of strangeness and charm; and hard scattering processes which serve as good primordial probes such as high p_T leptons, high p_T photons, and high p_T jets identified by their leading particle.

Debye screening due to the presence of a QGP can be systematically studied by measurements of J/ψ, ψ', and Υ production due to their different respective bound state radii: $R_\Upsilon(0.13\text{ fm}) < R_{J/\psi}(0.29\text{ fm}) < R_{\psi'}(0.56\text{ fm})$. In a sense, the Υ serves as an "experimental control" since it is unlikely to be screened at energy densities attainable at RHIC collision energies.

During the first year of RHIC operations, the luminosity will initially be 1% and grow to 10% of the design luminosity of 2×10^{26} cm^{-2}s^{-1}. The experiment expects to record 20 μb^{-1} to tape of $Au + Au$ collisions at a center-of-mass energy of 200 GeV per nucleon pair. Physics measurements that will be performed in the first year include measurement of global event properties such as N_{ch}, $< p_T >$, $dN_{ch}/d\eta$, E_T, and $dE_T/d\eta$, and fluctuations in these quantities. PHENIX will study the energy density and geometry of the collisions. Other early measurements include hadronic spectra and inclusive photon and π^0 measurements.

As the RHIC luminosity increases, feasible measurements next include open charm production (via single high p_T leptons as well as unlike-flavor lepton pairs), the Drell-Yan continuum, $\phi \to e^+e^-$, and $J/\psi \to e^+e^-$.

3. SPIN PHYSICS PROGRAM

RHIC will be able to produce longitudinally or transversely spin-polarized $p + p$ collisions at center-of-mass energies ranging from 50 to 500 GeV. It will provide two months of spin physics running per year beginning with the year 2000. The expected luminosity at $\sqrt{s} = 200$ GeV is 8×10^{31}cm^{-2}s^{-1}. The luminosity at $\sqrt{s} = 500$ GeV is 2×10^{32}cm^{-2}s^{-1}. PHENIX's initial request is for an integrated luminosity of 800 pb^{-1} collected at $\sqrt{s} = 500$ GeV and 320 pb^{-1} at $\sqrt{s} = 200$ GeV.

The primary goal of the PHENIX Spin Physics program is to use polarized proton-proton collisions to measure the helicity distributions of flavor separated quarks and antiquarks and gluon polarization in the nucleon. This information is obtained by studying the polarized Drell-Yan process, vector boson (J/ψ, W, Z) production, polarized gluon fusion, and polarized gluon Compton scattering (leading to direct photon production). Antiquark structure function measurements rely on analyzing Drell-Yan and vector-boson production data. Gluon polarization measurements rely on analyzing heavy quark, J/ψ, and prompt photon data. Efforts to understand the contributions of the spin of sea quarks and the polarization of gluons to the total nucleon spin may help explain the lack of agreement between experimental data and the Ellis-Jaffe sum rule. A further major goal of the PHENIX spin program is to perform precise tests of fundamental symmetries such as parity violation.

RHIC is expected to make important contributions to this field in ways that have not been possible historically via polarized deep inelastic scattering experiments. In essence, those experiments have used virtual photons to probe the nucleon. Such photons couple to the square of the electric charge and can not be used to distinguish quarks from antiquarks or to separate flavors. By using a proton beam to probe the nucleon, one can, through an appropriate program of measurements of helicity and transversity-dependent cross sections, untangle these quark-spin distributions as well as measure the gluon polarization in the nucleon.

An early study of particular interest is to study W boson production via a single μ tag. Fig. 9.1 demonstrates that this is very feasible for muons with $p_T > 20$ GeV/c. This allows measurement of a parity violating longitudinal spin asymmetry which can be optimized by selecting particular kinematic regions. This measurement is important to address the antiquark helicity distributions in the nucleon. These, and other, aspects of the spin program represent a very different, and fundamental, physics topic addressed using data from PHENIX. Work done commissioning, understanding, calibrating, and simulating the performance of PHENIX in support of the heavy ion program has immediate benefits to these spin studies.

4. EXPERIMENTAL OVERVIEW

PHENIX consists of eleven different detector subsystems. Fig. 9.2 provides a schematic drawing of PHENIX. The design philosophy of the experiment is to identify and trigger on exclusive leptons, photons, and high p_T hadrons with excellent momentum and energy resolution. This requires flexibility, multiple technologies, and the ability to handle

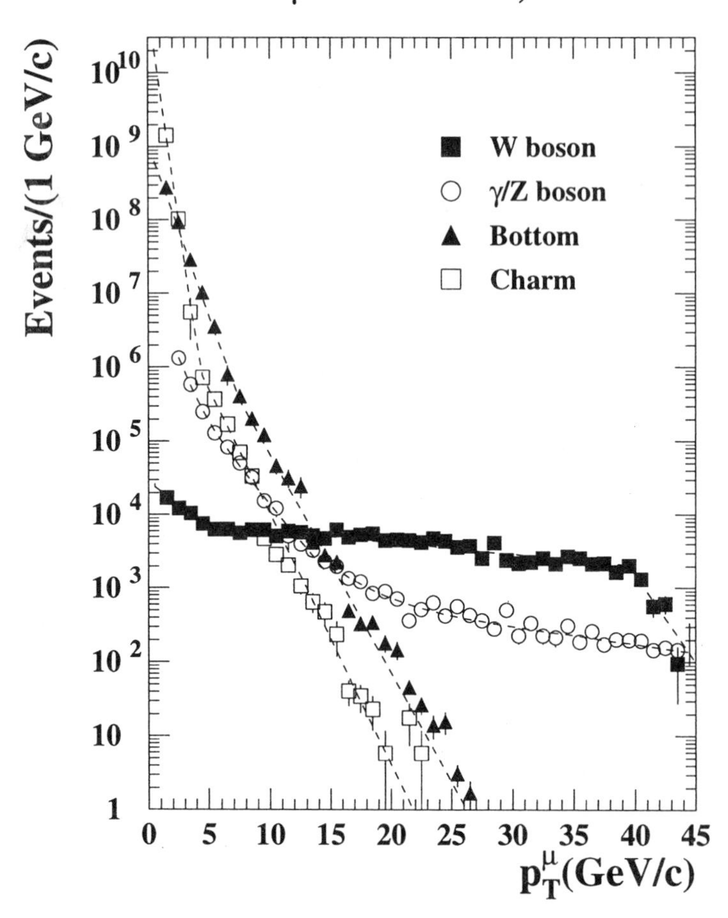

Figure 9.1 Simulated single muon transverse momentum distribution due to various decay processes. Distribution corresponds to an integrated luminosity of 2 fb^{-1} (one year of running at design luminosity) of $p+p$ collisions at $\sqrt{s} = 500$ GeV.

high data rates. The experiment is divided into central subsystems, east and west central arms, and north and south muon arms with each arm covering roughly one steradian. The magnetic field is axial in the central region and radial in the muon arms. PHENIX has a deadtime-less first-level trigger system to cope with background processes while measuring rare signal processes.

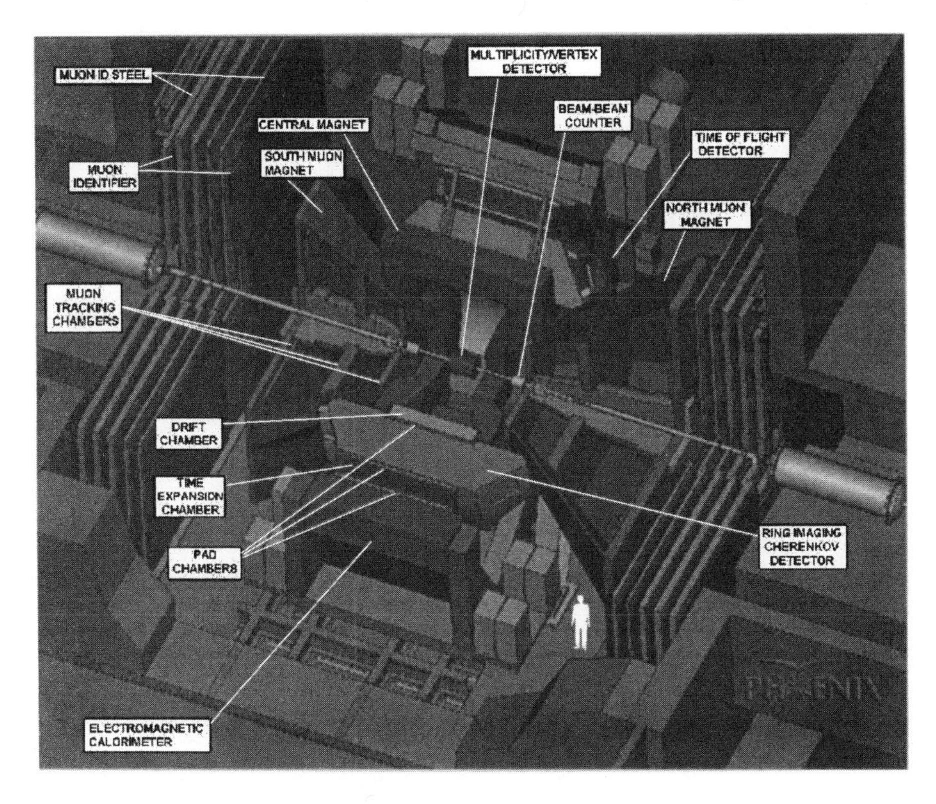

Figure 9.2 Schematic drawing of the PHENIX Experiment.

The central subsystems consist of the beam-beam counter and the multiplicity vertex detector. The beam-beam counter provides the time and longitudinal position of the interaction. It consists of two arrays of quartz Cerenkov telescopes. The multiplicity vertex detector is a 64 cm long silicon strip detector with silicon pad endcaps covering pseudorapidity from -2.5 to 2.5. It has approximately 35 thousand channels and a position resolution of 200 μm. It measures the collision vertex position in three dimensions as well as N_{ch} and $d^2N_{ch}/d\eta d\phi$.

5. PHENIX CENTRAL ARMS

A schematic of one of the two PHENIX central arms is shown in Fig. 9.3. The central arms cover pseudorapidity $-0.35 < \eta < 0.35$. Tracking in a central arm is provided by means of the drift chambers, pixel-pad chambers, and time expansion chambers (TEC). The drift chambers provide projective measurement of particle trajectories (with

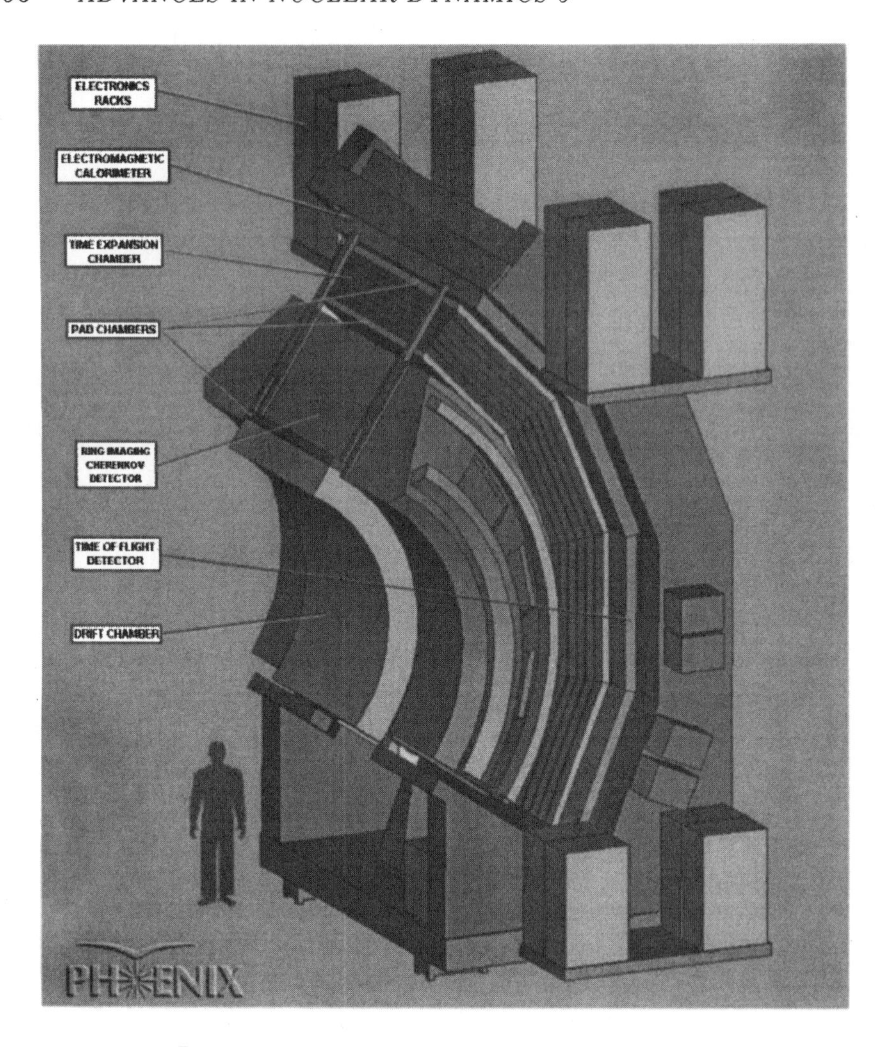

Figure 9.3 One of the two PHENIX Central Arms.

a position resolution of 150 μm) to yield a transverse momentum measurement. The pad chambers provide 3-dimensional space points which are critical for pattern recognition. Particle identification is provided by means of the time of flight (TOF) and Ring Imaging Cerenkov (RICH) subdetectors supplemented by the TEC. The TOF consists of 1056 scintillator slats. It provides 85 ps resolution time-of-flight information and can separate pions from kaons up to 2.5 GeV/c. This π/K particle identification separation is available for 30° of azimuth. The separation extends up to 1.4 GeV/c for the rest of the azimuthal coverage (using time-of-flight information provided by the electromagnetic calorimeter). The mass resolution for $\phi \to e^+e^-$ is 0.5% for $p_T < 2$ GeV/c. The π/e rejection factor for identified electrons is 10^{-4}.

The electromagnetic calorimeter has high resolution, high granularity, and provides timing information. It consists of approximately 25 thousand cells located 5 m from the interaction diamond. Each is about 15 to 18 radiation lengths long with a cross sectional area of about one Moliere radius square. It has a position resolution of a few mm/$\sqrt{E}$ and an energy resolution of 6–8%/$\sqrt{E}$, which has been tested extending up to 70 GeV. It can reliably separate π^0 mesons from single photons up to $p_T = 20$ GeV/c.

6. PHENIX MUON ARMS

The muon arms are a major contributor to the overall PHENIX physics program. The muon arms will be used to measure the production of vector mesons decaying into dimuons in heavy-ion collisions for masses ranging from that of the ϕ to the Υ. Measurement of the differential suppression of J/ψ, ψ', $\Upsilon(1S)$, and $\Upsilon(2S+3S)$ production will provide information concerning "deconfinement," *i.e.*, the Debye screening of the QCD potential. The muon arms also allow studies of the continuum dilepton spectrum in a much broader region of rapidity and mass than is accessible for e^+e^- in the central arms alone. In addition $e-\mu$ coincidences, using electrons detected by the central arms, will probe heavy-quark production and aid in the understanding of the shape of the continuum dielectron spectrum. This is feasible because unlike-sign $e-\mu$ pairs result primarily from $D\overline{D}$, while like-sign pairs are due to the combinatorial background arising predominantly from meson weak decays.

The muon arms consist of a muon tracker to track muons and provide sufficient mass resolution to separate the $\Upsilon(1S)$ from the $\Upsilon(2S+3S)$ and a muon identifier to provide an additional 100:1 $\pi:\mu$ rejection (beyond the factor of approximately 100 due to absorber material located before the muon identifier which filters out non-muons.) The muon tracker consists of three stations of cathode strip chambers per arm. The position resolution of the chambers is 100 μm. The muon identifier consists of 5 gaps per arm filled with planes of transversely oriented plastic proportional (Iarocci) tubes interleaved with layers of steel.

The muon arms provide coverage from $\eta = -2.3$ to -1.1 and 1.1 to 2.4. This corresponds to $\theta = 168°$ to $145°$ and $\theta = 10° - 35°$. The mass resolution for reconstruction $J/\psi \to \mu^+\mu^-$ is about 100 MeV/c^2. Table 9.1 lists the mass resolution, geometric acceptance times recon-

Table 9.1 Vector meson mass resolutions, acceptances, and expected number of particles per RHIC-year of minimum bias $Au + Au$ collisions (running at the design luminosity of 2×10^{26} cm^{-2}s^{-1}).

Particle	*Mass Resolution* *(MeV/c^2)*	*Acceptance* *(%)*	*Events/year*
ϕ	60	0.58	4×10^4
J/ψ	105	4.3	6.7×10^5
ψ'	105	4.3	1.2×10^4
Υ	180	3.0	382

struction efficiency, and estimated number of respective particles collected per arm per year at design luminosity (2.2×10^{26}cm^{-2}s^{-1}). For example, an estimated 0.67 million J/ψ are expected to be collected per arm per RHIC-year of $Au + Au$ minimum bias collisions.

Not included in the table are the estimated approximately 290 Υ per year for which one muon goes into each arm. This corresponds to acceptance for Υ production at $y = 0$. This acceptance extends down to approximately 5 GeV/c^2 providing good coverage for Drell-Yan pairs at $y = 0$ (with a yield varying as $1/M^3$). Thus, the presence of *two* muon arms represents more than a factor of two in acceptance because they additionally provide coverage for production at $y = 0$. It is critical to have *two* muon arms in order to study Z production as part of the spin program.

7. CONSTRUCTION STATUS

The PHENIX experiment is currently under construction. The highlights of the current status of construction and installation as of early 1999 are as follows. The detector magnets have been installed and tested. All of the steel for the north and south muon arms is installed. The first drift chamber is mechanically complete, with the second chamber underway. The inner and outer pixel pad chambers have working prototypes and are in fabrication. The first ring imaging Cerenkov vessel is installed on the west arm, with the second in fabrication. The third sector of the time expansion chamber is in fabrication. The TOF counters are ready for installation. Four electromagnetic calorimeter sectors are installed on the west arm and those for the east arm are ready for installation. The muon identifier panels are all installed. The muon tracker cathode strip chambers are being fabricated. Working prototypes exist for all central arm front end electronics, global first-level trigger electronics, data acquisition systems, and online control systems. Advanced simulation, reconstruction, and analysis code has been developed and tested.

8. SUMMARY

The PHENIX experiment has been described both in terms of its physics goals and its technological implementation. The heavy ion and spin physics programs represent two broad programs of experimental endeavor to address fundamental physics questions. The experiment is comprised of multiple subsystems and detector technologies to address these physics goals. Important physics will be available beginning with the first year. The highlights of the present status of construction were presented.

Acknowledgments

ORNL is managed by Lockheed Martin Energy Research Corporation under contract DE-AC05-96OR22464 with the U.S. Department of Energy. This work has also been supported by the U.S. Department of Energy under contract DE-FG02-96ER40982 with the University of Tennessee.

References

[1] "The PHENIX Experiment at RHIC," D. P. Morrison for the PHENIX Collaboration, Y. Akiba *et al.,* Nucl. Phys. A638 (1998) 565c-569c.

[2] "Spin Physics with the PHENIX Detector System," N. Saito for the PHENIX Collaboration, Y. Akiba *et al.,* Nucl. Phys. A638 (1998) 575c-578c.

[3] T. Matsui and H. Satz, Phys. Lett. B178 (1986) 416.

Chapter 10

THE COULOMB DISSOCIATION OF ^8B – HOW FAR CAN WE GO AND STILL BE RIGHT?

Moshe Gai
Dept. of Physics, U46, University of Connecticut,
2152 Hillside Rd., Storrs, CT 06269-3046, USA;
gai@uconnvm.uconn.edu; http://www.phys.uconn.edu

1. INTRODCUTION– THE COULOMB DISSOCIATION OF ^8B

The solar neutrino puzzle [1, 2] has now been upgraded to a status of a problem with solution(s) that may allow for new break through in our understanding of stars and the standard model [3, 4]. The precise knowledge of the astrophysical S_{17}-factor of the ^{7}Be(p,γ)^{8}B reaction at the Gamow peak, about 20 keV, is of crucial importance for interpreting terrestrial measurements of the solar neutrino flux [5]. This is particularly true for the interpretation of results from the Homestake, Kamiokande and SuperKamiokande experiments [6, 7, 8] which measured high energy solar neutrinos mainly or solely from ^{8}B decay.

Seven direct measurements were reported for the ^{7}Be(p,γ)^{8}B reaction [9, 10, 11, 12, 13, 14, 15]. Since the cross section at $E_{\rm cm} \approx 20$ keV is too small to be measured, the S_{17}-factors at low energies have to be extrapolated from the experimental data with the help of theoretical models (see e.g. Johnson *et al.* [16], Jennings *et al.* [17]). But also in the energy range above $E_{\rm cm} \approx 100$ keV, where measurements can be performed, experimental difficulties, mainly connected with the determination of the effective target thickness of the radioactive ^{7}Be target, hamper the derivation of accurate results. These difficulties are reflected in the fact that the five precise measurements [10, 11, 12, 14, 15] can be grouped into two distinct data sets which agree in their energy dependence but disagree in their absolute normalization by about 30%. Since this discrepancy is larger than the error of adopted S_{17} in the standard solar

Advances in Nuclear Dynamics, 5,
Edited by Bauer and Westfall, Kluwer Academic / Plenum Publishers, New York, 1999.

model, experimental studies with different methods are highly desirable for improving the reliability of the input to the standard solar model [18].

The Coulomb Dissociation (CD) [19, 20] is a Primakoff [21] process that could be viewed in first order as the time reverse of the radiative capture reaction. In this case instead of studying for example the fusion of a proton plus a nucleus (A-1), one studies the disintegration of the final nucleus (A) in the Coulomb field, to a proton plus the (A-1) nucleus. The reaction is made possible by the absorption of a virtual photon from the field of a high Z nucleus such as ^{208}Pb. In this case since π/k^2 for a photon is approximately 1000 times larger than that of a particle beam, the small cross section is enhanced. The large virtual photon flux (typically 100-1000 photons per collision) also gives rise to enhancement of the cross section. Our understanding of the Coulomb dissociation process [19, 20] allows us to extract the inverse nuclear process even when it is very small. However in Coulomb dissociation since αZ approaches unity (unlike the case in electron scattering), higher order Coulomb effects (Coulomb post acceleration) may be non-negligible and they need to be understood [23, 24]. The success of CD experiments [22] is in fact contingent on understanding such effects and designing the kinematical conditions so as to minimize such effects.

Hence the Coulomb dissociation process has to be measured with great care with kinematical conditions carefully adjusted so as to minimize nuclear interactions (i.e. distance of closest approach considerably larger then 20 fm, or very small forward angles scattering), and measurements must be carried out at high enough energies (many tens of MeV/u) so as to maximize the virtual photon flux.

2. THE RIKENII RESULTS AT 50 AMEV

The RIKENII measurement [25] of detailed angular distributions for the Coulomb dissociation of ^8B allowed us to extract the E2 amplitude in the CD of ^8B. The same data also allowed us to extract values for S_{17} as recently published [26]. The ^{208}Pb target and ^{8}B beam properties in this experiment were as in Ref. [27], but the detector system covered a large angular range up to around 9° to be sensitive to the E2 amplitude. The E1 and E2 virtual photon fluxes were calculated [25] using quantum mechanical approach. The nuclear amplitude is evaluated based on the collective form factor where the deformation length is taken to be the same as the Coulomb one. This nuclear contribution results in possible uncertainties in the fitted E2 amplitude. Nevertheless, the present results lead to a very small E2 component at low energies, below 1.5 MeV, of the order of a few percent, even smaller than the low value predicted by Typel and Baur [24] and considerably below the upper limit on the E2 component extracted by Gai and Bertulani [28] from the RIKENI data. These low values of the E2 amplitude were however recently challenged by the MSU group [29] that used interference between E1 and E2 components to observe an asymmetry of the longitudinal-momentum distribution of ^{7}Be [30]. They [29] claim: $S_{E2}/S_{E1} = 6.7^{+2.8}_{-1.9} \times 10^{-4}$ at 0.63 MeV. These results are considerably larger than reported by Kikuchi *et al.* [25], and just below the upper limit reported by Gai and Bertulani [28]. A recent analysis of the RIKENII data [25] by Bertulni and Gai [31] confirmed the small E2 extracted by Kikuchi et al. [25] as well as the negligible nuclear contribution. Note that in this analysis [31] the acceptance of the RIKENII detector is taken into account using the matrix generated by Kikuchi *et al.* [25]. Recently a possible mechanism to reduce the E2 dissociation amplitude was proposed by Esbensen and Bertsch [30].

3. THE GSI RESULTS AT 254 A MEV

An experiment to measure the Coulomb dissociation of ^{8}B at a higher energy of 254 *A* MeV was performed at GSI [32]. The material presented in this section is in fact as submitted for publication by the GSI collaboration. The experimental setup is shown in Fig. 1. The present experimental conditions have several advantages: (i) forward focusing allows us to use the magnetic spectrometer KaoS [33] at GSI for a kinematically complete measurement with high detection efficiency over a wider range of the p-^{7}Be relative energy; (ii) because of the smaller influence of straggling on the experimental resolution at the higher energy, a thicker target can be used for compensating the weaker beam intensity, (iii) effects that obscure the contribution of E1 multipolarity to the Coulomb dissociation like E2 admixtures and higher-order contributions are reduced [23, 24]. The contribution of M1 multipolarity is expected to be enhanced at the higher energy, but this allows to observe the M1 resonance peak and determine its γ width.

A ^{8}B beam was produced by fragmentation of a 350 *A* MeV ^{12}C beam from the SIS synchrotron at GSI that impinged on a beryllium target with a thickness of 8.01 g/cm^2. The beam was isotopically separated by the fragment separator (FRS) [34] by using an aluminum degrader with a thickness of 1.46 g/cm^2 with a wedge angle of 3 mrad. The beam was transported to the standard target-position of the spectrometer KaoS [33]. The average beam energy of ^{8}B in front of the breakup target was 254.5 *A* MeV, a typical ^{8}B intensity was 10^4 /spill (7s/spill). Beam-particle identification was achieved event by event with the TOF-

ΔE method by using a beam-line plastic scintillator with a thickness of 5 mm placed 68 m upstream from the target and a large-area scintillator wall discussed later placed close to the focal plane of KaoS. About 20 % of the beam particles were ^{7}Be, which could however unambiguously be discriminated from breakup ^{7}Be particles by their time of flight.

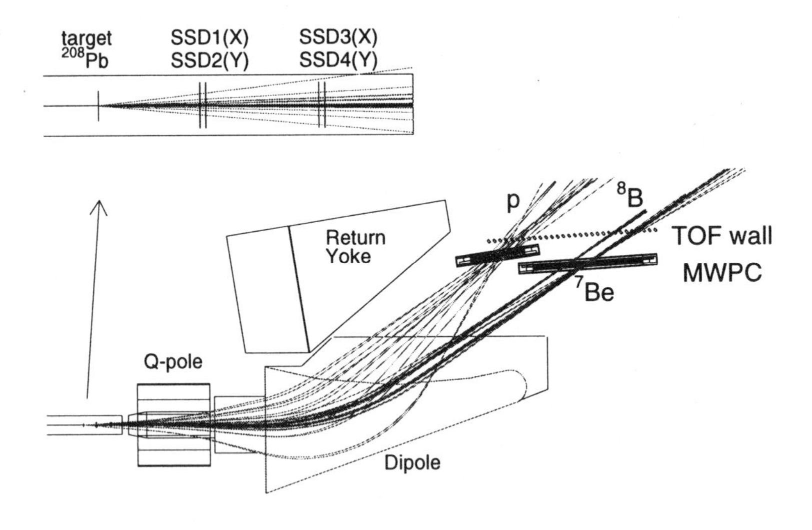

Figure 10.1 The experimental setup of the GSI experiment.

An enriched ^{208}Pb target with a thickness of 199.7 ($\pm$ 0.2) mg/cm^2 was placed at the entrance of KaoS. The average energy at the center of the target amounted to 254.0 A MeV. The reaction products, ^{7}Be and proton, were analyzed by the spectrometer which has a large momentum acceptance of $\Delta p/p \approx 50$ % and an angular acceptance of 140 and 280 mrad in horizontal and vertical directions, respectively. For scattering-angle measurement or track reconstruction of the two reaction products, two pairs of silicon micro-strip detectors were installed at about 14 and 31 cm downstream from the target, respectively, measuring either x- or y-position of the products before entering the KaoS magnets. Each strip detector had a thickness of 300 μm, an active area of 56 $\times$ 56 mm^2, and a strip pitch of 0.1 mm.

The energy deposit on each strip was recorded using an analog-multiplexing technique. This configuration enabled us to measure emission angles of the reaction products with a 1σ-accuracy of 0.9 mrad which is sufficiently smaller than the angular straggling in the target (3.4, 3.1, and 5.5 mrad for ^{8}B, ^{7}Be, and proton, respectively). The vertex of the two reaction products could be determined with a 1σ-accuracy of 5.5 mm, 0.4 mm, and 0.4 mm for in-beam, horizontal, and vertical directions, respectively. The vertex information was used to eliminate a large amount of background events produced in layers of matter other than the target.

Momenta of the reaction products were analyzed by trajectory reconstruction using position information from the micro-strip detectors and two two-dimensional multi-wire proportional chambers (MWPC) which were optimized for detecting the protons or the Be ions and located close to the focal plane of KaoS. The MWPC for protons and the heavier ions had active areas of 60×30 cm^2 and 120×35 cm^2, respectively, and a wire pitch of 5 mm. Signals from each wire were individually read by a transputer system [33], yielding a position resolution of about 1 mm by determining the centroid of the hit distribution. The momentum reconstruction was based on Monte-Carlo simulations with the code GEANT [35]. The procedure was checked successfully by sweeping a primary ^{12}C beam of 94 A MeV over the two MWPC and a secondary ^{8}B beam of 261 A MeV over the MWPC for the heavy ions.

A large-area (180×100 cm^2) scintillator wall [33] consisting of 2×15 plastic scintillator paddles with a thickness of 2 cm each was placed just behind the MWPC. It served as a trigger detector for the data acquisition system as well as a stop detector for time-of-flight (TOF) measurements. A breakup event was identified by coincident hits in the left and right 15 paddles of the scintillator wall. For counting the number of ^{8}B beam particles, single hits in only one half of the scintillator wall, downscaled by a factor of 1000, were added to the trigger.

The measured complete kinematics of the breakup products allowed us to reconstruct the p-^{7}Be relative energy and the scattering angle θ_8 of the center-of-mass of proton and ^{7}Be (excited ^{8}B) with respect to the incoming beam from the individual momenta and angles of the breakup products.

To evaluate the response of the detector system, Monte-Carlo simulations were performed using the code GEANT[35]. The simulations took into account the measured ^{8}B beam spread in energy, angle, and position at the target, as well as the influence of angular and energy straggling and energy loss in the layers of matter. Losses of the products due to limited detector sizes were also accounted for. Further corrections in the simulation are due to the feeding of the excited state at 429 keV in ^{7}Be. We used the result by Kikuchi *et al.* [25] who measured the γ-decay in coincidence with Coulomb dissociation of ^{8}B at 51.9 A MeV.

The Monte Carlo simulations yielded relative-energy resolutions from the energy and angular resolutions of the detection system to be 0.11 and 0.22 MeV (1σ) at $E_{\rm rel} = 0.6$ and 1.8 MeV, respectively. The total efficiency calculated by the simulation was found to be larger than 50% at $E_{\rm rel} = 0 - 2.5$ MeV. due to the large acceptance of KaoS.

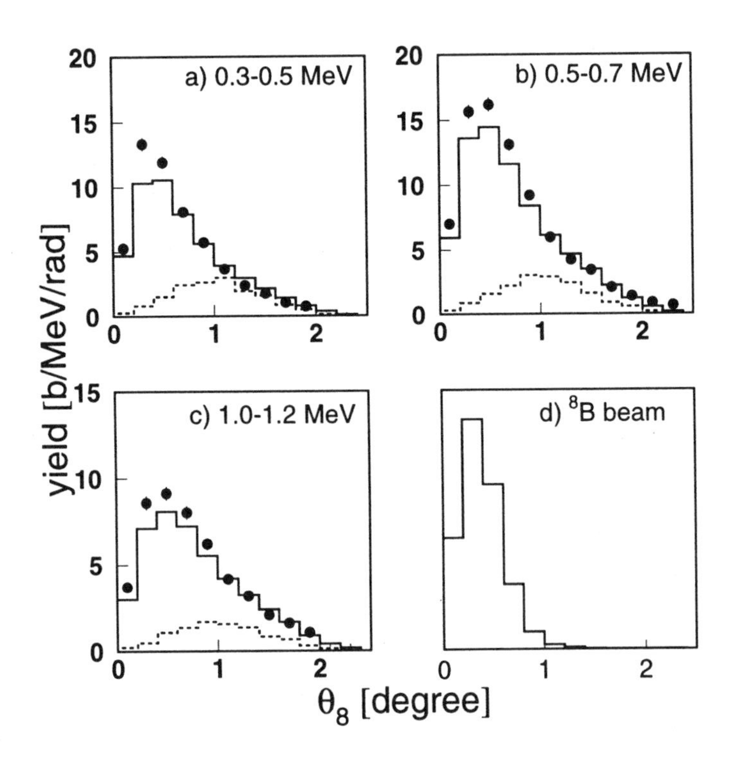

Figure 10.2 (a,b,c) Yields of breakup events plotted against θ_8, the scattering angle of the excited ^{8}B (center of mass of the p-^{7}Be system), for three relative energy bins. The full histograms show the result of a simulation taking into account the angular spread of the incident ^{8}B beam (shown in d) and assuming E1+M1 multipolarity. The dashed histograms show the effect of including an E2 contribution according to the calculations of Bertulani and Gai.

In Fig. 2, the experimental yield is plotted against the scattering angle of the excited ^{8}B for three relative-energy bins, 0.3–0.5 (a), 0.5–0.7 (b), and 1.0–1.2 (c) MeV, together with results of the Monte-Carlo simulations assuming E1 excitation. The M1 transition also contributes to the 0.5–0.7 MeV bin with the same angular dependence. Since we did not measure the incident angle of ^{8}B at the target, the experimental angular-distributions represent the θ_8 distributions folded with the angular spread of the incident beam shown in Fig. 3d. The angular resolution was estimated to be 0.35° (1σ), smaller than the observed widths. As seen in Fig. 3 the experimental distributions are well reproduced by the simulation using only E1 multipolarity, in line with the results of Kikuchi *et al.* [25].

To study our sensitivity to a possible E2 contribution, we have added the simulated E2 angular distribution from the E2 Coulomb dissociation cross section calculated by Bertulani and Gai [31]. Note that nuclear breakup effects are also included in the calculation. By fitting the experimental angular distributions to the simulated ones, we obtained 3σ upper limits of the ratio of the E2- to E1-transition amplitude of the ^{7}Be(p,γ)^{8}B reaction, $S_{\mathrm{E2}}/S_{\mathrm{E1}}$ of 0.06×10^{-4}, 0.3×10^{-4} and 0.6×10^{-4} for $E_{\mathrm{rel}} = 0.3-0.5$, $0.5-0.7$ and $1.0-1.2$ MeV, respectively. These numbers agree well with the results of Kikuchi *et al.* [25], and suggest that the results of the model dependent analysis of Davids *et al.* [29] need to be checked.

4. CONCLUSION

In conclusion we demonstrated that the Coulomb dissociation (when used with "sechel") provides a viable alternative method for measuring small cross section of interest for nuclear-astrophysics. First results on the CD of ^8B are consistent with the lower measured values of the cross section and suggest the extracted $S_{17}(0) \approx 19 \pm 1\ eV - b$. The accuracy of the extracted S-factors are now limited by our very understanding of the Coulomb dissociation process, believed to be approx. ±10%. The value of the E2 S-factor as extracted from both the RIKEN and GSI experiments are consistent and shown to be very small, S_{E2}/S_{E1} of the order of 10^{-5} or smaller.

Acknowledgments

I would like to acknowledge the work of N. Iwasa, T. Kikuchi, K. Suemmerer, F. Boue and P. Senger on the data analyses. I also acknowledge discussions and encouragements from Professors J.N. Bahcall, C.A. Bertulani, G. Baur, and Th. Delbar. Work Supported by USDOE Grant No. DE-FG02-94ER40870.

References

[1] John N. Bahcall, Neutrino Astrophysics, Cambridge University Press, New York, 1989.

[2] J.N. Bahcall, F. Calaprice, A.B. McDonald, and Y. Totsuka; Physics Today**30,#7**(1996)30 and references therein.

[3] L. Wolfenstein, Phys. Rev. **D17**(1978)2369, ibid **D20**(1979)2634.

[4] S.P. Mikheyev, and A.Yu. Smirnov, Yad. Fiz. **44**(1985)847.

[5] E.G. Adelberger, S.A. Austin, J.N. Bahcall, A.B. Balantekin, G. Bertsch, G. Bogaert, L. Buchmann, F.E. Cecil, A.E. Champagne, L. de Braeckeleer, C.A. Duba, S.R. Elliott, S.J. Freedman, M. Gai, G. Goldring, C.R. Gould, A. Gruzinov, W.C. Haxton, K.M. Heeger, E. Henley, M. Kamionkowski, R.W. Kavanagh, S.E. Koonin, K. Kubodera, K. Langanke, T. Motobayashi, V. Pandharipande, P. Parker, R.G.H. Robertson, C. Rolfs, R. Sawyer, N. Shaviv, T.D. Shoppa, K. Snover, E. Swanson, R.E. Tribble, S. Turck-Chiez, J.F. Wilkerson.; Rev. Mod. Phys. **70**(1998)1265.

[6] R. Davis, Prog. Part. Nucl. Phys. **32**(1994)13.

[7] Y. Fukuda *et al.*, Phys. Rev. Lett. **77**(1996)1683.

[8] Y. Fukuda *et al.*; Phys. Rev. Lett. **81**(1998)1158; Err.-ibid **81**(1998)4279.

[9] R.W. Kavanagh, Nucl. Phys. **15**(1960)411.

[10] P.D. Parker *et al.*, Astrophys. J. **153**(1968)L85.

[11] R.W. Kavanagh *et al.*, Bull. Am. Phys. Soc. **14**(1969)1209.

[12] F.J. Vaughn *et al.*, Phys. Rev. C **2**(1970)1657.

[13] C. Wiezorek *et al.*, Z. Phys. A **282**(1977)121.

[14] B. W. Filippone *et al.*, Phys. Rev. C **28**(1983)2222.

[15] F. Hammache *et al.*, Phys. Rev. Lett. **80**(1998)928.

[16] C.W. Johnson *et al.*, Astrophys. J. **392**(1992)320.

[17] B.K. Jennings, S. Karataglidis, and T.D. Shoppa, preprint nucl-th/9806067 v2 (1998)

[18] J.N. Bahcall and M.H. Pinsonneault, Rev. Mod. Phys. **64**(1992)885.

[19] G. Baur, C.A. Bertulani, and H. Rebel; Nucl. Phys. **A458**(1986)188.

[20] C.A. Bertulani and G. Baur; Phys. Rep. **163**(1988)299.

[21] H. Primakoff; Phys. Rev. **81**(1951)899.

[22] Moshe Gai, Nucl. Phys. **B(Sup.)38**(1995)77.

[23] C.A. Bertulani; Phys. Rev. **C49**(1994)2688.

[24] S. Typel and G. Baur; Phys. Rev. **C50**(1994)2104.

[25] T. Kikuchi, T. Motobayashi, N. Iwasa, Y. Ando, M. Kurokawa,, S. Moriya, H. Murakami, T. Nishio, J. Ruan (Gen), S. Shirato, S. Shimoura, T. Uchibori, Y. Yanagisawa, M. Ishihara, T. Kubo,

Y. Watanabe, M. Hirai, T. Nakamura, H. Sakurai, T. Teranishi, S. Kubono, M. Gai, R.H. France III, K.I. Hahn, Th. Delbar, C. Michotte, and P. Lipnik; Phys. Lett. **B391**(1996)261.

[26] T. Kikuchi, T. Motobayashi, N. Iwasa, Y. Ando, M. Kurokawa, S. Moriya, H. Murakami, T. Nishio, J. Ruan (Gen), S. Shirato, S. Shimoura, T. Uchibori, Y. Yanagisawa, H. Sakurai, T. Teranishi, Y. Watanabe, M. Ishihara, M. Hirai, T. Nakamura, S. Kubono, M. Gai, R.H. France III, K.I. Hahn, Th. Delbar,P. Lipnik, and C. Michotte; Eur. Phys. J. **A3**(1998)213.

[27] T. Motobayashi, N. Iwasa, M. Mourakawa, S. Shimoura, Y. Ando, H. Murakami, S. Shirato, J. Ruan (Gen), Y. Watanabe, N. Inabe, T. Kubo, M. Ishihara, M. Gai, R.H. France III, K.I. Hahn, Z. Zhao, T. Teranishi, T. Nakamura, Y. Futami, K. Furataka, and T. Delbar; Phys. rev. Lett. **73**(1994)2680. and N. Iwasa *et al.*; Jour. Phys. Soc. Jap. **65**(1996)1256.

[28] Moshe Gai, and Carlos A. Bertulani; Phys. Rev. **C52**(1995)1706.

[29] B. Davids *et al.*, Phys. Rev. Lett. **81**(1998)2209.

[30] H. Esbensen and G.F. Bertsch; Phys. Lett. **B359**(1995)13 ibid 531 Nucl. Phys. **A600**(1996)37.

[31] Carlos A. Bertulani and Moshe Gai; Nucl. Phys. **A636(1998)227**.

[32] N. Iwasa, F. Boue, G. Surowka, T. baumann, B. Blank, S. Czajkowski, A. Forster, M. Gai, H. Geissel, E. Grosse, M. Hellstrom, P. Koczon, B. Khlmeyer, R. Kulessa, F. Laue, C. Marchand, T. Motobayashi, H. Oeschler, A. Ozawa, M.S. Pravikoff, E. Schwab, P. Senger, J. Speer, A. Surowiec, C. Sturm, K. Summerer, T. Teranishi, F. Uhlig, A. Wagner, and W. Walus; GSI Scientific Report, 1998.

[33] P. Senger *et al.*, Nucl. Instr. Meth. A **327**, 393 (1993).

[34] H. Geissel *et al.*, Nucl. Instr. Meth. B **70**, 286 (1992).

[35] Detector description and simulation tool by CERN, Geneva, Switzerland.

Chapter 11

TWO-PROTON CORRELATIONS FROM PB+PB CENTRAL COLLISIONS

F. Wang (for the NA49 collaboration)
Nuclear Science Division, Lawrence Berkeley National Laboratory
One Cyclotron Road, Berkeley, CA 94720, USA
FQWang@lbl.gov

Abstract The two-proton correlation function at midrapidity from Pb+Pb central collisions at 158 AGeV has been measured by the NA49 experiment. The preliminary results are compared to model predictions from proton source distributions of static thermal Gaussian sources and the transport models of RQMD and VENUS. We obtain an effective proton source size $\sigma_{\text{eff}} = 4.0 \pm 0.15(\text{stat.})^{+0.06}_{-0.18}(\text{syst.})$ fm. The RQMD model underpredicts the correlation function ($\sigma_{\text{eff}} = 4.41$ fm), while the VENUS model overpredicts the correlation function ($\sigma_{\text{eff}} = 3.55$ fm).

Keywords: two-proton correlation, heavy ion

Nuclear matter under extreme conditions of high energy density has been extensively studied through high energy heavy ion collisions. The baryon density plays an important role in the dynamical evolution of these collisions. To measure the baryon spatial density, one needs information on the space-time extent of the baryon source. The space-time extent of the proton source at freeze-out can be inferred from two-proton correlation functions. The peak in the correlation function at $q_{\text{inv}} = \sqrt{-q_\mu q^\mu}/2 \approx 20$ MeV/c is inversely related to the effective size of the proton source [1, 2]. In the above, q_μ is the difference of the proton 4-momenta, and q_{inv} is the momentum magnitude of one proton in the rest frame of the pair.

We report the first, preliminary results on the two-proton correlation function in the midrapidity region from Pb+Pb central collisions at 158 GeV per nucleon. The measurement was done by the NA49 experi-

Advances in Nuclear Dynamics, 5,
Edited by Bauer and Westfall, Kluwer Academic / Plenum Publishers, New York, 1999.

ment [3] at the SPS, using the 5% most central events, corresponding to collisions with impact parameter $b \leq 3.3$ fm.

Two independent analyses were performed on the data: dE/dx analysis which used particles in the rapidity range $2.9 < y < 3.4$ (assuming proton mass) with at least 70% probability to be a proton obtained from their ionization energies deposited in the time projection chambers (TPCs), and TOF analysis which used identified protons by combining the time of flight information and the dE/dx. Both analyses used protons up to a transverse momentum $p_T = 2$ GeV/c.

The proton samples are contaminated by weak decay protons ($\Lambda + \Sigma^0$ and Σ^+) which are incorrectly reconstructed as primary vertex tracks. From the measured single particle distributions [4, 5] and model calculations of RQMD and VENUS, we estimate the contamination to be $15^{+15}_{-5}\%$. This results in about $30^{+20}_{-10}\%$ of the proton pairs having at least one proton from weak decays. These protons do not have correlation with protons from the primary interactions. In the dE/dx analysis, further contamination is present from kaons on the lower tail of their dE/dx distribution merging into the region where particles have at least 70% probability to a proton. This resulted in 25% K^+p pairs and fewer than 2% K^+K^+ pairs in the proton pair sample. The coulomb hole of K^+p pairs does not affect the resulting two-proton correlation function in the interested low q_{inv} region below 50 MeV/c.

The two-proton correlation function is obtained as the ratio of the q_{inv} distribution of true proton pairs to that of mixed-event pairs with protons from different events. The number of mixed-event pairs was large enough so that the statistical error on the correlation function is dominated by the statistical uncertainty in the number of true pairs. To eliminate the effect of close pair reconstruction inefficiency, a cut of 2 cm was applied on the pair distance at the middle plane of the TPC for both true and mixed-event pairs.

The correlation functions obtained from the two analyses can be directly compared because of the nearly symmetric acceptances about midrapidity $y_{\text{c.m.}} = 2.9$. The correlation functions (with the K^+p contamination corrected in the dE/dx analysis) are consistent. In the results reported below, the true pairs and the respective mixed-event pairs from the two analyses were combined. The combined sample had about 10^5 pairs with $q_{\text{inv}} < 120$ MeV/c, 75% of which were from the TOF analysis. The q_{inv} distributions of the true pairs and the mixed-event pairs are shown in the top panel of Fig. 11.1. The number of mixed-event pairs was normalized to that of true pairs in the range $q_{\text{inv}} > 500$ MeV/c. The resulting correlation function is shown in the bottom panel of Fig. 11.1. The prominent peak at $q_{\text{inv}} \approx 20$ MeV/c is evident. There is a statistically significant structure in the correlation function at $q_{\text{inv}} \approx 70$ MeV/c. Many systematic effects have been studied; none has been identified that can account for the structure. There have been suggestions that a sharp edge in the two-proton density distribution of the source can produce such an effect [6].

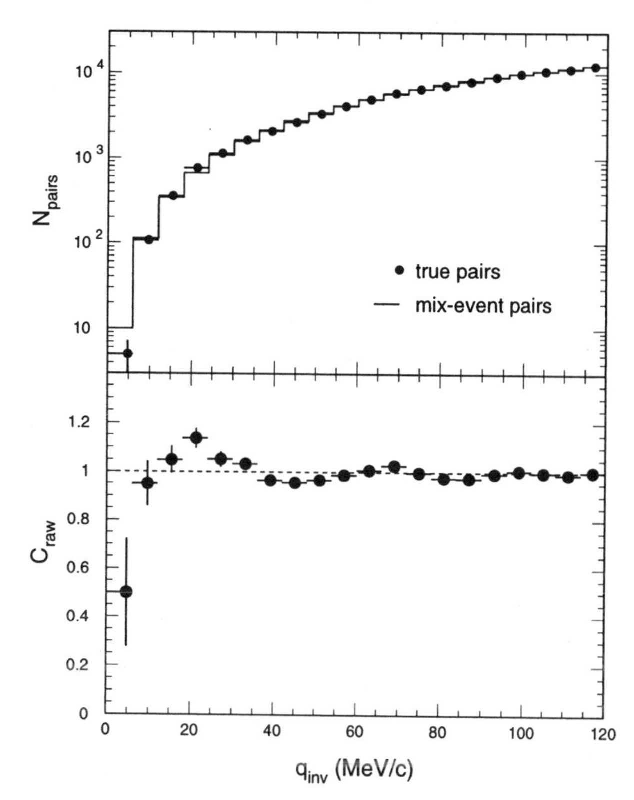

Figure 11.1 Top: q_{inv} distributions of true proton pairs (points) and mixed-event proton pairs (histogram). Bottom: the measured two-proton correlation function. The contamination from weak decay protons and the finite momentum resolution were not corrected. The errors shown are statistical only.

We correct the measured correlation function for the contamination from weak decay protons and for the finite momentum resolution. The effect of the momentum resolution is only significant in the first two data points of the measured correlation function. The corrected correlation function is plotted in Fig. 11.2 as filled points.

In order to assess the proton freeze-out conditions, we compare the measured two-proton correlation function to theoretical calculations. Given the proton phase space density distribution, the two-proton correlation function can be calculated by the Koonin-Pratt Formalism [1, 7]. Two types of proton source were used:

(I) Isotropic Gaussian sources of widths $\sigma_{x,y}, \sigma_z$ and σ_t for the space and time coordinates of protons in the source rest frame, and thermal momentum distribution of temperature T. No correlation between space-time and momentum of the protons is present. Following combinations of parameters were used in the calculations: $\sigma_{x,y} = \sigma_z = \sigma$, $\sigma_t = 0$ and σ, and $T = 120$ MeV (as derived in [8]), 300 MeV (measured inverse slope of proton transverse mass spectrum [4]) and 70 MeV (inverse slope observed at low energy, as an extreme).

(II) Protons generated for Pb+Pb central collisions ($b \leq 3.3$ fm) at 158 AGeV by two microscopic transport models: the Relativistic Quantum Molecular Dynamics (RQMD) model (version 2.3) [9] and the VENUS model (version 4.12) [10]. Both models describe a variety of experimental data on single particle distributions reasonably well. Protons at freeze-out have correlations between space-time and momentum intrinsic to the dynamical evolution in the models.

Only protons in the experimental acceptance are used to calculate the two-proton correlation functions, the results of which are shown in Fig. 11.2 for RQMD, VENUS, and the Gaussian source with $\sigma_{x,y} = \sigma_z = \sigma_t = 3.8$ fm and $T = 120$ MeV, respectively. While VENUS overpredicts the amplitude of the correlation function, RQMD slightly underpredicts the amplitude.

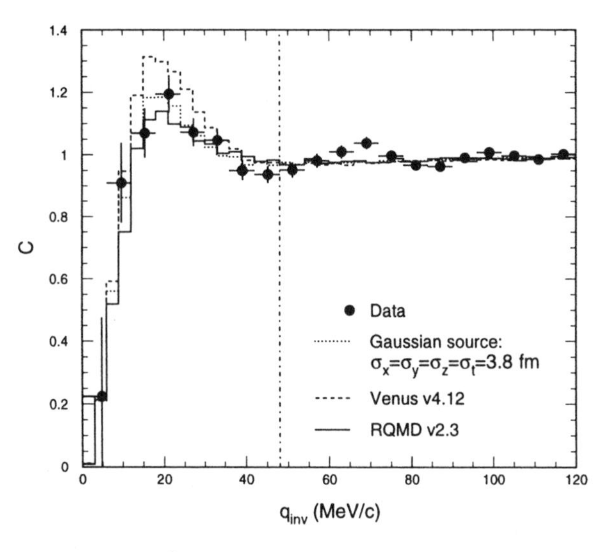

Figure 11.2 The two-proton correlation function after corrections for the 30% contamination due to weak decay protons and the finite momentum resolution (points), compared to calculations for a Gaussian source (dotted), and for freeze-out protons from RQMD v2.3 (solid) and VENUS v4.12 (dashed). The errors shown on the data points are statistical only.

We use χ^2/ndf, the normalized mean square of the point-to-point difference between the data and the calculation in the range $q_{\text{inv}} < 48$ MeV/c (*i.e.*, 8 data points or ndf = 8), to quantify how well the calculations agree with data. We characterize the effective size of the proton source from each model by $\sigma_{\text{eff}} = \sqrt[3]{\sigma_{\Delta x} \cdot \sigma_{\Delta y} \cdot \sigma_{\Delta z}}/\sqrt{2}$, where $\sigma_{\Delta x}, \sigma_{\Delta y}$ and $\sigma_{\Delta z}$ are the Gaussian widths fitted to the distributions in $\Delta x, \Delta y$ and Δz, distance between the protons of close pairs with $q_{\text{inv}} < 48$ MeV/c. The distance is evaluated in the pair rest frame at the time when the later particle freezes out [11]. Respectively for RQMD and

VENUS, the χ^2/ndf values are 1.47 and 2.29; the fitted Gaussian widths are $(\sigma_{\Delta x}, \sigma_{\Delta y}, \sigma_{\Delta z}) = (5.91, 6.00, 6.83)$ fm and $(4.57, 4.57, 6.08)$ fm, where z is the longitudinal coordinate; the resulting effective sizes are $\sigma_{\text{eff}} =$ 4.41 fm and 3.55 fm.

In Fig. 11.3, we study the χ^2/ndf as a function of σ_{eff}. The χ^2/ndf values from all three models follow roughly the same solid line, drawn through the points for the Gaussian sources with $T = 120$ MeV to guide the eye [12]. ¿From the minimum χ^2/ndf point and the points where χ^2/ndf has increased by 0.125 (note ndf = 8), we extract $\sigma_{\text{eff}} = (4.0 \pm 0.15)$ fm, where 0.15 fm is the statistical error [13]. By applying a correction to the measured correlation function using a proton pair contamination of 20% and 50%, we obtain a systematic error of $^{+0.06}_{-0.18}$ fm on σ_{eff}. We note that $\sigma_{\text{eff}} = 4.0$ fm corresponds to a uniform density hard sphere of radius $\sqrt{5}\sigma_{\text{eff}} = 8.9$ fm.

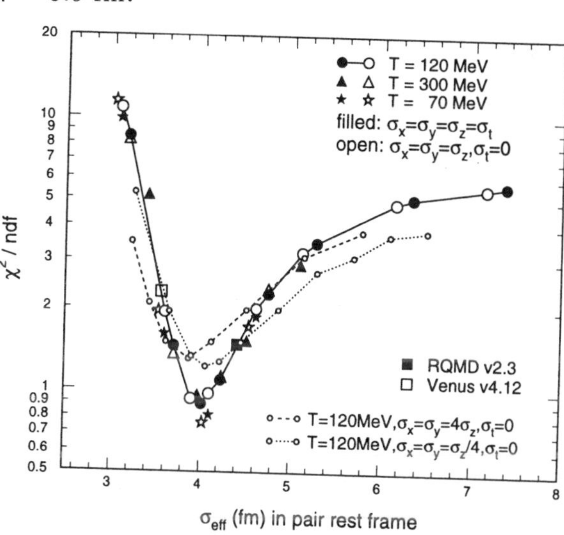

Figure 11.3 The χ^2/ndf values as function of the effective source size σ_{eff} for various model calculations with respect to the measured correlation function. The model calculations are for Gaussian sources (circles, triangles, and stars) and for freeze-out protons generated by RQMD v2.3 (filled square) and VENUS v4.12 (open square).

We have also studied Gaussian sources with extreme shapes: oblate $\sigma_{x,y} = 4\sigma_z$ and prolate $\sigma_{x,y} = \sigma_z/4$ (both with $\sigma_t = 0$). The corresponding χ^2/ndf versus σ_{eff} curves are shown in Fig. 11.3 as dashed and dotted lines, respectively. They do not fall along the solid line, implying that the two-proton correlation function has certain sensitivity to the shape of the proton source. In principle, multi-dimensional two-proton correlation functions could reveal the shape of the proton source. However, the lack of statistics has prevented us from drawing conclusions.

We note that there is no simple relation between σ_{eff} and the proton source in RQMD and VENUS model. ¿From single proton distributions at freeze-out, we obtain the following Gaussian widths in the source rest frame: $(\sigma_x, \sigma_y, \sigma_z, \sigma_t) = (7.6, 7.7, 6.4, 7.0)$ fm for RQMD and $(3.6, 3.6, 4.3, 1.9)$ fm for VENUS. The two-proton correlation function, therefore, appears to measure a smaller region of the source, *i.e.*, $\sigma_{\Delta i} < \sqrt{2}\sigma_i$ $(i = x, y, z)$, due to space-time-momentum correlation. The effect is more dramatic in RQMD than in VENUS, which is consistent with the expectation that more secondary particle interactions in RQMD result in a stronger correlation between space-time and momentum of freeze-out protons. The fact that the χ^2/ndf values versus σ_{eff} for RQMD and VENUS lie on the curves obtained for the isotropic Gaussian sources, in which no space-time-momentum correlation is present, suggests that the effect of the space-time-momentum correlation is small in σ_{eff}.

Finally, we comment on our two-proton correlation function in the context of other measurements. The pion source size measured by interferometry increases with the pion multiplicity [14], which increases steadily with bombarding energy in similar colliding systems [15]. Due to the large pion-nucleon cross-section, one would expect that protons and pions freeze-out under similar conditions, therefore, the proton source size would increase with bombarding energy as well. However, our measurement, in conjunction with preliminary results obtained at GSI [16] and AGS [17] energies, shows that the peak height is rather insensitive to the bombarding energy. This implies that the effective sizes of the freeze-out proton sources are similar in heavy ion collisions over a wide energy range. More detailed studies are needed to understand the possible acceptance and instrumental effects in these measurements.

In summary, the NA49 experiment has measured the two-proton correlation function at midrapidity from Pb+Pb central collisions at 158 GeV per nucleon. ¿From comparisons between the data and the calculations, we extract an effective proton source size of $\sigma_{\text{eff}} = 4.0 \pm 0.15(\text{stat.})^{+0.06}_{-0.18}(\text{syst.})$ fm. The RQMD model underpredicts the amplitude of the correlation function ($\sigma_{\text{eff}} = 4.41$ fm), while the VENUS model overpredicts the amplitude ($\sigma_{\text{eff}} = 3.55$ fm). Due to the space-time-momentum correlation, the two-proton correlation function is sensitive only to a limited region of the proton source. Our measurement together with the measurements at lower energies suggest a very weak dependence of the two-proton correlation function on bombarding energy.

Acknowledgment: I would like to thank M. Cristinziani who did part of the data analysis. Fruitful discussions with Drs. P.M. Jacobs, R. Lednicky, S. Panitkin, A.M. Poskanzer, H.G. Ritter, P. Seyboth, S. Voloshin and N. Xu are greatly acknowledged. This work was partially supported by the Director, Office of Energy Research, Division of Nuclear Physics of the Office of High Energy and Nuclear Physics of the US Department of Energy under Contract DE-AC03-76SF00098.

References

[1] S.E. Koonin, Phys. Lett. **70B**, 43 (1977).

[2] D.H. Boal, C. Gelbke, and B.K. Jennings, Rev. Mod. Phys. **62**, 553 (1990); R. Lednicky and V.L. Lyuboshits, Sov. J. Nucl. Phys. **35**, 770 (1982).

[3] S. Wenig (NA49 Coll.), NIM **A409**, 100 (1998); S. Afanasiev *et al.* (NA49 Coll.), CERN-EP/99-001 (1999), NIM in press.

[4] H. Appelshäuser *et al.* (NA49 Coll.), nucl-ex/9810014 (1998), Phys. Rev. Lett. in press.

[5] C. Bormann *et al.* (NA49 Coll.), J. Phys. **G23**, 1817 (1997).

[6] M.I. Podgoretsky, Sov. J. Part. Nucl. **20**, 266 (1989); A. Makhlin and E. Surdutovich, hep-ph/9809278 (1998); D. Brown and E. Shuryak, private communications.

[7] S. Pratt and M.B. Tsang, Phys. Rev. C **36**, 2390 (1987).

[8] H. Appelshäuser *et al.* (NA49 Coll.), Eur. Phys. J. **C2**, 661 (1998).

[9] H. Sorge, H. Stocker, and W. Greiner, Ann. Phys. **192**, 266 (1989); H. Sorge *et al.*, Phys. Lett. B **289**, 6 (1992); H. Sorge, H. Stöcker, and W. Greiner, Nucl. Phys. **A498**, 567c (1989); H. Sorge, Phys. Rev. C **52**, 3291 (1995).

[10] K. Werner, Phys. Rep. **232**, 87 (1993).

[11] For a chaotic Gaussian source, σ_{eff} is close to the non-relativistic approximation of the effective size in the source rest frame, $\sqrt[6]{(\sigma_{x,y}^2 + \frac{T}{2m}\sigma_t^2)^2 \cdot (\sigma_z^2 + \frac{T}{2m}\sigma_t^2)}$, where $\frac{T}{m}$ is the one-dimensional mean squared velocity of the thermal protons, and $\frac{T}{2m}$ is that of the close pairs.

[12] The shape of the χ^2/ndf versus σ_{eff} curve can be understood as follows: the χ^2/ndf approaches an asymptotic value for large sources to which the two-proton correlation function is not sensitive to the source size any more, whereas for small sources, the two-proton correlation function is very sensitive and its strength decreases rapidly with increasing source size.

[13] Particle Data Group, Eur. Phys. J. **C3**, 172–177 (1998).

[14] K. Kaimi *et al.* (NA44 Coll.), Z. Phys. C **75**, 619 (1997); I.G. Bearden *et al.* (NA44 Coll.), Phys. Rev. C **58**, 1656 (1998); M.D. Baker (E802 Coll.), Nucl. Phys. **A610**, 213c (1996).

[15] M. Gazdzicki and D. Roehrich, Z. Phys. C **65**, 215 (1995).

[16] C. Schwarz *et al.* (ALADIN Coll.), nucl-ex/9704001 (1997); R. Fritz *et al.* (ALADIN Coll.), nucl-ex/9704002 (1997).

[17] S. Panitkin *et al.* (E895 Coll.), BNL/E895 preliminary.

Chapter 12

NUCLEAR PHASE TRANSITION STUDIED WITHIN AMD-MF

Yoshio Sugawa and Hisashi Horiuchi
Physics Department
Kyoto University
Kyoto, Japan
sugawa@ruby.scphys.kyoto-u.ac.jp, horiuchi@ruby.scphys.kyoto-u.ac.jp

Abstract Liquid-gas phase transition of finite nucleus is studied by means of microscopic reaction theory, AMD-MF. By caluculating time development of hot system in a potential well, thermodynamic variables such as temperature and pressure and their relationship to the excitation energy of the system are obtained. We see clearly the existence of three regions in calculated caloric curve; namely liquid-dominant, plateau and gas regions. The transition from liquid-dominant phase to gas-phase begins with the cracking of hot liquid nucleus and disintegration into fragments. Property of plateau region strongly depends on the pressure of the system. Gas phase is well reproduced by van der Waals equation.

Keywords: Nuclear Phase Transition, Molecular dynamics, Thermodynamics

1. INTRODUCTION

The idea of nuclear phase transition and the possibility of its observation have been one of the major concerns in nuclear physics. Its origin is the similarity of inter-nucleon force to the inter-molecular force, the combination of repulsive and attractive forces. Theoretically, phase transition of nuclear matter and finite nuclei have been studied by using several models with various effective forces. In order to investigate the possible nuclear phase transition, the collision of two heavy ions has been utilized. When two nuclei collides, the incident energy is converted to the thermal energy of nuclei and the heated nuclei break into pieces.

Advances in Nuclear Dynamics, 5,
Edited by Bauer and Westfall, Kluwer Academic / Plenum Publishers, New York, 1999.

This nuclear multi-fragmentation reflects the information of the thermodynamic property of such high temperature stage. Therefore nuclear collision and observation of multi-fragmentation is one of the important means to study the nuclear phase transition.

Recently, the experimental result presented by ALADIN collaboration[1] called much interest in this field. The extracted caloric curve, or temperature - excitation energy diagram, consists of three typical regions. When the temperature is not so high, the temperature rises as excitation energy increases as is usually described by the Fermi gas model. They called this energy region as evaporation region. At certain excitation energy, 3 MeV/nucleon, the temperature doesn't go up even if the excitation energy becomes higher. This anomaly is called as plateau of temperature and they insisted on the correspondence of this plateau to the liquid-gas phase transition. Finally, when the excitation energy exceeds 10 MeV, the temperature begins to rise steeply. This energy region is called the vapor region.

Since there are several difficulties in those experiments such as definition of the temperature, impact parameter dependence of excitation energy of points on caloric curves, there is not yet concrete conclusion whether the caloric curve they show indicates the occurence of liquid-gas phase transition or not.

One of the difficulties underlying in analysis of this kind of heavy ion collision is that the observation is limited to the fragment from excited nucleus and that we don't know the detailed information of collision dynamics: how the excitation energy of and the temperature of transient colliding system is reflected to the observables. From this point of view, we believe it is important to deal with the heavy ion colision directly within microscopic molecular dynamics framework and to have the detailed knowledge of excited nucleus. At the same time, it is significant to make it clear how a finite nucleus is excited and change its property from Fermi gas model-like liquid to classical gas.

To this end, we explored the thermodynamical property of finite nuclear system within AMD(antisymmetrized molecular dynamics framework). By calculating long time evolution of the various heated finite nuclear system, we will show the thermodynamical variables and the dependence of liquid gas nuclear phase transition on the volume of the system.

2. FRAMEWORK AND CALCULATIONS

Here we briefly explain the framework of AMD and AMD-MF. We refer to [2, 3] for the detailed information. In AMD, the wave function of the total nuclear system is represented by one Slater determinant,

$$|\Phi\rangle = \frac{1}{\sqrt{A!}} \det[\varphi_i(j)]. \tag{12.1}$$

Each nucleon wave function is the multiplication of spatial part that is Gaussian wave packet and spin-isospin part that is constant in time,

$$\varphi_i = \phi_{\mathbf{Z}_i} \chi_{\alpha_i}, \tag{12.2}$$

where

$$\langle \mathbf{r} | \phi_{\mathbf{Z}_i} \rangle = \left(\frac{2\nu}{\pi} \right)^{\frac{3}{4}} \exp\left\{ -\nu \left(\mathbf{r} - \frac{\mathbf{Z}_i}{\sqrt{\nu}} \right)^2 + \frac{1}{2} \mathbf{Z}_i^2 \right\}. \tag{12.3}$$

The time development of the system is obtained by using the time-dependent variational principle,

$$\delta \int_{t_1}^{t_2} dt \frac{\langle \Phi(\mathbf{Z}) | \left(i\hbar \frac{d}{dt} - H \right) | \Phi(\mathbf{Z}) \rangle}{\langle \Phi(\mathbf{Z}) | \Phi(\mathbf{Z}) \rangle} = 0, \tag{12.4}$$

which leads to

$$i\hbar \sum_{j\tau} C_{i\sigma,j\tau} \dot{Z}_{j\tau} = \frac{\partial \langle H \rangle}{\partial Z_{i\sigma}^*}, \tag{12.5}$$

where

$$C_{i\sigma,j\tau} = \frac{\partial^2}{\partial Z_{i\sigma} \partial Z_{j\tau}^*} \ln \langle \Phi(\mathbf{Z}) | \Phi(\mathbf{Z}) \rangle, \tag{12.6}$$

$$\langle H \rangle = \frac{\langle \Phi(\mathbf{Z}) | H | \Phi(\mathbf{Z}) \rangle}{\langle \Phi(\mathbf{Z}) | \Phi(\mathbf{Z}) \rangle}. \tag{12.7}$$

The equation (12.5) is the equation of motion that determines the time development of the complex vectors $\mathbf{Z}$.

It is shown[3] that if we only follow the time evolution of the central value of wave packets $\mathbf{Z}$, thermodynamical property such as nuclear temperture is not correctly refrected to the kinetic energy of nucleons that are emitted from excited nucleus, Therefore we apply the momentum fluctuation procedure [3] at the emission. We call this extention AMD-MF(momentum fluctuation).

Within AMD-MF, we put 36 nucleons into a container potential with has the following shape,

$$V_{\text{wall}} = \begin{cases} 0 & \text{r} < r_{\text{wall}}, \\ \infty & \text{r} \geq r_{\text{wall}}. \end{cases} \tag{12.8}$$

As r_{wall}, we take 5,6,...,12 fm and total excitation energies of the system are taken from $E^*/A = 2$MeV to $E^*/A = 38$MeV. For each pair of r_{wall} and E^*/A, we calculated up to about 15000 fm/c to calculate the property of equilbrated system. Cluster of the system at each time step is defined from the chain cluster method(nucleon i and j are connected if $|\mathbf{R}_i - \mathbf{R}_j| \leq 2.5$fm.) The heaviest cluster in the system at the time step is refered as 'liquid' nucleus in the followings. Unconnected nucleons are regarded as 'gas' nucleons and the temperature of the system at the moment is defined from their kinetic energies, $\frac{3}{2}T = \frac{1}{N_{\text{gas}}}\sum_{i=1}^{N_{\text{gas}}} K_i$. Pressure is defined from the momentum given to the wall from the nucleons when they hit the container wall. One have to keep it in mind that these physical variables are defined at each time step and they always have fluctuations around the central values.

3. CALORIC CURVES

In fig.12.1, we show the caloric curves for the three cases of the radius of the system, $r_{\text{wall}} = 5, 8$ and 12. In this figure, we can wee that the

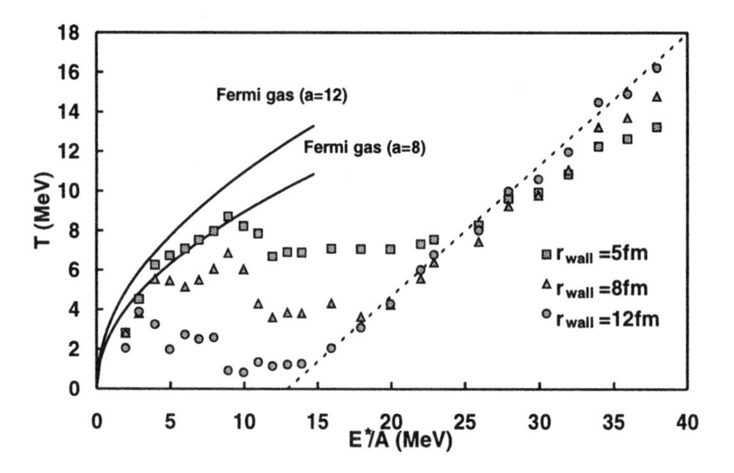

Figure 12.1 Caloric curves for the case of $r_{\text{wall}} = 5, 8$ and 12. Thick curves are the caloric curve for Fermi-gas model $E^*/A = T^2/a$ with $a = 8$ and $a = 12$. Dotted line is that of ideal gas.

liquid-gas phase transition takes place in all case of volumes. For low excitation energy, caloric curves follow that of Fermi gas model expectation for all case of r_{wall}. In this energy region, the system consists of one heavy excited nucleus and several gas nucleons. Therefore the thermodynamic property of the total system is ruled by that of heavy nucleus(Fermi gas-like). For much higher energies, the caloric curves follows that of ideal gas. In this energy region, the system is considered to be in gas phase. There are somewhat flat reigon midst of these two phases. In this region, tempareture doesn't rise considerably so we call this plateau. From these obervations, we can say that the system experiences the liquid-gas phase transition in all cases of volumes.

When we look at the caloric curves more precisely, we find there are fall of the temperature just before the plateau region, when the excitation energy grows.

The reason for this fall should be attributed to the sudden increase of the number of clusters in the system. In fig.12.2, we show the corresponding change in the number of clusters for the case of $r_{\text{wall}} = 8\text{fm}$ When the excitation energy is low ($E^*/A \leq 9\text{MeV}$), number of cluster

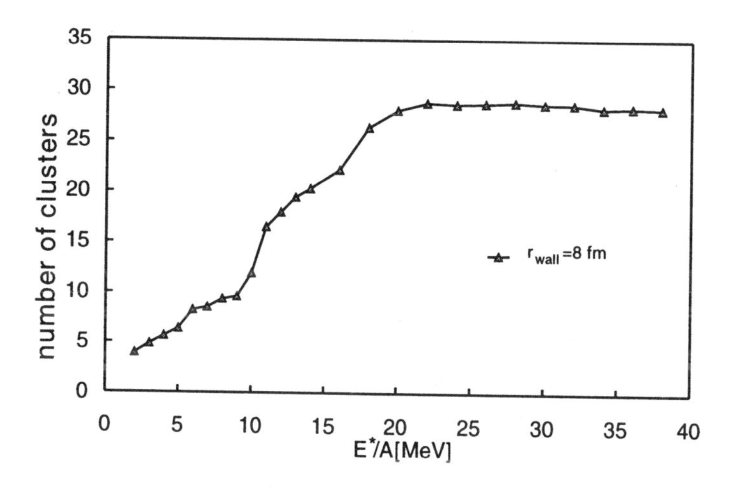

Figure 12.2 Average number of clusters in the system for each point of the caloric curve. The curve experiences sudden increase of the number(see text.)

gradually increases. However, at around 10MeV/A, it shows a sudden rise. This excitation energy corresponds to the point where the temparature suddenly falls in the caloric curve. At this energy, system enters into the plateau region.

4. CRACKING—MULTIPLICITY DISTRIBUTIONS

This understanding can also be certified from the change of multiplicity distributions(Fig. 12.3).

In the upper panel, three typical multiplicity distribution are shown for each case of radius. When the excitation energy is low, there is a

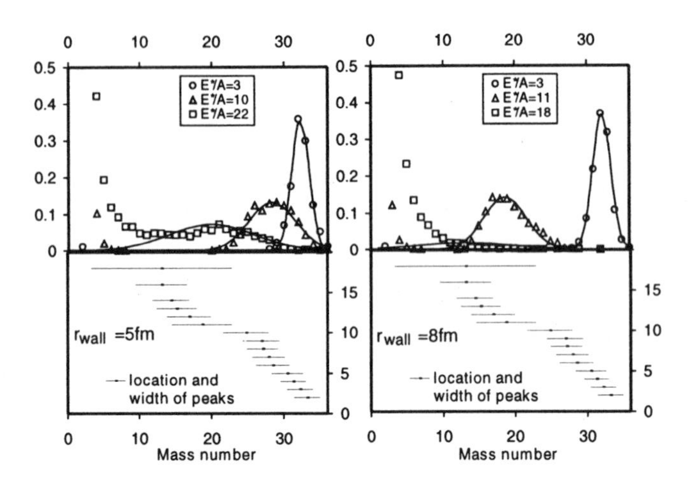

Figure 12.3 Change in the multiplicity distribution in the case of $r_{\mathrm{wall}} = 5\mathrm{and}8\mathrm{fm}$. Multiplicity peaks for the liquid nucleus are fitted by Gaussian(Upper panels.) Diversity of the center and width of the Gaussian are shown in the lower panels.

peak in the distriution, whose central value means the most probable mass number of liquid nucleus. As the excitation energy grows, the peak moves to left and width gradually becomes larger. At the excitation energy that corresponds to the end of plateau region, distribution loses its peak shape. The lower panel show the change of the location of such peaks and the width. In the case of $r_{\mathrm{wall}} = 8\mathrm{fm}$, sudden jump we mentioned above is clear. For lower excitation energy, the mass number of liquid cluster gradually decreases up to certain excitation energy. At $E^*/A = 10\mathrm{MeV}$, the mass number suddenly decreases. Therefore, the way the system changes from liquid dominant phase to plateau region can be depicted as follows: As the excitation energy rises, liquid nucleus becomes excited and it emits nucleons. At certain energy, liquid nucleus can't hold excitation energy any more and the energy put into the system is only poured into the separation energy of disintegrated clusters and emitted nucleons and their kinetic energies. Therefore, the number of cluster suddenly increases at this excition energy. We call this feature 'cracking', which is mentioned in [4]. On the other hand, in the case of $r_{\mathrm{wall}} = 5\mathrm{fm}$, there is no such a jump in the lower panel of Fig.12.3. The reason for this lies in the radius of the system. In this case, there is not enough room for hot nucleus to disintegrate or emit too much nucleons. Therefore, in this volume ($r_{\mathrm{wall}} = 5\mathrm{fm}$), cracking is gradual. From the density snapshot(Fig. 12.4), the difference between the volumes is obvious. Thus, the way system enters plateau region, or how cracking takes places, is also dependent on the volume of the system.

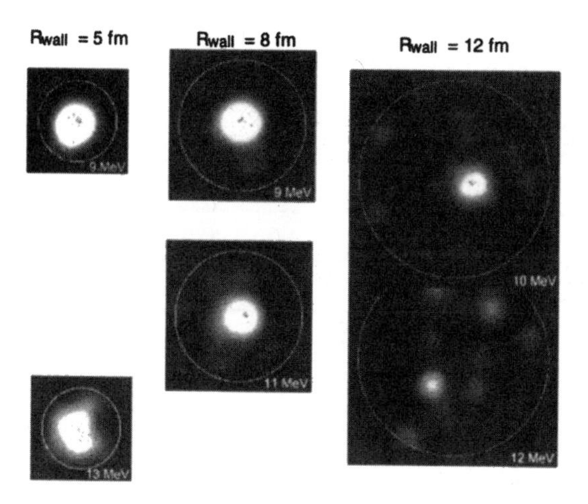

Figure 12.4 Density snapshots of several excitation energies in the plateau region. In the case of $r_{\text{wall}} = 12\text{fm}, E^*/A = 12\text{MeV}, \alpha$ clusters are formed besides liquid nucleus.

5. PLATEAU REGION—CORRELATION BETWEEN THE HEAVIEST AND THE SECOND HEAVIEST CLUSTER

We next show the property of the nuclear phase transition by looking at the favored cluster configuration during the transition. In Fig. 12.5, the correlation between the heaviest cluster and the second heaviest cluster that exist during the time development at several excitation energies are shown.

When the system is in the smallest volume ($r_{\text{wall}} = 5\text{fm}$ in Fig.12.5), there are various sizes of clusters as the 1st and 2nd heaviest clusters. Transitional phase from the liquid-dominant to the gas phase consists of clusters with various sizes. Therefore, the transition is somewhat violent as the boiling water.

When the system is in the middle volume ($r_{\text{wall}} = 8\text{fm}$ in Fig.12.5), clusters with $A > 2$ are hardly formed as the second heaviest cluster and the equilibration is established between the heaviest nucleus and the mixture of gas nucleons and deuterons in the plateau region. This description is closer to the naive picture of phase transition where mixed phase contains only liquid nucleus and gas nucleons.

When the system is in the largest volume ($r_{\text{wall}} = 12\text{fm}$ in Fig.12.5, during the plateau, many α particles are produced in the space outside of the liquid nucleus. In this volume, the plateau temperature is very low (around 1 MeV/nucleon); therefore, the emitted nucleons don't have

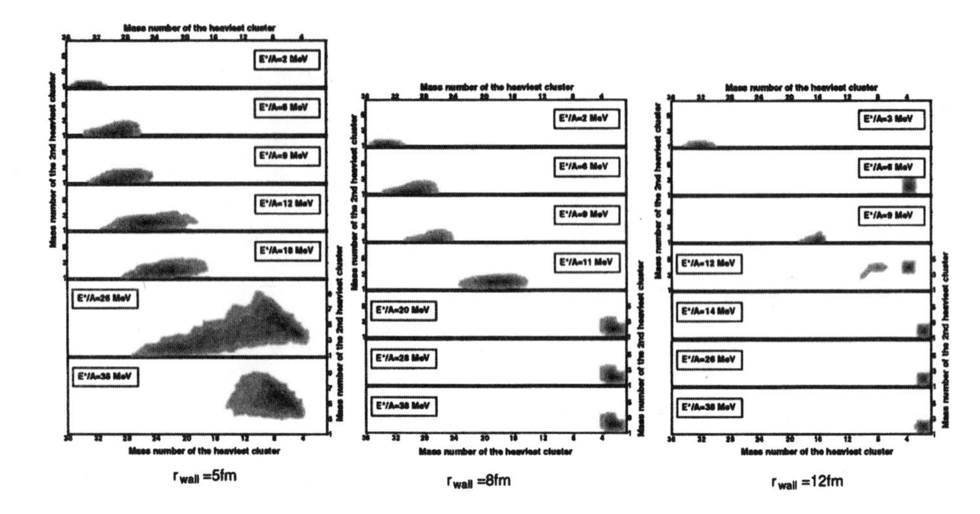

Figure 12.5 Correlation between the heaviest cluster and the second heaviest cluster at each excitation energy in the case of $r_{\rm wall} = 5, 8 \text{and} 12$ fm.

much kinetic energy and are easily recombined to form heavier cluster in the 'gas' space.

From these figures, one finds that the transitional state between the liquid-dominant phase and gas-phase is not simple mixture of these two, but it gives qualitatively different pathways from liquid phase to gas phase depending on the volume of the system. Our figures are drawn by changing the total excitation energy and keeping the volume constant. The nuclear system that is formed during the nuclear collision has more complex feature, since the volume changes during the time evolution. Therefore, verification of the nuclear phase transition might require much more delicate treatment.

6. GAS PHASE

In Fig.12.6, the relationship of these temperature and pressure is shown. Each square represents the averaged value of p and T for the entire time development under fixed value of excitation energy and radius. Solid line starting from the origin is the relationship of p and T at the same radius for the ideal gas and the other line is obtained by the best fit of the two parameters, a and b, in van der Waals equation:

$$\left(p + \frac{a}{v^2}\right)(v - b) = T \tag{12.9}$$

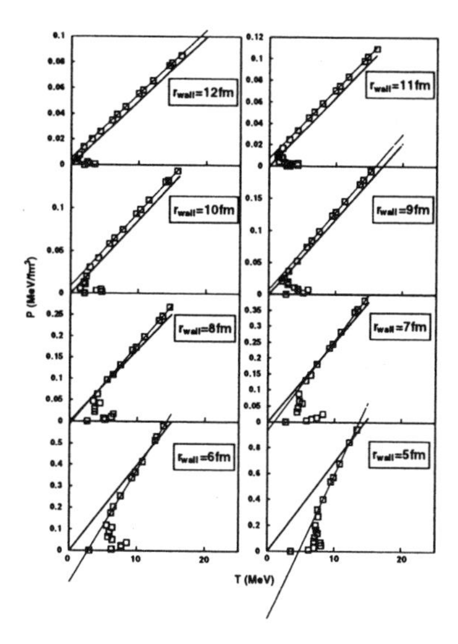

Figure 12.6 Relationship between temperature and pressure. Squares are calculated points at each excitation energy. Thick lines are caloric curves for ideal gas. Thin lines are van der Waals fits.

where v is volume per nucleon,

$$v = V/A \tag{12.10}$$

The characteristics common for all the cases of radius are that the pressure can be expressed as a linear function of temperature when the system is in high excitation and that it ceases to follow the line at certain lower temperatures.

In the case of large volume, squares corresponding to high temperature are close to both lines of ideal gas and van der Waals gas. For the smaller volume, the gas squares are not close to the ideal gas any more but they are still well fitted by van der Waals equation.

When one looks at Table 12.1, the parameter a constantly increases from negative value to positive value when the volume decreases. Since a represents the strength of the attractive part of van der Waals force, the nuclear force of the gas phase works repulsively when the volume is big, and turns to attractive for the smaller volume.

Parameter b corresponds to the volume (per gas nucleon) of the effective repulsive core where particles can't go inside. In our calculation, not only the repulsive part of nuclear force but also the momentum fluctuation procedure acts repulsively. It is because succeeding absorption of nucleon and emission of another nucleon by a cluster is equivalent as a reflection collision between cluster and nucleon.

Table 12.1 Parameters a and b of van der Waals equation that is fitted to the calculated relation between pressure and temperature of the gas phase

r_{wall}	12	11	10	9	8	7	6	5
a (MeV/fm^3)	-172	-109	-51.0	-19.9	25.9	50.4	93.4	102
b (fm^3)	5.42	5.75	7.33	5.80	5.68	5.13	5.95	5.17

In Table 12.1 the best-fit values of the parameter b are also shown. Although it looks like the value of b doen't change so much for different value of r_{wall}, we should not take the values of b for large r_{wall} so seriously. The reason is as follows. For large r_{wall}, v is much larger than the value of b given in Table 12.1. Since b is contained in the van der Waals equation only in the form of $(v - b)$, even if we make the choice of $b = 0$ it fits still for the case of large r_{wall}. On the other hand, for small r_{wall} the values of b given in Table 12.1 should be regarded as being determind uniquely. If the value of b is 5.8 fm^3, it means that the volume wich gas nucleons should avoid amounts to 185.6 fm^3, which corresponds to the volume of the nucleus of mass number $A = 29 - 37$ with normal density.

The fact that the gas nucleon obeys van der Waals equation in all case of r_{wall} jusitifies our definition of temperature, since the kinetic energy part of the internal energy of van der Waals gas is the same as that of ideal gas and is independent on parameters a and b.

7. DISCUSSION AND SUMMARY

By means of the microscopic theory AMD-MF, we studied the change in the behavior of nuclear system when its excitation energy and volume are varied. Transitional stage that connects liquid-dominant phase and gaseous phase is shown not to be the mere mixture of the gas nucleus and liquid nucleus but complicated compounds of clusters with various sizes. This consequence qualitatively agrees with the result which is predicted in the statistical multi-fragmentation studies[4, 5]. The most important point is that in the present study, the Fermi-gas like property of the liquid nucleus, van der Waals like behavior of gas nucleons, the existence of cracking and the property of plateau region are all obtained from microscopic calculations *without any assumption*. This feature makes it possible to explore the collision dynamics with AMD.

In AMD-MF (and its superset, AMD-V[6]), quantum statistics is automatically taken into account and one can safely compare the calculation with the experimental results. We don't need any assumption on the reaction stage like equilibration, reaction geometry or switching from dynamical to quantum statistical stage as statistical framework and quantm molecular dynamics do. With AMD-MF we will be able to check whether the agreement of the isotopic temperature to other temperatures in this ideal condition.

Our study revealed the importance to treat the volume of the system explicitly, since the volume of the system determines the property of the plateau region even qualitatively. The temperature of plateau is dependent on the volume and hence so is the way the liquid-dominant system turns into totally gaseous phase. From the observation that the volume is the key parameter to determine the behavior of the transition, we realize the importance of the estimation, in the experimental condition, of the equivalent volume in which the system can be regarded as equilibrated. Each point on experimental caloric curve may correspond to different effective volume and might be a representative of different path of the phase transition.

References

[1] J. Pochodzalla, T. Mohlenkamp, T. Rubehn, A. Schuttauf, A. Worner, E. Zude, M. Begemann-Blaich, T. Blaich, C. Gross, H. Emling, A. Ferrero, G. Imme, I. Iori, G.J. Kunde, W.D. Kunze, V. Lindenstruth, U. Lynen, A. Moroni, W.F.J. Muller, B. Ocker, G. Raciti, H. Sann, C. Schwarz, W. Seidel, V. Serfling, J. Stroth, A. Trzcinski, W. Trautmann, A. Tucholski, G. Verde, B. Zwieglinski, Phys. Rev. Lett. **75**, 1040 (1995).

[2] A. Ono, H. Horiuchi, T. Maruyama and A. Ohnishi, Prog. Theor. Phys. **87**, 1185 (1992).

[3] A. Ono and H. Horiuchi, Phys. Rev. **C53**, 845 (1996), Phys. Rev. **C53**, 2341 (1996)

[4] J. P. Bondorf, R. Donangelo, I. N. Mishustin, C. J. Pethick, H. Schulz, K. Sneppen, Nucl. Phys. **A443**, 321 (1985).

[5] D. H. E. Gross, Z. Xiao-ze, and X. Shu-yan Phys. Rev. Lett. **56**, 1544 (1986).

[6] A. Ono and H. Horiuchi, Phys. Rev. **C53**, 2958 (1996).

Chapter 13

RESULTS FROM THE E917 EXPERIMENT AT THE AGS

Birger B. Back
Argonne National Laboratory
Argonne, IL 60439, USA

for the E917 collaboration
B. B. Back[1], R. R. Betts[1,6], H. C. Britt[5], J. Chang[3], W. Chang[3], C. Y. Chi[4], Y. Y. Chu[2], J. Cumming[2], J. C. Dunlop[8], W. Eldredge[3], S. Y. Fung[3], R. Ganz[6,9], E. Garcia-Soliz[7], A. Gillitzer[1,10], G. Heintzelman[8], W. Henning[1], D. J. Hofman[1], B. Holzman[1,6], J. H. Kang[12], E. J. Kim[12], S. Y. Kim[12], Y. Kwon[12], D. McLeod[6], A. Mignerey[7], M. Moulson[4], V. Nanal[1], C. A. Ogilvie[8], R. Pak[11], A. Ruangma[7], D. Russ[7], R. Seto[3], J. Stanskas[7], G. S. F. Stephans[8], H. Wang[3], F. Wolfs[11], A. H. Wuosmaa[1], H. Xiang[3], G. Xu[3], H. Yao[8], C. Zou[3]
[1] *Argonne National Laboratory, Argonne, IL 60439, USA*
[2] *Brookhaven National Laboratory, Upton, NY 11973, USA*
[3] *University of California at Riverside, Riverside, CA92521, USA*
[4] *Columbia University, Nevis Laboratories, Irvington, NY10533, USA*
[5] *Department of Energy, Division of Nuclear Physics, Germantown, MD 20874, USA*
[6] *University of Illinois at Chicago, Chicago, IL 60607, USA*
[7] *University of Maryland, College Park, MD 20742, USA*
[8] *Massachusetts Institute of Technology, Cambridge, MA 02139, USA*
[9] *Max Plank Institute für Physik, D-80805 München, Germany*
[10] *Technische Universität München, D85748 Garching, Germany*
[11] *University of Rochester, Rochester, NY14627, USA*
[12] *Yonsei University, Seoul 120-749, Korea*

Abstract Collisions of Au+Au have been studied at beam kinetic energies of 6, 8, and 10.8 GeV/nucleon at the AGS facility at Brookhaven National Laboratory. Particles emitted from the collisions were momentum ana-

Advances in Nuclear Dynamics, 5,
Edited by Bauer and Westfall, Kluwer Academic / Plenum Publishers, New York, 1999.

lyzed and identified in a magnetic spectrometer system. Measurements were made at spectrometer angles in the range 14° - 59°. m_t-spectra of protons from central collisions were analyzed to derive integrated rapidity distributions and inverse slope as a function of rapidity. The results are compared with a thermal model and it is concluded that there is either substantial transparency or longitudinal expansion at all three beam energies.

Keywords: AGS, Protons, Rapidity distributions, Thermal models

1. INTRODUCTION

One of the expected ways of achieving the Quark Gluon Plasma phase of hadronic matter is to increase the matter density to 5-10 times that of normal nuclear matter. It has been thought that these high densities may be achieved in central collisions of heavy nuclei at AGS energies, provided that the relative motion is stopped. The question of the degree of stopping in head-on collisions is thus of central importance at these energies. Direct information pertaining to this issue may be obtained by studying the transverse mass spectra of protons, the majority of which are primordial, over a wide rapidity range to assess whether the observed rapidity distribution is consistent with the initial momentum of the projectile being converted to isotropic emission from a source at rest in the center-of-mass system. We find that complete stopping, defined in this way, is not achieved for the central Au-Au collisions at any of the energies studied in the E917 experiment.

2. EXPERIMENTAL ARRANGEMENT

The experimental arrangement is illustrated in Fig. 13.1. Beams of ^{197}Au with momenta of 6.84, 8.86 and 11.69 GeV/c per nucleon corresponding to kinetic energies of 6.0, 8.0 and 10.8 GeV per nucleon were obtained from the AGS at Brookhaven National Laboratory and focused onto a Au-target of 1 mm thickness, which corresponds to $\sim$ 3% interaction probability in the target. The trajectory of each beam particle was determined by a beam vertex detector consisting of four planes of scintillating fibers read out by position sensitive photo-multiplier tubes arranged in orthogonal pairs and located at 5.84 m and 1.72 m upstream from the target position. Further beam characterization was performed using beam-time-zero and halo counters also placed upstream from the target. Triggers for beam interactions with the target were obtained by requiring that a signal of less than 75% of that expected for the full energy loss of a Au-beam nucleus was registered in a circular "bulls-eye" Čerenkov detector placed 11 m downstream from the target.

The centrality of each beam-target interaction was derived from the multiplicity of particles (mostly pions) registered in a multiplicity detector array subtending a solid angle of about 6.85 sr around the target and/or by the total energy of the projectile remnant measured in the zero degree calorimeter.

In order to determine the reaction plane orientation in peripheral collisions a hodoscope consisting of two orthogonal planes of 1 cm wide plastic scintillator slats was placed in front of the zero degree calorimeter. The azimuthal angle of the reaction plane determined from the average position of the charged projectile remnants in the hodoscope, relative to the beam

axis will be used to study collective flow characteristics of peripheral collisions and the reaction plane dependence of the apparent source size obtained in a Hanbury-Brown Twiss analysis of pion pairs. Such an analysis is the subject of B. Holzman's talk at this workshop [1].

Particle spectra were obtained by momentum analysis in a movable magnetic spectrometer consisting of a 0.4-Tesla magnet (Henry Higgins) and a number of multi-wire ionization chambers used to determine the straight line trajectories of particles entering and exiting the magnetic field. In addition, a plastic scintillator wall located behind the spectrometer provided particle identification by time-of-flight measurements relative to the Čerenkov start detector located in front of the target.

This arrangement is capable of identifying charged pions, kaons, protons, anti-protons and heavier nuclei. The separation of pions and kaons is effective up to an energy of ~1.75 GeV. Protons are separated from pions up to ~3.4 GeV, although there is a negligible kaon contamination above ~2.9 GeV.

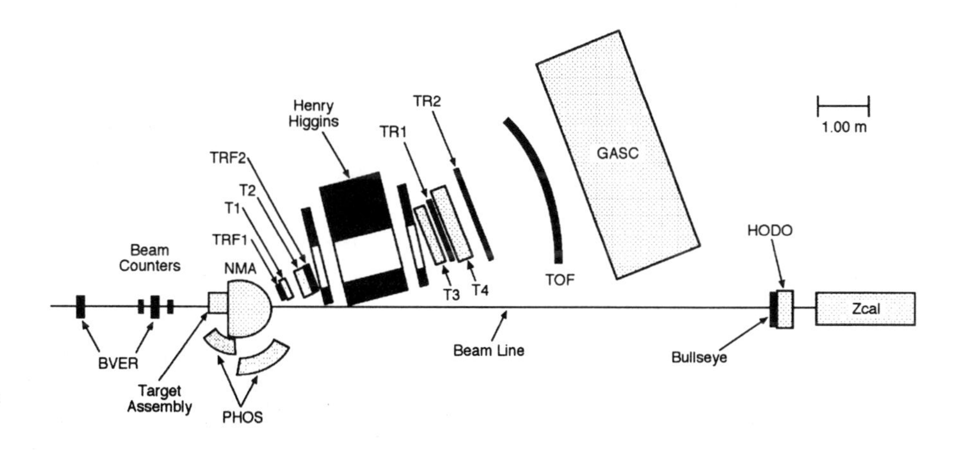

Figure 13.1 Experimental arrangement.

The acceptance of the spectrometer is also sufficient to detect correlated decay products of $\Phi \to K^+K^-$, $\Lambda \to p\pi^-$, and $\overline{\Lambda} \to \overline{p}\pi^+$. The analysis of Φ-mesons and $\overline{\Lambda}$ is the subject of W.-C. Chang's talk at this workshop[2].

The present talk will concentrate on the proton spectra obtained at the three beam kinetic energies of 6, 8, and 10.8 GeV/nucleon, and an analysis to the resulting rapidity distributions and fitted inverse slopes obtained from the measured m_t-spectra.

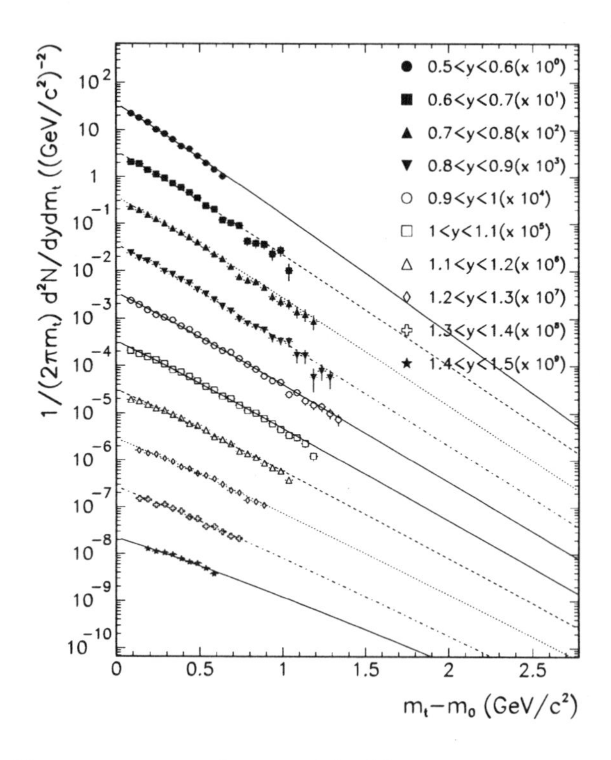

Figure 13.2 Proton m_t-spectra for central (<5%) Au-Au collisions at a beam kinetic energy of 10.8 GeV/nucleon for different rapidity bins (laboratory system). Curves are Boltzmann distributions fitted to the data. Spectra for adjacent rapidity bins are offset by factors of ten to avoid overlapping.

3. RESULTS

In Fig. 13.2, spectra of the invariant probability for proton emission per trigger event are plotted for the 10.8 GeV/nucleon beam energy as a function of the transverse mass $m_t - m_0$ for the 5% most central collisions as determined from the energy deposition in the zero-degree calorimeter.

The spectra are shown for different rapidity bins as indicated. Only statistical error bars are shown. Note that adjacent spectra are offset by factors of ten to avoid overlapping. The range in m_t reflects the acceptance of the spectrometer in the different rapidity bins ranging from backwards to near mid-rapidity at $y_{pp} = 1.613$. The curves represent the best fits to the spectra using a Boltzmann distribution, *i.e.*

$$\frac{1}{2\pi m_t}\frac{d^2N}{dydm_t} = Cm_t\exp(-m_t/T), \tag{13.1}$$

where C is a normalization constant and T is the inverse slope of the spectrum, both of which are determined from the fit to the data. We observe that the experimental m_t-spectra are in excellent agreement with this shape although they could also be described almost equally well by a pure exponential function.

From these fits we derive the total probability for proton emission per unit of rapidity, dN/dy, which is plotted as a function of rapidity in the center-of-mass frame, $y - y_{cm}$, in the left panels of Fig. 13.3. The derived inverse slopes T are shown in the right hand panels. Data are shown for 5% central collisions at all three beam energies, where the centrality for the 6 and 8 GeV/nucleon data are obtained from the the multiplicity array at the target position. The measured points are represented by solid circles, whereas reflection around mid-rapidity results in the open points. Error bars on the fit parameters are purely statistical and do not include possible systematic errors.

We note that all three dN/dy distributions are quite flat over the measured rapidity range. The inverse slopes, T, show, however, a distinct peaking at mid-rapidity. The solid curves in Fig. 13.3 represent the expectation for isotropic emission from a thermal source with temperature, T_0, at rest in the center-of-mass system. For such a source one expects that the y-dependence of the inverse slope is:

$$T = T_0/\cosh y \tag{13.2}$$

and a dN/dy distribution of [3]

$$\frac{dN}{dy} \propto T_0\left(m_0^2 + 2m_0\frac{T_0}{\cosh y} + 2\frac{T_0^2}{\cosh^2 y}\right)\exp(-m_0\cosh y/T_0), \tag{13.3}$$

where m_0 is the proton rest mass. In the right hand panels of Fig. 13.3 we compare the rapidity dependence of the inverse slope, T, with those predicted by this model. The source temperature, T_0, was adjusted to account for the observed inverse slope at mid-rapidity. We note that this naive model gives a rather good representation of the observed inverse slopes.

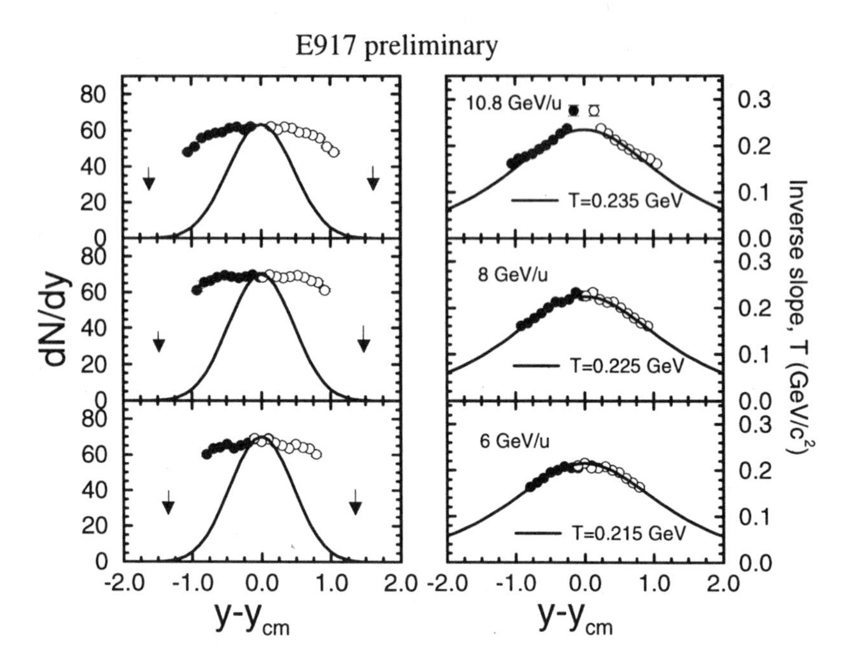

Figure 13.3 Proton rapidity distributions in the center-of-mass system (left panels) and inverse slopes (right panels) are compared to a simple thermal source prediction for the three beam energies. The arrows indicate target and beam rapidities.

On the other hand, a comparison of the predicted distribution in rapidity dN/dy with the measurements (left hand panels in Fig. 13.3) reveals a discrepancy which clearly demonstrates that the observed proton spectra are inconsistent with isotropic emission from a single source at rest in the center-of-mass system. Rather, we note that the rapidity distributions for protons are flat over a wide range of rapidities indicating a significant degree of either incomplete stopping or longitudinal expansion at all three beam energies.

Of course, the naive model shown here also disregards the possible effects of radial expansion in the fireball. It has been shown[3], however, that, within a wide range of parameters, there is a strong anti-correlation between the radial expansion velocity, v_0, and the source temperature, T_{source}, such that it is impossible to disentangle their relative values from fits to spectra of a single particle species *e.g.* protons. In the present analysis we have therefore chosen to use only a single parameter, namely the *apparent* source temperature, T_0, keeping in mind that its value does not necessarily represent the true temperature of the source formed in a central Au-Au collision.

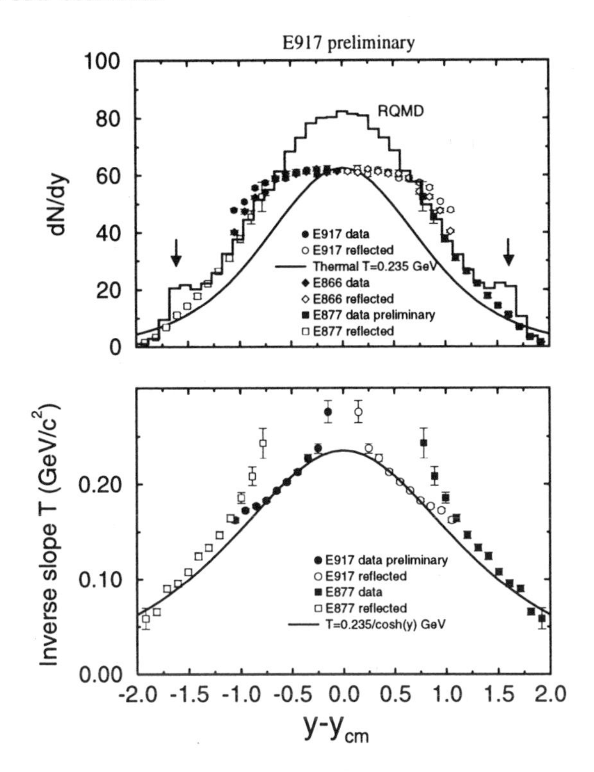

Figure 13.4 Proton rapidity distributions (top) and inverse slopes (bottom) for Au-Au collisions at 10.8 GeV/nucleon beam kinetic energy compared with previously published results from the E866 and E877 experiments at the AGS. The arrows indicate target and beam rapidities.

In Fig. 13.4 we compare the dN/dy (top panel) and inverse slopes T (bottom panel) from our experiment to results from the E866 [4] and E877[5] experiments at the AGS. At mid-rapidity there is good agreement between the present data for the dN/dy distribution and that from the E866 experiment, but a relatively small discrepancy between the two data sets is apparent at less central rapidities. The source of this discrepancy is presently being investigated. The E877 data were measured in the extreme forward rapidity region but overlap with the present (reflected) data in the rapidity region $y - y_{cm}$ = 0.8 - 1.1. Here, there appear to be substantial discrepancies between the data sets. The general trend of the three data sets is, however, clear. There is a wide range -0.7< $y - y_{cm}$ < 0.7 around mid-rapidity, where the dn/dy distribution is essentially flat, followed by a monotonic decrease on either side. We note that this observed range of essentially constant dN/dy is inconsistent with thermal emission from a stopped source (solid curve), as well as early predictions of the Relativistic Quantum Molecular Dynamics model[6] (RQMD v1.08, [5]) (solid histogram), both of which are too sharply peaked at mid-rapidity.

The slopes, however, appear to be in reasonable agreement with the $1/\cosh y$ dependence expected from the thermal model (solid curve) although this may be fortuitous since the rapidity distribution clearly show that there is either a significant amount of transparency or longitudinal expansion at these energies which violates the assumption of a thermal source at rest in the center-of-mass system.

4. CONCLUSION

An analysis of m_t-spectra for protons emitted in central Au-Au collisions at beam kinetic energies of 6, 8 and 10.8 GeV/nucleon in terms of Boltzmann distributions has been carried out, and the resulting dN/dy-distributions and inverse slopes derived. They are compared to a simple thermal model assuming isotropic emission from a source at rest in the center-of-mass system corresponding to complete stopping of the colliding Au nuclei in central collisions. We find that the rapidity distributions dN/dy are substantially wider and essentially constant around mid-rapidity although the inverse slopes exhibit the expected $1/\cosh y$ dependence on rapidity. We interpret this as a manifestation of incomplete stopping or longitudinal expansion of the entrance channel momenta at all three beam energies.

Acknowledgments

This work was supported by the Department of Energy, the National Science Foundation, (USA), and KOSEF (Korea).

References

[1] Holzman, B. *et al.* (1999) Contribution to these proceedings.

[2] Chang, W.-C. *et al.* (1999) Contribution to these proceedings.

[3] Schnedermann, E. *et al.* (1993) Phys. Rev. **C48**, 2462.

[4] Ahle, L. *et al.* (1999) Phys. Rev. **C57**, R466.

[5] Lacasse, R. *et al.* (1996) Nucl. Phys. **A610**, 153c.

[6] Sorge, H. *et al.* (1989) Ann. Phys. (NY) **192**, 266.

Chapter 14

SIZE MATTERS: WHAT REDUCIBILITY CAN TELL US

Wolfgang Bauer
National Superconducting Cyclotron Laboratory
and Department of Physics
Michigan State University
East Lansing, MI 48824-1321, USA
bauer@nscl.msu.edu; http://www.nscl.msu.edu/~bauer/

Abstract We show that the experimentally observed binomial multiplicity distributions for intermediate mass fragments can be explained as mainly due to the effects of the finite size of the emitting system. The dependence of the binomial parameters p and m on the total transverse energy, as well as the scaling and reducibility properties, can be explained by examining the acceptance of the detector used in the fragmentation experiments that found these properties.

Keywords: Scaling, reducibility, heavy-ion reactions, fragmentation, intermediate-mass fragments (IMF), Poisson distribution, Binomial distribution, finite-size effects, detector acceptance.

1. INTRODUCTION

One of the most interesting and challenging problems in many-body physics is the determination of the phase diagram of nuclear matter. Because of the presence of the short-ranged nuclear force and the long-ranged Coulomb interaction, because of the large variety of fermionic and bosonic constituents, because of the small numbers of constituents involved, and because of the lack of a separation of scales this area of study has generated an astounding collection of suggestions for different phase transitions: quark-gluon plasma, chiral restoration, Lee-Wick matter, Delta matter, pion condensation, kaon condensation, pion- and/or sigma laser and Bose-Einstein condensate, superconducting phase of QCD,

Advances in Nuclear Dynamics, 5,
Edited by Bauer and Westfall, Kluwer Academic / Plenum Publishers, New York, 1999.

shape transitions, pairing transitions, liquid-gas transition, and the percolation transition.

Experimentally, this field is equally challenging, because the duration of the heavy ion reactions considered here is on the order of 10^{-22} to 10^{-20} s – too short by many orders of magnitude for direct observation – and because the size of the system is only of the order of 10^{-14} m – too small by many orders of magnitudes for direct observation. The system cannot be prepared at one point in the nuclear phase diagram, but will migrate along a path. If equilibrium (thermal and/or chemical) was established along the way is not clear *a priori*. All of this information has to be extracted from the asymptotic momentum states of the particles (primary and secondary baryons and mesons, as well as composite fragments) [1].

In the intermediate energy regime, 20 to 100 MeV per nucleon, there is credible evidence for a phase transition between the nuclear Fermi liquid and a hadron gas. This is, of course, motivated by the character of the nucleon-nucleon interaction, which is of the van der Waals type. Credible experimental evidence for a first [2] and second [3, 4, 5] order phase transition has been presented. The picture of a first-order transition terminating in a second-order transition at the critical point is consistent with a liquid-gas type of universality class. However, that question is still open. Another class of models that are consistent with the experimental observables are those that are based on percolation theory [6, 7].

2. BINOMIAL FRAGMENT MULTIPLICITY DISTRIBUTIONS

Moretto and co-workers recently have published a series of papers [8] in which they study systematics of the event-by-event multiplicity distributions of intermediate mass fragments (IMFs), i.e. fragments with charges $3 \leq Z \leq 20$, as a function of the total transverse kinetic energy of all detected particles,

$$E_t \equiv \sum_{\ell} E_\ell \sin^2 \theta_\ell \tag{14.1}$$

They found that the probability P_n of emitting n intermediate mass fragments (IMFs) follows a binomial distribution

$$P_n(m,p) = \frac{m!}{n!\,(m-n)!} p^n (1-p)^{m-n} \tag{14.2}$$

The parameters m and p are related to the average and variance of the distribution,

$$\langle n \rangle = \sum_{n=0}^{\infty} n\, P_n(m,p) = m \cdot p \tag{14.3}$$

$$\sigma^2 = \sum_{n=0}^{\infty} (n - \langle n \rangle)^2\; P_n(m,p) = m\, p\, (1-p) \tag{14.4}$$

and, consequently:

$$p = 1 - \frac{\sigma^2}{\langle n \rangle} \quad \text{and} \quad m = \frac{\langle n \rangle^2}{\langle n \rangle - \sigma^2} \tag{14.5}$$

This suggest that p could be considered the elementary probability for the emission of one fragment, and the parameter m as the total number of tries, indicating that the problem of multi-fragment emission is *reducible* to that of multiple one-fragment emission. The claim for reducibility and its interpretation as the consequence of a simple barrier penetration phenomenon was further strengthened by the observation that $\ln(p^{-1})$ has a linear dependence on $1/\sqrt{E_t}$. Finally, the same *scaling* was found for different beam energies and different projectile-target combinations.

While many explanations for the observed scaling were proposed and inconsistencies in the analysis were pointed out [9], a complete explanation has lacked so far. It is, however, clear that the interpretation of thermal reducibility and scaling contradicts the picture of a fragmentation phase transition in nuclear matter. We have thus recently studied the origins of the observed binomial distributions and their scaling behavior [10, 11, 12, 13], and here we summarize the outcome of these studies.

3. FINITE-SIZE EFFECTS

We will show here that the effects of the finite size of the emitting system and the finite detector acceptance force the fragment multiplicity distributions towards the binomial limit of sub-poissonian statistics. In Ref. [13] we show that one can start with a super-poissonian distribution and achieve this effect. He we start with a poissonian multiplicity distribution in the limit of infinite source size, as generated by, for example, and event generator that produces power-law distributed fragment mass spectra, $N(Z) \propto Z^{-\tau}$. The exact Poisson distribution has an infinite tail that is cut off by the finite size of the emitting system. This is demonstrated in Fig. 14.1. The solid lines represent the value of $\tau = 2.5$, which is the infinite size limit. The circles represent the result of the simulations for $Z_{sys} = 100$ and 10, respectively. Clearly, one can see the deviation from the power-law that is caused by the finite size.

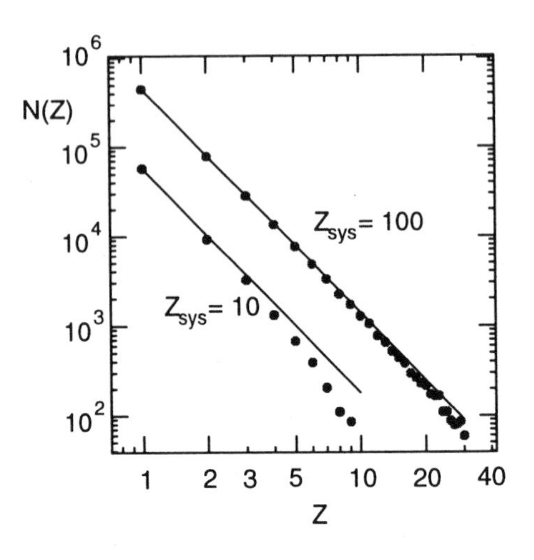

Figure 14.1 Fragment size distribution output of our event generator for finite system sizes. Here the value of τ was chosen as 2.5.

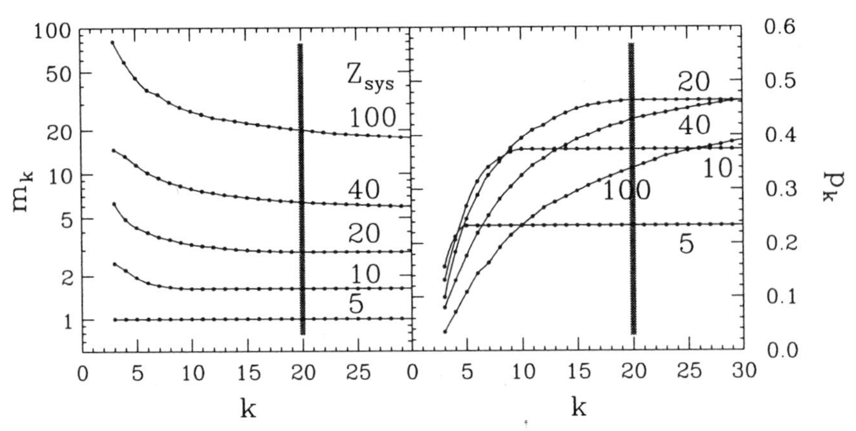

Figure 14.2 Dependence of the binomial parameters p and m of the fragment multiplicity distribution on the upper mass number limit, k, of the summation over intermediate mass fragments, for different total charges of the fragmenting system, Z_{sys}. Again the value of τ was chosen as 2.5

This immediately has the effect that the infinite size limit Poisson distribution for each fragment species is cut off and shifted towards a binomial distribution. For the resulting distribution, one can extract the parameters p and m via Eq. 14.5 individually for each fragment species. Since IMFs are defined as fragments with charges $3 \leq Z \leq 20$, we need to produce a combined distribution for all fragments. The extracted parameter p for the combined distribution is shown in Fig. 14.2. The index k is the upper summation index for the combined fragment multiplicity distributions of fragments of charge 3 to k.

We can extract several important observations from this figure. First, let us look at one of the lines, each of which represent a fixed system size, $Z_{\rm sys}$. One can see that p_k is a monotonously rising function of the summation index k, and m_k is a monotonously falling one. This is a consequence of the fact that when one combines two sub-poissonian distributions the combined distribution is further away from the Poisson limit, because the mean of the combined distribution is the sum of the mean value of the individual distributions, but the variance is smaller than the sum of the individual variances. Each of the lines for fixed $Z_{\rm sys}$ saturates for $k \geq Z_{\rm sys}$, because there are no additional contributions to the combined distribution from fragment sizes that exceed the system size.

The second important information that can be extracted from this figure is that the curves $p_k(k)$ for different system size cross each other. Thus, even though the multiplicity distributions for individual fragments are closer to the poissonian limit for larger system size (see $k = 3$, where the summation only extends over the Lithium fragments), the combined IMF multiplicity distributions (see $k = 20$) are further away from the Poissonian limit for larger system size. This is a direct consequence of the saturation property discussed in the previous paragraph.

4. CORRELATION BETWEEN TRANSVERSE ENERGY AND NUMBER OF RECOVERED CHARGES

The finite-size effects discussed so far provide a large part of the solution of the reducibility puzzle. The second part of the solution is contained in the correlation between the total charge detected in an event, $Z_{\rm sys}$, and the total measured transverse energy, E_t. This relation is shown in Fig. 14.3 for the reaction 55 AMeV Kr+Au and the MSU Miniball 4πdetector, which was used in the experiments analyzed by Moretto et al. [14]. It should be stressed though, that this correlation is not unique to that particular detector, but is shared by most 4πdetectors used with fixed-target reactions. The most notable exceptions are presently TPCs with internal targets, where a more-or-less complete event reconstruction is possible.

This correlation is due to the fact that projectile remnants travel down the beam pipe and target remnants may remain below the detection threshold energy or may not even escape from a solid target within a

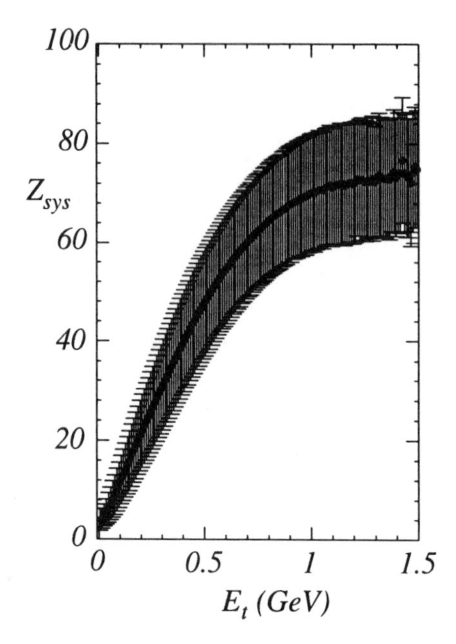

Figure 14.3 Dependence of the number of detected charges, Z_{sys}, on the total measured transverse energy, E_t, for the reaction 55 *A*MeV Kr+Au [14]. The error bars indicate the width (standard deviation) of the distribution on an event-by-event basis.

fixed-target experiment. These are the reasons that the total transverse energy is considered an excellent measure for impact parameter – at least at higher energies.

The essential feature visible from Fig. 14.3 is the linear dependence of E_t on the system size, $Z_{\rm sys}$, for small to intermediate value of the transverse energy. With this information, we can now relate the extracted binomial parameters to the total transverse energy. This is shown on the left side of Fig. 14.4 for a value of $\tau = 3.0$. It is clear that we can see a linear relationship between p and $1/\sqrt{E_t}$ already. (Other values of τ in the physically reasonable space of values between 2.0 and 4.0 [3] yield very similar results, with more-or-less straight lines and different slope values [11].) This linear relationship was previously considered the primary indication for reducibility and thermal scaling in the work of Moretto et al. [8]. Here, however, this linear dependence is caused exclusively by the finite size effects in source emitting the fragments and by the linear relationship between the total measured transverse energy and the number of charges recovered in each event.

In the right side of Fig. 14.4, we allow the power-law exponent τ to vary as a function of the impact parameter – and with it as a function of the transverse energy. We chose the functional dependence

$$\tau(E_t) = 3.5 - E_t/(0.5 \text{ GeV}) , \tag{14.6}$$

which yields a value of 3.5 for peripheral and 2.0 for central collisions. If we adopt this procedure, then we obtain more-or-less complete agree-

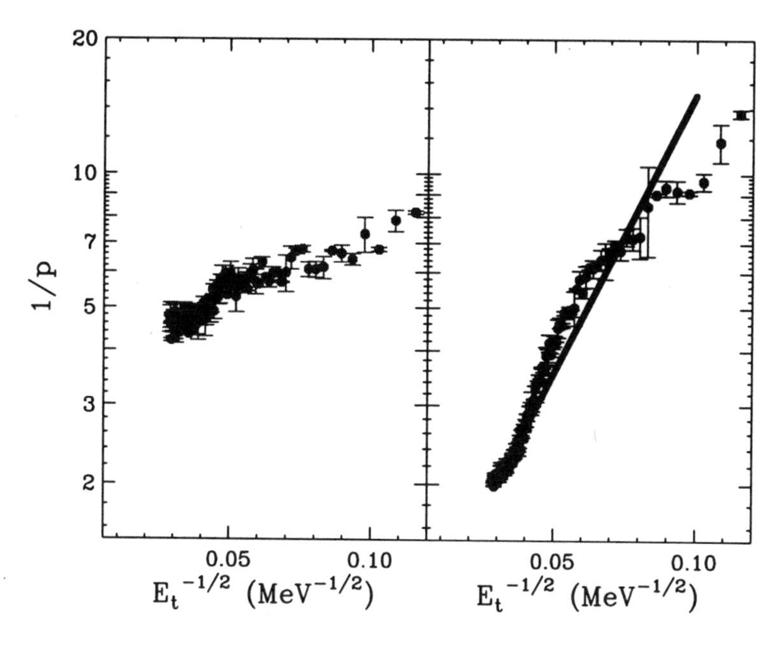

Figure 14.4 Values of the extracted binomial parameter p vs. $1/\sqrt{E_t}$. Left panel: fixed value of $\tau = 3$ in the simulation; right panel: variable τ according to Eq. 14.6. The straight line in the right panel represents the experimental results of Moretto et al.

ment with the data (shown by the thick straight gray line in the right panel of Fig. 14.4).

5. CONCLUSIONS

We have shown that only minimal assumptions about the fragment emission process are needed to reproduce the binomial fragment multiplicity distributions and their dependence on the measured transverse energy. These minimal assumptions are only that of a power-law distributed fragment spectrum and are fulfilled by almost all multifragmentation models on the market. In particular, it does not matter, if these models deliver Poissonian or sub- or super-Poissonian multiplicity distributions in infinite nuclear matter. The constraints imposed by the finite size of the emitting source and the correlations between the total charge recovered in the detector and the total transverse energy measured are the determining factors for the observed scaling and apparent reducibility. Thus the observations of the regularity in the binomial character of the multiplicity distributions hardly tell us anything about the fundamental fragment emission process in the nuclear matter limit, but mainly about the finiteness of the emitting source: size matters!

Acknowledgments

The author would like to thank Scott Pratt and Tarek Gharib for fruitful collaborations that led to the results reported on here. This research was supported by the U.S. National Science Foundation under grant number PHY-9605207

References

[1] For a review and extensive list of references please see: W. Bauer, C.-K. Gelbke, and S. Pratt, Annu. Rev. Nuc. Part. Sci. **42**, 77 (1992).

[2] J. Pochodzalla et al., Phys. Rev. Lett. **75**, 1040 (1995); J. Pochodzalla, Prog. Part. Nucl. Phys. **39**, 443 (1997).

[3] T. Li et al., Phys. Rev. Lett. **70**, 1924 (1993); Phys. Rev. C **49**, 1630 (1994); C. Williams et al., Phys. Rev. C **55**, R2132 (1997).

[4] M.L. Gilkes et al., Phys. Rev. Lett. **73**, 1590 (1994); J.B. Elliott et al., Phys. Rev. C **49**, 3185 (1994); H.G. Ritter et al., Nucl. Phys. A **583**, 491c (1995).

[5] W. Bauer and W.A. Friedman, Phys. Rev. Lett. **75**, 767 (1995); W. Bauer and A. Botvina, Phys. Rev. C **52**, R1760 (1995); Phys. Rev. C **55**, 546 (1997).

[6] W. Bauer et al., Phys. Lett. **B150**, 53 (1985); W. Bauer et al, Nucl. Phys. A **452**, 699 (1996); W. Bauer, Phys. Rev. C **38**, 1297 (1988).

[7] X. Campi, J. Phys. A **19**, L917 (1986).

[8] L.G. Moretto et al., Phys. Rev. Lett. **71**, 3935 (1993); Phys. Rev. Lett. **74**, 1530 (1995); Phys. Rep. **287**, 249 (1997); L. Phair et al., Phys. Rev. Lett. **77**, 822 (1996).

[9] A. Del Zoppo *et al.*, Phys. Rev. Lett. **75**, 2288 (1995); J. Toke, D.K. Agnihotri, B. Djerroud, W. Skulski and W.U. Schroeder, Phys. Rev. C **56**, R1686 (1997); M.B. Tsang and P. Danielewicz, Phys. Rev. Lett. **80**, 1178 (1998).

[10] T. Gharib, W. Bauer, and S. Pratt, Phys. Lett. **B444**, 231 (1998).

[11] W. Bauer and S. Pratt, Phys. Rev. C, in print; Los Alamos preprint archive nucl-the/9808068 (1998).

[12] W. Bauer, T. Gharib, and S. Pratt, Prog. in Part. and Nucl. Phys. **42** (1999), in print.

[13] W. Bauer, in Proceedings of the XXXVII International Winter Meeting on Nuclear Physics, Bormio, Italy.

[14] M.B. Tsang, private communication.

Chapter 15

THE BOMBARDING ENERGY DEPENDENCE OF π^- INTEFEROMETRY AT THE AGS

M.A. Lisa[j], N.N. Ajitanand[m], J. Alexander[m], D. Best[a], P. Brady[e], T. Case[a], B. Caskey[e], D. Cebra[e], J. Chance[e], I. Chemakin[d], P. Chung[m], V. Cianciolo[i], B. Cole[d], K. Crowe[a], A.C. Das[j], J. Draper[e], S. Gushue[b], M. Gilkes[l], M. Heffner[e], H. Hiejima[d], A. Hirsch[l], E. Hjort[l], L. Huo[g], M. Justice[h], M. Kaplan[c], J. Klay[e], D. Keane[h], J. Kintner[f], D. Krofcheck[k], R. Lacey[m], J. Lauret[m], E. LeBras[m], H. Liu[h], Y. Liu[g], R. McGrath[m], Z. Milosevich[c], D. Olson[a], S. Panitkin[h], C. Pinkenburg[m], N. Porile[l], G. Rai[a], H.-G. Ritter[a], J. Romero[e], R. Scharenburg[l], L. Schroeder[a], R. Soltz[i], B. Srivastava[l], N.T.B. Stone[b], T.J. Symons[a], S. Wang[h], R. Wells[j], J. Whitfield[c], T. Wienold[a], R. Witt[h], L. Wood[e], X. Yang[d], W. Zhang[g], Y. Zhang[d]

[a] *Lawrence Berkeley Lab,* [b] *Brookhaven National Lab,* [c] *Carnegie Mellon University,* [d] *Columbia University,* [e] *U.C. Davis,* [f] *St. Mary's College,* [g] *Harbin Institute, China,* [h] *Kent State University,* [i] *Lawrence Livermore National Lab,* [j] *The Ohio State University,* [k] *University of Auckland, NZ,* [l] *Purdue University,* [m] *SUNY at Stony Brook*

Abstract We present a preliminary excitation function of π^- intensity inteferometry at AGS energies (2-8 AGeV). The sensitivity of the multidimensional correlation functions to both the geometry and dynamics of the hot pion-emitting system provide a stringent test of transport models of heavy ion collisions. Detailed comparisons are performed with the RQMD transport model, both with and without an explicit nuclear meanfield. These comparisons suggest that the evolution in the reaction dynamics with beam energy is different in the model and the data. A suggested signal of the onset of the transition to quark-gluon plasma in the inteferometry excitation function is not observed.

Intensity inteferometry (a.k.a. HBT) techniques have been used extensively to probe the space-time structure of heavy ion collisions (see, e.g. [1]). Multidimensional analysis of two-pion correlation functions has been shown to be sensitive to the geometry of the pion-emitting source [1], the duration of pion emission [2, 3, 4, 5], and the details of resonance decay contributions [4, 6, 7, 8]. The dependence of the correlations on kinematic variables displays sensitivity to the dynamics of the collision, including collective flow [4, 10, 11]. In particular, flow generates space-momentum correlations in the freeze-out configuration, resulting in an apparent (or effective) source size that is smaller than the true size [4, 10, 11]. Therefore, correlation functions– and especially HBT excitation functions– are a powerful tool to stringently test the dynamics generated in microscopic transport models [7, 12].

Here, we compare the preliminary E895 analysis to the RQMD model (v2.3) [13], which incorporates known and calculated hadronic cross sections and resonances, relativistic kinematics, rescattering effects, and a parameterized nuclear mean field. This model has reproduced measured observables, including pion correlations, reasonably well at AGS and SPS energies [12, 7, 14]. The predictions of such models represent the current knowlege of effects arising from "normal" hadronic physics in a heavy ion collision. As such, one would like to use these models at higher collision energies– e.g. at Brookhaven's nearly complete Relativistic Heavy Ion Collider (RHIC)– to describe a normal collision, and attribute (perhaps subtle) discrepancies with the data to "new" physics that may occur at those energies. Confidence in such an interpretation would be greatly enhanced if the model agrees with the evolution of the measured systematics over a large range of lower collision energies.

The most intriguing "new" physics sought in heavy ion collisions is the phase transition from hadronic matter to a quark gluon plasma (QGP), in which partons are deconfined. If such a transition takes place, the timescale for pion emission is expected to increase significantly, to tens of fm/c, detectable via 2-particle inteferometry as an apparent source geometry extended in the direction of the pion momentum [4]. Heavy ion reaction studies at Brookhaven's Alternating Gradient Synchrotron (AGS) probe nuclear matter at very high baryon density ($\rho \sim$6-8ρ_0) [15, 16]. Some models [15, 16, 17, 18, 19] and equilibrium-based meta-analyses of data [20] suggest that sufficient energy density may be generated in collisions below maximum AGS energies to trigger QGP formation. Recent hydrodynamic calculations suggest an HBT excitation function as an excellent method to detect the transition to QGP through a sudden increase of emission timescale [21].

The E895 collaboration has measured about 0.75 million Au+Au collisions at E_{beam} = 2, 4, 6, and 8 $A{\cdot}GeV$ at the Brookhaven AGS using the EOS Time Projection Chamber (TPC) [22]. The TPC, located in the

MPS magnet operated at 0.75 or 1.0 T, provides nearly full acceptance for pions with rapidity $y_{lab} \geq 0.5$ with no lower threshold in transverse momentum (p_T).

Particle identification was achieved by correlating the magnetic rigidity of a track with its specific ionization (dE/dx) in the P10 gas. While $\bar{p}$ and K^- production are negligible compared to π^- at these energies, the electron and π^- bands overlap somewhat in the dE/dx-rigidity space. After track-quality cuts, we estimate the level of e^- contamination at $\sim 5\%$ for the highest beam energy, with less contamination at the lower beam energies and higher p_T.

Momentum resolution effects can distort the measured correlation functions, leading to reduced λ and radii fit parameters. After refitting primary tracks with the event vertex, the momentum resolution from the TPC tracking software is $\delta p/p \sim 1\%$. Equally important is the contribution from multiple Coulomb scattering and straggling in the target (3% interaction probability) and trigger scintillator, which worsens the overall resolution to 1.5-3%. All contributions depend on the total π^- momentum and emission angle, and have been studied in detail. We correct the correlation functions for resolution effects with an iterative technique similar to that used by the NA44 collaboration [28]. This correction typically increases the λ and radii fit parameters by 15% and 5%, respectively. When comparing the measured correlation functions to those from the model, we use the corrected data, and do not smear the model results by the momentum resolution.

Correlation functions are generated from the offline event reconstruction after a series of quality cuts. Requiring that a track projects back to the interaction vertex eliminates "ghost" tracks originating from unused hit combinatorics. This cut also eliminates many π^- from decays of long-lived particles (e.g. Λ). However, detailed simulations with the RQMD model reveal that roughly 10% (5%) of the π^- from the 8 $A{\cdot}GeV$ (2 $A{\cdot}GeV$) collisions that pass this cut, originate from long-lived particles; these pions are typically at low p_T and their effect is to lower the strength of the correlation at zero relative momentum (quantified by the λ fit parameter, see below). As in the data, these pions are included when generating correlation functions for the model.

The correlation function is constructed by dividing the two-particle yield as a function of relative momentum vector, q, by a "background" generated by mixing pions of the current event with those of the previous 10 events, following the standard event-mixing technique [29]. The large acceptance of the TPC eliminates several complications associated with correlation measurements with "keyhole" spectrometers [23]. Firstly, the measured pion multiplicity in each event is much greater than unity, eliminating possible ambiguitites about differing classes of events used in the correlated distribution and the background. Second, the effect of residual correlations in the background are greatly reduced. Finally, an unrestricted acceptance in q provides greater sensitivity to the shape of the effective pion source in coordinate space.

To eliminate track-merging effects in the TPC, which suppress the correlation function at low q, we require that the two tracks forming a pair are well separated over a distance of at least 18 cm in the beam direction. In applying this cut to the mixed event background, we account for event-to-event variation in primary vertex position. Further increasing the required hit separation distance or number of padrows does not affect the measured correlations. Simulations show that This cut also obviates the need for a further track-splitting cut; false "pairs" from split tracks are also eliminated by this cut.

The Coulomb correction applied to the raw correlation functions is obtained by averaging the square of the Coulomb wave function [4, 30] over a spherical Gaussian source of 5 fm radius. Changing this radius by ± 1 fm has negligible effect on the fit parameters. Our correction is the same as that used by several groups [31, 26]. Identical Coulomb corrections were applied to the data and to RQMD model predictions.

The momentum resolution of the TPC ($\delta p/p \sim$1.5-3%, including multiple Coulomb scattering and straggling effects) artificially broadens the experimental correlation functions. Using detailed studies of our resolution, we correct the correlation functions with an iterative technique similar to that used by the NA44 collaboration [28]. This correction typically increases the λ and radii fit parameters by 15% and 5%, respectively. In our model comparisons, we use the corrected data, and do not smear the model results by the momentum resolution.

Multidimensional correlation functions were generated by decomposing the relative momentum in the Au+Au c.m. frame using the Bertsch-Pratt (BP) parameterization [3, 4]. Here, q_{long} is the component of the relative momentum parallel to the beam, q_{side} is perpendicular to the beam and to the total momentum of the pair, and q_{out} is perpendicular to q_{long} and q_{side}. The quality of the data is clear from Figure 15.1, where projections of the three-dimensional correlation functions are plotted along all three components of the relative momentum. The correlation functions are constructed from midrapidity ($y = y_{cm} \pm 0.25$) pions with $p_T = 0.1 - 0.8\ GeV/c$ from central (11% σ_T as determined by event multiplicity) Au+Au collisions. This lower p_T cut was selected to reduce the effects of e^- contamination [9] and to allow comparison with correlation measurements at midrapidity by the E866 collaboration at 10.6 $A{\cdot}GeV$ [40]. The projections shown for a given component are integrated over the range ± 30 MeV/c in the other two components. For these cuts, $< p_T >_{\pi^-}$ changes from 239 MeV/c for the lowest bombarding energy to 305 MeV/c at the highest.

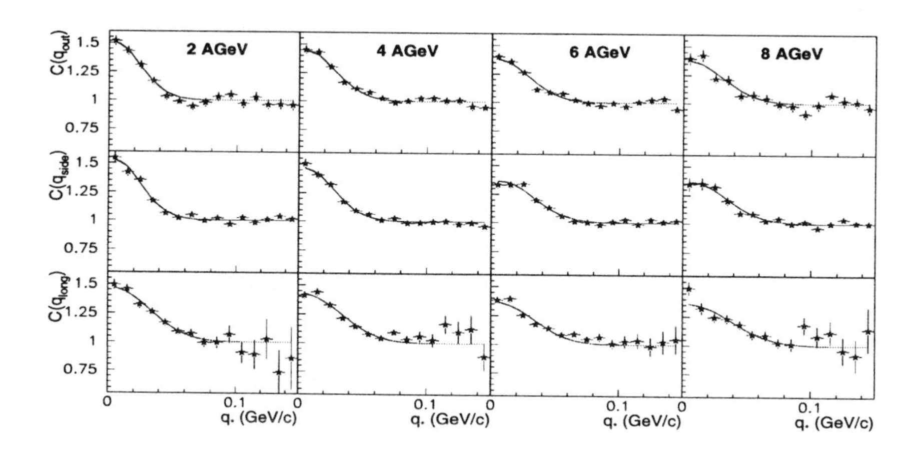

Figure 15.1 Projections of the measured two-pion in the out-side-long components of relative momentum (integrating over $\pm$ 30 MeV/c in the other components), for the 2, 4, 6, and 8 AGeV bombarding energy. All correlation functions were taken at $y = y_{cm} \pm 0.25$ and $p_T = 0.1 - 0.8$ GeV/c .

Also shown are projections of maximum log-likelihood fits of the 3-dimensional correlation functions to the form

$$C(q_{out}, q_{side}, q_{long}) = 1 + \lambda e^{-R_{out}^2 q_{out}^2 - R_{side}^2 q_{side}^2 - R_{long}^2 q_{long}^2 - 2R_{ol}^2 q_{out} q_{long}} \tag{15.1}$$

The cross-term [32] R_{ol}^2 (which may be negative) was included in all fits, but always came out consistent with zero with very large uncertainty. At midrapidity, this term is expected to vanish identically [32]. It was verified for all energies that fixing $R_{ol}^2 = 0$ in the fits of the midrapidity correlation functions has negligible effect on the other parameters. The correlation functions are well described by this Gaussian form ($\chi^2/n.d.f. \leq 1$) even down to very small relative momentum; hence the fit parameters may be used to summarize the correlations [4].

The excitation function of the BP fit parameters is shown in the left panels of Figure 15.2. Near unity at the lowest energy, the λ parameter falls sharply with beam energy, due mostly to increased production of long-lived π^--emitting particles at the higher energies [7]. Values of $\lambda \approx$

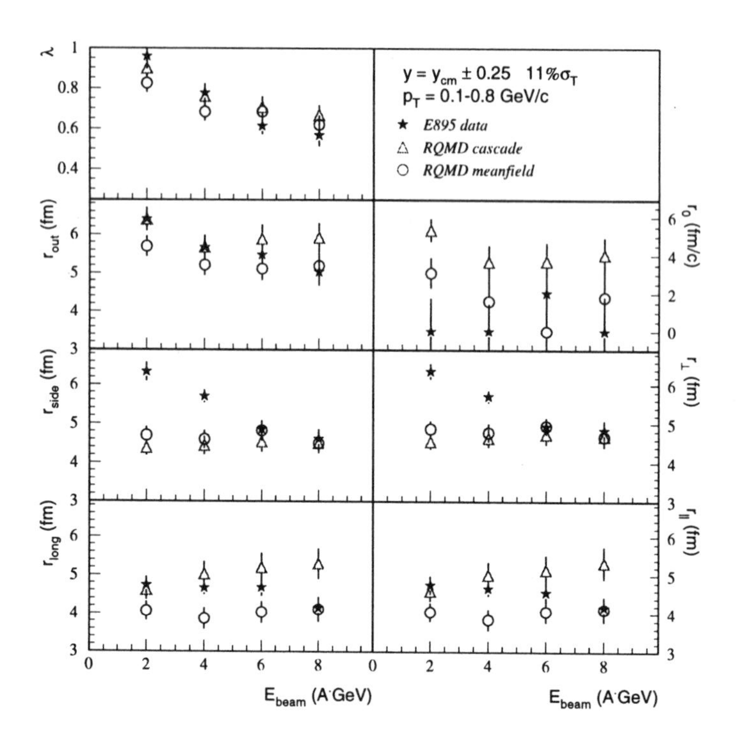

Figure 15.2 Fit parameters for the measured correlation functions decomposed with the BP (left panels) and YKP (right panels) decompositions are shown as stars. Triangles and circles, respectively, represent the results of fits to correlation functions predicted by the RQMD model in cascade and meanfield modes.

1 have been reported by some (e.g. [9]) but not all [23] groups measuring correlation functions at similar energies at the Bevalac. The small (¡ 5%) e^- contamination is estimated to reduce λ by up to 10% for the highest beam energy [27]. Purposely increasing the e^- contamination by relaxing particle cuts, affects only λ, not the extracted radii, consistent with observations from higher energies [26].

R_{side}, which is most directly related to source size transverse to the beam direction and R_{out}, which is sensitive to temporal as well as spatial extent [3, 4], are also seen to decrease with increasing bombarding energy, while R_{long} remains almost constant.

Also shown in Fig. 15.2 are results of fits to correlation functions from the RQMD transport model. At each energy, sets of events with impact parameter b=0-5 fm (geometrically weighted) were generated with and without the meanfield turned on in the model. Model correlation functions were generated by correlating π^- from the RQMD freeze-out phasespace (including pions from long-lived decays) with the correlator code of Pratt [12].

Good agreement in the λ fit parameter is observed between the data and the model. This suggests that the fraction of π^- originating from long-lived decays (e.g. Λ, η) is well reproduced in the model, independent of meanfield. The parameters R_{out} and R_{long} from the model fits agree reasonably well with the data, and exhibit some sensitivity to the action of the meanfield during the collision. However, the model predicts a flat beam energy dependence of R_{side}, with or without the meanfield, in strong disagreement with the data. R_{side} is underpredicted by about 1.5 fm for 2 $A{\cdot}GeV$ collisions, while the agreement between data and model improves with increasing beam energy.

These discrepancies suggest that the effective source generated by the model is too small, and with too large an effective lifetime. Such an interpretation is more clearly illustrated by constructing the correlation functions using the Yano-Koonin-Podgoretskiĭ (YKP) decomposition [33] of relative momentum, in which the effective lifetime is fit more directly. In this parameterization, one fits the correlation function to the form

$$C(q_0, q_\perp, q_\|) = 1 + \lambda e^{-R_\perp^2 q_\perp^2 - R_\|^2(q_\|^2 - q_0^2) - (R_o^2 + R_\|^2)(q \cdot U)^2} \qquad (15.2)$$

where $U = \gamma \cdot (1, 0, 0, v)$ and v is the actual source velocity in the beam direction, relative to the analysis frame; $\gamma = (1 - v^2)^{-1/2}$. For all energies, v is consistent with zero, which is expected since the analysis is performed in the system c.m. frame, and the correlations are constructed from midrapidity pions. It was verified that fixing $v = 0$ in the fits has no significant effect on the other parameters [34].

R_0 measures the effective lifetime of the π^--emitting source, and $R_\perp$ and $R_\|$ the effective size perpendicular to and parallel to the beam axis. As seen in the right panels of Figure 15.2, the data indicate π^- emission from an effective source with zero effective lifetime (although the true time duration is almost certainly nonzero, flow effects will reduce the apparent lifetime) and $\sim$ 6 fm Gaussian size at the lowest energy. As the beam energy increases, the source size decreases. As is clear from the right panels of Figure 15.2, the effective lifetime is indeed overpredicted by the RQMD, especially in cascade mode, while the predicted source size is too small, especially at the lower beam energies.

Given this energy dependence of the discrepancy between RQMD and the data, it is interesting to note that for Pb+Pb collisions at 158 AGeV, the effective π^- source predicted by RQMD is significantly ($\sim$35%) larger than the measured size [35]. (e.g. R_{side}=4.45 fm for data vs 6.23 fm for the model.)

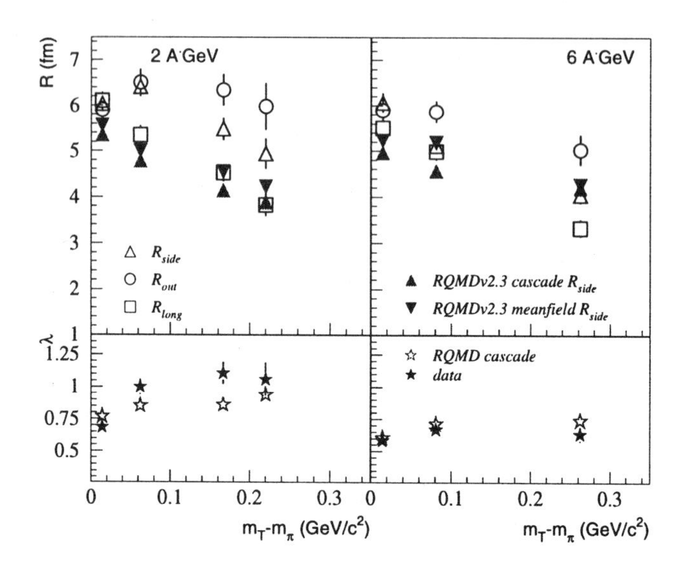

Figure 15.3 The m_T dependence of the measured BP HBT radii and λ parameter for midrapidity ($y = y_{cm} \pm 0.4$) π^- are shown for central (13 % σ_T) 2 $A{\cdot}GeV$ (left) and 6 $A{\cdot}GeV$ Au+Au collisions. Also shown is the transverse radius R_{side} from RQMD calculations with and without meanfield, and the λ value from RQMD without meanfield. (λ from RQMD changes inperceptably when the meanfield is on.)

Some insight into the discrepancies may be gained by examining the dependence of the correlations on $m_T = \sqrt{p_T^2 + m^2}$. Collective flow effects that induce position-momentum correlations [11, 35], as well as source expansion coupled with a finite emission timescale [4], both tend to decrease the effective source size at higher m_T. Differences in the m_T spectra of direct pions and pions from long-lived decay has also been suggested as a cause of the decrease of R_{side} with m_T [6].

As is clear from figure 15.3, the magnitude of these effects evolves with beam energy. At all energies, R_{long} decreases strongly with m_T. However, only for the higher beam energies do the transverse radii fall quickly with m_T, similar to observations at much higher energy [35, 36, 14]. At 2 $A{\cdot}GeV$, R_{side} and R_{out} show a smaller and more complicated decrease with m_T. Prior reports on the presence [24] or absence [25] of a momentum dependence of transverse radii correlation analyses for heavy systems at the Bevalac at slightly lower energy are mutually inconsistent.

In contrast, the m_T dependence predicted by the RQMD model is almost independent of collision energy, as may be seen by the predictions for R_{side} shown in Figure 15.3. While in fair agreement with the correlations at 8 AGeV, the model predicts a much stronger decrease of R_{side} with m_T than seen in the data at 2 AGeV; the underprediction of R_{side} observed in Figure 15.2 is due in large part to this strong fall-off.

As expected due to the increasing fraction of π^- from particle decay at low p_T, λ increases with m_T; similar trends are seen at higher energy [35]. Since the model reproduces the trends of λ reasonably well (c.f. Figs 15.2 and 15.3), it is unlikely that the details of resonance production dominate the discrepancy in m_T systematics [6].

Because flow and lifetime effects both influence the m_T dependence of R_{side} similarly, it is difficult to isolate the physical origin of the difference between the model and data. However, the discrepancy appears somewhat worse when the meanfield is switched off. It is known that collective flow at these energies is better reproduced when the meanfield is included in the model [37, 38, 39], and space-momentum correlations are actually stronger with the meanfield on. Thus, the reduced emission timescale may be the main reason for better agreement with the meanfield. Further study on this point is required.

Finally, we study whether it is possible to observe a possible non-azimuthally-symmetric pion source in non-central collisions. At finite impact parameter, the geometrical overlap between the two ions transverse to the beam is elongated in a direction perpendicular to the reaction plane. Naively, this overlap region is the dominant source of the pions, and one may observe an oscillation in R_{side} (and perhaps R_{out}) as the emission angle of the pair with respect to the reaction plane varies.

The E895 collaboration has studied reaction plane reconstruction extensively [38], and achieves a measured resolution on the order of 17° (36°) for midcentrality collisions at 2 (8) AGeV. Only baryons are used to determine the reaction plane, so autocorrelation effects with the pions are not an issue. Importantly, to construct the event-mixed background of the correlation functions, only those pions from events whose reconstructed reaction plane are within 5° of each other are mixed. Statistics limit the cuts on emission angle with respect to the reaction plane (ϕ_{rp}) to be 90° wide, centered at 0°, 90°, 180°, and 270°. Preliminary results for 2, 4, and 6 AGeV collisions with impact parameter b$\approx 5 - 7$ fm (as determined from charged particle multiplicity) are shown in Fig. 15.4.

At the lower energy, we observe the oscillation we naively expect from geometric considerations. The magnitude of the oscillation ($\sim$0.5 fm) is consistent with simulations with the RQMD model. At the higher energy, the effect is not observed. However, this is due at least in part to

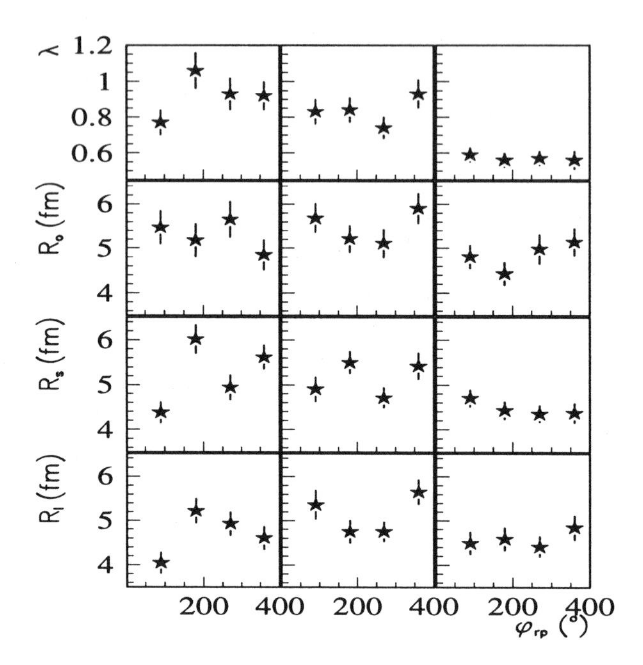

Figure 15.4 Fits to correlation functions in the Bertsch-Pratt parameterization for semicentral Au+Au collisions at 2 (left), 4 (middle), and 6 (right) AGeV bombarding energy. l and radii parameters are plotted against the pion emission angle with respect to the reconstructed reaction plane.

the worsening resolution with which the reaction plane is measured; the radii are not corrected for the finite dispersion. Ideally, the parameters should take identical values at 90° and 270°, and, since these data are taken around midrapidity, they should be the same at 0° and 180°. The observed deviations from this expectation are being invesitgated.

In summary, we have presented a preliminary excitation function of pion inteferometry at the AGS. The correlations at midrapidity for central collisions show no sharp increase in emission timescale that would signal QGP formation. A detailed comparison of measured multidimensional correlation functions to those from the popular transport model RQMD, revealed systematic discrepancies. In particular, the effective pion sources generated by the model are too small in size and too large in temporal extent for the lower bombarding energies. The discrepancy diminishes with increasing energy, finally reversing at CERN energies, where the predicted source is too large. The m_T dependence of the correlations reveals an evolution in the strength of dynamical and/or lifetime effects at AGS energies; this evolution with collision energy appears to be absent in the model. It is important to understand energy-dependent discrepancies in more detail before extrapolating transport models to the still higher energies of RHIC. We have also begun investigation into a novel combination of interferometry and flow measurements, capable of imaging the non-azimuthally-symmetric pion source. It is hoped that these more refined measurements will further enhance the power of interferometric studies as a probe of collision geometry and dynamics.

Acknowledgments

For enlightening discussions, MAL gratefully acknowledges enlightening discussions with Drs. D. Hardtke, U. Heinz, T. Humanic, S. Pratt, R. Soltz, H. Sorge, and J. Sullivan We thank Drs. S. Pratt and H. Sorge for the use of their codes. This work supported by the U.S. National Science Foundation under grants PHY-9722653, PHY-9601271, PHY-9225096; by the U.S. Department of Energy under contract DE-AC03-76SF00098 and grants DE-FG02-89ER40531, DE-FG02-88ER40408, DE-FG02-87ER40324; and by the University of Auckland Research Committee, NZ/USA Cooperative Science Programme CSP 95/33.

References

[1] W. Bauer, C.K. Gelbke, and S. Pratt, Ann. Rev. Nucl. Part. Sci. **42**, 77 (1992).

[2] S. Pratt, Phys. Rev. **C49** 2772 (1994)

[3] G. Bertsch, M. Gong, and M. Tohyama, Phys. Rev. **C37** 1896 (1988)

[4] S. Pratt, T. Csörgő, and J. Zimányi, Phys. Rev. **C42**, 2646 (1990)

[5] S. Pratt, Phys. Rev. **D33**, 72 (1986).

[6] B.R. Schlei and N. Xu, Phys. Rev. **C54**, R2155 (1996).

[7] J.P. Sullivan *et al.*, Nucl. Phys. **A566**, 531c (1994).

[8] M. Gyulassy, S.S. Padula, Phys. Rev. **C41**, R21 (1990).

[9] D. Beavis, *et al.*, Phys. Rev. **C27**, R910 (1983)

[10] S. Pratt, Phys. Rev. Lett. **53**, 1219 (1984).

[11] A.N. Mahklin, Y.M. Sinyukov, Z. Phys. **C39**, 69 (1988)

[12] S. Pratt, Nucl. Phys. **A566**, 103c (1993).

[13] H. Sorge, Phys. Rev. **C52**, 3291 (1995)

[14] G. Roland *et al.* (NA35), Nucl. Phys. **A566**, 527c (1994).

[15] B.A. Li and C.M. Ko, Nucl. Phys. **A601**, 457 (1996).

[16] P. Danielewicz *et al.*, Phys. Rev. Lett. **81**, 2438 (1998).

[17] J.I. Kapusta, A.P. Vischer, R. Venugopalan, Phys. Rev. C**51**, 901 (1995).

[18] D. Rischke *et al.*, J. Phys. G14, 191, (1988); Phys. Rev. D**41**, 111 (1990)

[19] N.K. Glendenning, Nucl. Phys. **A512**, 737 (1990).

[20] P. Braun-Munzinger and J. Stachel, Nucl. Phys. **A606**, 320 (1996).

[21] D.H. Rischke, Nucl. Phys. **A610**, 88c (1996)

[22] H.G. Pugh, *et al.*, LBL-22314 (1986); G. Rai, *et al.*, IEEE Trans. Nucl. Sci. **37**, 56 (1990).

[23] W.A. Zajc, *et al.* Phys. Rev. **C29**, 2173 (1984)

[24] W.B. Christie, *et al.*, Phys. Rev. **C47**, 779 (1993)

[25] A.D. Chacon, *et al.*, Phys. Rev. **C43**, 2670 (1991)

[26] K. Kadija, *et al.* (NA49), Nucl. Phys. **A610**, 248c (1996).

[27] T. Csörgő, B. Lörstad, and J. Zimányi, Z. Phys. **C71**, 491 (1996).

[28] H. Bøggild, *et al.* (NA44), Phys. Lett. **B302**, 510 (1993).

[29] G.I. Kopylov, Phys. Lett. **B50**, 472 (1974).

[30] A. Messiah, **Quantum Mechanics**, North-Holland (1961)

[31] J. Barrette, *et al.* (E877), Nucl. Phys **A610**, 227c (1996); J. Barrette, *et al.*, Phys. Rev. Lett **78**, 2916 (1997).

[32] S. Chapman, P. Scotto, and U. Heinz, Phys. Rev. Lett. **22**, 4400 (1995).

[33] U. Heinz, Nucl. Phys. **A610**, 264c (1996).

[34] Heinz et al. (Phys. Lett. **B382**, 181, (1996)) provide consistency relations between the YKP and BP fit parameters. The consistency of the fit parameters for all correlation functions in this work has been verified.

[35] I.G. Bearden, *et al.* (NA44), Phys. Rev. **C58**, 1656 (1998)

[36] S. Pratt, Nucl. Phys. **A638**, 125c (1998).

[37] H. Sorge, Phys. Rev. Lett. **78**, 2309 (1997).

[38] C. Pinkenburg, *et al.*, (E895) submitted to PRL, and R. Lacey, contribution to these proceedings.

[39] H. Liu *et al.*, (E895) Nucl. Phys. **A638**, 451c (1998).

[40] M.D. Baker, *et al.* (E866), Nucl. Phys. **A610**, 213c (1996)

Chapter 16

PROBING THE QCD PHASE BOUNDARY WITH FINITE COLLISION SYSTEMS

T. A. Trainor
Nuclear Physics Laboratory 354290
University of Washington
Seattle, WA 98195, USA
trainor@hausdorf.npl.washington.edu

1. INTRODUCTION

This article treats three closely-related topics. The common theme is the degree to which classical thermodynamic and statistical concepts are applicable to small-number dynamical systems near a bulk-matter phase boundary.

I begin with an examination of finite-size effects and the observability of a proposed QCD tricritical point. In contrast to macroscopic bulk-matter behavior a finite system cannot exhibit dramatic variation of order-parameter dependence on system control parameters (*e.g.*, first-order phase transition). Predictions of novel structure on a phase boundary should include an estimate of observability based on available system size and statistical realities.

I then consider the relationship between linear response coefficients (*e.g.*, heat capacity) and fluctuations in local extensive quantities. Non-statistical fluctuations are proposed as a signature of the QGP phase transition. Gaussian fluctuation theory implies that fluctuation amplitudes can be represented by linear response coefficients. This description fails near a phase boundary.

Finally, I consider application of the Central Limit Theorem to event-by-event distributions of global variables. Detailed examination of a proposed differential measure based on the CLT reveals that there are a number of contributions to such a measure, making interpretation difficult. Direct study of more elementary differential measures provides a better path to the underlying physics.

Advances in Nuclear Dynamics, 5,
Edited by Bauer and Westfall, Kluwer Academic / Plenum Publishers, New York, 1999.

2. OBSERVABILITY OF THE QCD TRICRITICAL POINT

General arguments based on universality have recently been put forward in support of a localized structure on the QCD phase boundary: a tricritical point or critical endpoint located at some intermediate position in the (T, μ_B) plane [1]. Detection and study of such a structure would be an important contribution to our understanding of full QCD. From an experimental viewpoint the question immediately arises what affect finite system size or scale interval has on the observability of such a structure?

The relevant scale interval is bounded by the characteristic size a of a typical momentum-carrying object and the characteristic size L of a 'causally connected' or 'equilibrated' region. The ratio L/a then determines the maximum sharpness of structures near critical points. The effective scale interval is a function of space and time in the collision space-time volume. We can use 'system size' informally to mean some average or effective L/a where this is not confusing.

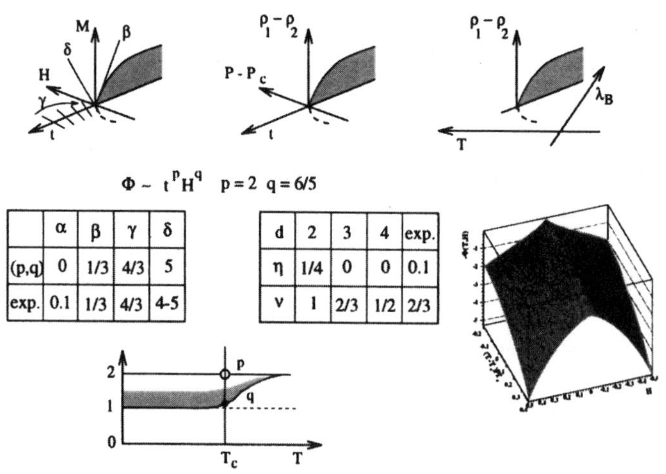

	α	β	γ	δ
(p,q)	0	1/3	4/3	5
exp.	0.1	1/3	4/3	4-5

d	2	3	4	exp.
η	1/4	0	0	0.1
ν	1	2/3	1/2	2/3

Figure 16.1 Three instances of a critical point - Ising model, liquid-gas phase boundary and proposed QCD critical endpoint (top). Critical exponents corresponding to parameter pair $(p, q) = (2, 6/5)$ (middle). Schematic temperature dependence of exponent parameters and corresponding free-energy surface (bottom).

The tricritical-point proposal suggests varying 'control parameters' (state variables) as a method of study. Practically accessible control parameters are collision geometry, energy and nuclear collision partners. These parameters map in some loose way onto thermodynamic equivalents (T, μ_B). However, the relationship between these two parameter systems is itself a subject of intense study and may always be poorly defined due to lack of complete system equilibration.

To explore the consequences of finite system size on critical phenomena the Ising magnet serves as a paradigm. Near the critical point the standard set of six critical exponents are linearly interdependent through

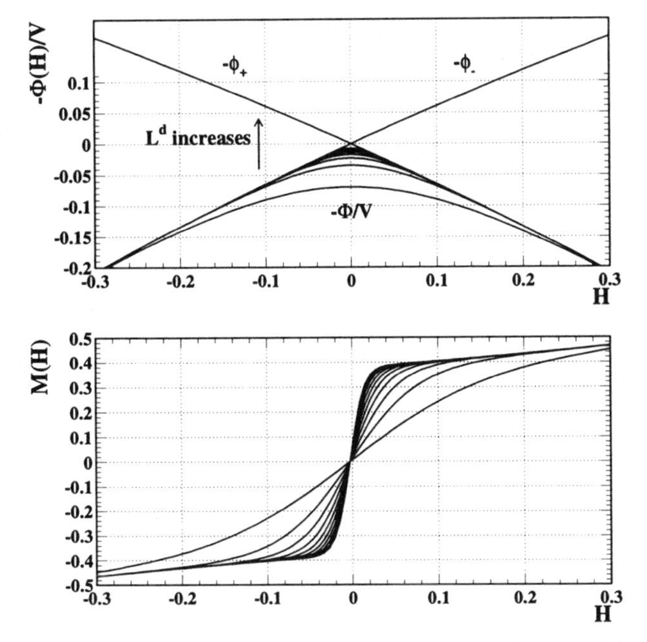

Figure 16.2 Sketch of free-energy dependence on external field for Ising model and various system sizes L. In the asymptotic limit $L \to \infty$ one recovers a first-order transition.

a system of scaling laws and can be reduced to two independent quantities (say (p, q)). These quantities represent the compound curvature of the free-energy surface in the neighborhood of the critical point along the control parameters T and H respectively. For this universality class the critical exponents are well represented by $(p, q) = (2, 6/5)$ as shown in Fig. 16.1.

How does finite system size affect this picture? Considering the discontinuity in $M(H)$ for bulk matter below the critical temperature, we can express the total free-energy density in terms of the individual free-energy densities ϕ_+ and ϕ_- of the two spin populations. The overall free-energy density can be written

$$\Phi(H) \approx -log\left\{e^{-\phi_+(H)\cdot(L/a)^d} + e^{-\phi_-(H)\cdot(L/a)^d}\right\}\cdot(a/L)^d \quad (16.1)$$

If $L/a \to \infty$ and given the nearly linear dependence of $\phi_\pm$ on H near the origin we have

$$\Phi(H) \approx |H|^1 \quad (16.2)$$

for $|H| \approx 0$, at and below the critical temperature, leading to an asymptotic (in scale interval) singularity in the magnetic susceptibility *vs* external field. The functional form of $M(H)$ therefore depends parametrically on the scale interval as $(L/a)^d$. This dependence is illustrated in Fig. 16.2. Thus, the consequence of finite system size is the general elimination of sharp structures on a phase boundary. If the QCD phase boundary is 'observed' with a finite collision system the manifestation of a tricritical point may be undetectable because of finite-scale-interval effects.

3. FLUCTUATION THEORY NEAR A CRITICAL POINT

Critical phenomena include nonstatistical fluctuations in local extensive quantities. Fluctuation characteristics are determined by the dependence of the free energy on relevant state variables. The standard treatment of fluctuations is the Einstein/gaussian theory or linear response theory [2]. This treatment cannot predict fluctuation properties in the neighborhood of a critical point.

The state of a large but bounded dynamical system in equilibrium can be characterized by a set of global intensives $\vec{\alpha} = (T, \lambda, P, ...)$, the system control parameters. If this system is randomly partitioned the partition elements can be characterized by local extensives $X_\alpha = (E, N, V, ...)$. An equilibrated system can be further characterized by a free energy $\Phi = (TS, A, G, ...)$. A local extensive can then be expressed in terms of its conjugate intensive and a free energy as $X_\alpha = \frac{\partial \Phi}{\partial \alpha}$. Linear response coefficients χ_α (C_V, κ_T,...) can be defined by

$$\begin{aligned} \chi_\alpha &= \frac{\partial X_\alpha}{\partial \alpha} \\ &= \frac{\partial^2 \Phi}{\partial \alpha^2} \end{aligned} \quad (16.3)$$

The basic consequence of the gaussian fluctuation model (free energy assumed gaussian in the neighborhood of its maximum) is that a fluctuation variance is directly proportional to a corresponding linear response coefficient

$$\begin{aligned} (\cdot)\sigma^2_{X_\alpha} &= (\cdot)\frac{\partial X_\alpha}{\partial \alpha} \\ &= N \cdot \frac{\partial log\{X_\alpha\}}{\partial log \alpha} \end{aligned} \quad (16.4)$$

where $(\cdot)$ is defined by requiring that $(\cdot)X^2_\alpha \approx N^2$.

If the density of DoF (degrees of freedom) near a phase boundary is represented as the product of a fugacity and a singular part $\rho(\alpha) =$

$\lambda \cdot g(\alpha)$ then a typical variance density can be written as

$$(\cdot)\sigma^2_{X_\alpha}/V \;=\; \rho(\alpha)\{1 + \hat{\alpha} \cdot \nabla_\alpha log[g(\alpha)]\} \tag{16.5}$$

where ∇_α is a logarithmic gradient with respect to the global intensives $\vec{\alpha}$ and $\hat{\alpha}$ is a unit vector in the intended direction of parameter variation. Far from a phase boundary, where gaussian theory is applicable

$$\{1 + \hat{\alpha} \cdot \nabla_\alpha log[g(\alpha)]\} \;\approx\; C_V/N,\ P \cdot \kappa_T,\ V \cdot \alpha_t, ... \tag{16.6}$$

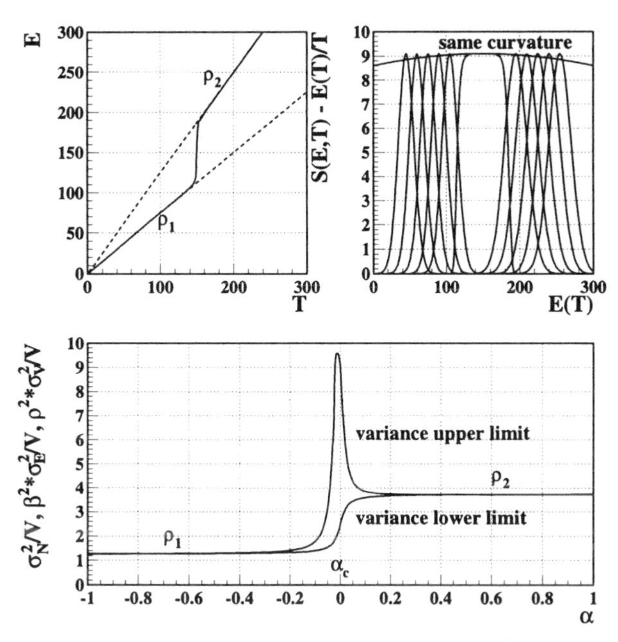

Figure 16.3 Sketch of E(T) dependence near a phase transition (top left), fluctuation distribution for various temperatures (top right) and plot of fluctuation amplitude vs temperature estimated by linear response coefficients (upper limit) and DoF density (lower limit) (bottom).

These relationships are represented in Fig. 16.3. The $E(T)$ dependence is piece-wise linear, corresponding to two different density regions separated by a phase boundary. At top right is a representation of energy fluctuation distributions at various temperatures. Near the critical temperature the fluctuations increase in amplitude but are bounded, whereas the linear response coefficient (slope of $E(T)$) may be arbitrarily large. The apparent paradox then is that linear response coefficients may become arbitrarily large in the neighborhood of a critical point but the observed fluctuation amplitude may not be large. Gaussian fluctuation theory would predict

$$\sigma^2_{X_\alpha} \propto \chi_\alpha \;\rightarrow\; \infty \tag{16.7}$$

when fluctuations are in fact bounded.

The paradox is resolved by noting that linear response coefficients measure the *curvature* of the free-energy surface near its maximum. If the free-energy surface is approximately gaussian in the neighborhood of its maximum then Eq. (16.4) is a good approximation. However, near a critical point the free-energy curvature and the width may have *no well-defined relationship.* The gaussian treatment breaks down and linear response coefficients cannot predict or represent fluctuation amplitudes. Near a phase boundary a separate determination must be made of the local variance density and the linear response coefficients, to the extent that the latter are meaningful or useful.

In the bottom panel of Fig. 16.3 the upper curve represents an estimate of fluctuation variance density based on linear response coefficients and serves as an upper limit in the neighborhood of a phase boundary. The lower curve is the density of DoF and serves as a lower limit to the variance density. Actual behavior near the phase boundary depends on the detailed structure of the free-energy surface. Both linear response coefficients and fluctuation amplitudes are in principle measurable. Instead of being redundant (as in gaussian theory) they may actually provide complementary information about the underlying system Hamiltonian.

4. ANALYSIS OF MEAN-P_T FLUCTUATIONS

Various event-wise global variables can be extracted from a collision final-state multiparticle momentum distribution. The most straightforward are the moments of the p_t distribution for charged hadrons, especially the first moment $\langle p_t \rangle$. Preliminary analysis of the $\langle p_t \rangle$ distribution for 100k central Pb-Pb events has been completed by the NA49 collaboration [3]. To summarize, it is found that this distribution is very close to a simple gaussian. The question arises whether this distribution is consistent with simple finite-number statistics modified by known hadronic physics, or whether the distribution contains evidence at some level for nonstatistical fluctuations that one might expect near a QCD phase boundary. How does one formulate this problem in a quantitative way?

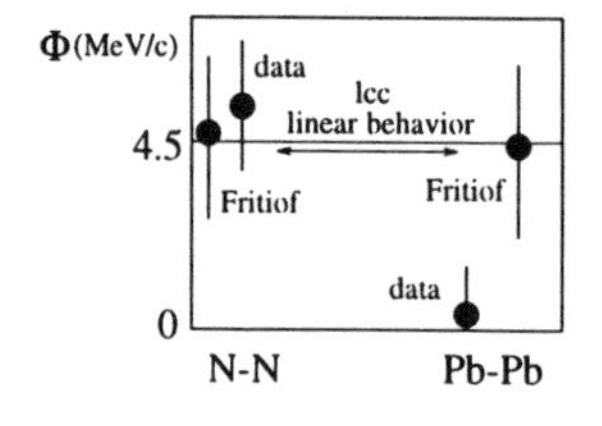

Figure 16.4 Summary of NA49 Φ analysis results.

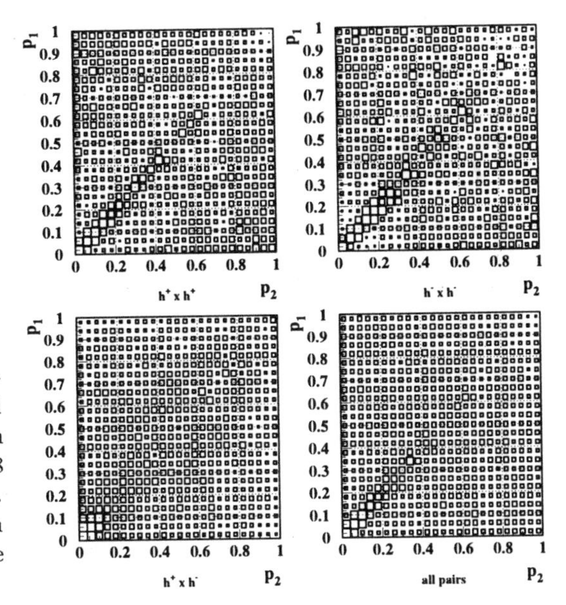

Figure 16.5 Distribution of sibling/mixed pair-density ratio on $p_t \otimes p_t$ space for 158 GeV/c Pb-Pb collisions. Density-ratio variation (box size) is within the interval $1 \pm .005$.

The basis for forming a differential measure applicable to this problem is the Central Limit Theorem (CLT). The CLT can be formulated in the following way. If events of N *independent* p_t samples with total momentum P_t are drawn from a *static* parent p_t distribution the distribution of event means $<\mathrm{p}_t> = \mathrm{P}_t/\mathrm{N}$ approximates a gaussian distribution with mean approaching the parent mean and variance $\sigma^2_{P_t}$ related to the parent variance $\sigma^2_{p_t}$ by

$$\sigma^2_{P_t} \quad = \quad N \cdot \sigma^2_{p_t} \tag{16.8}$$

A related differential width measure, the so-called Φ_{p_t} measure , has been proposed [4] and used to analyze the NA49 $<\mathrm{p}_t>$ distribution for 100k events [5]. It has the basic structure

$$\Phi_{p_t} \quad = \quad \sqrt{\sigma^2_{P_t}/N} - \sqrt{\sigma^2_{p_t}} \tag{16.9}$$

and is zero if the event population is consistent with the CLT.

Departures from the CLT ($\Phi_{p_t} \neq 0$) are due to 1) (anti)correlated samples and 2) variation of the parent during the sample history. Net positive correlation implies that $N_{eff} < N$, that is, the *effective* sample is less than N. We then expect $\Phi_{p_t} > 0$. The opposite is true for net anticorrelated samples. If the parent distribution varies during the sampling it can be shown that $\Phi_{p_t} \approx \sigma^2_{\overline{p_t}}$, where $\sigma^2_{\overline{p_t}}$ is a variance measure for the (slowly-varying) parent-distribution centroid.

Preliminary results of the NA49 Φ_{p_t} analysis which compares N-N and Pb-Pb collision events are summarized in Fig. 16.4 and seem to indicate that correlations observed in N-N collisions are reduced or eliminated, possibly by rescattering, in Pb-Pb collisions [6]. In order to provide a clearer understanding of the Φ_{p_t} measure and offer improved insight into the underlying physics we can decompose Φ_{p_t} in the following way

$$\Phi_{p_t} \approx \frac{A^2 + B^2}{2\sigma_{p_t} \cdot \overline{N}} \tag{16.10}$$

where

$$\begin{aligned} A^2 &= \sum_{i \neq j} p_i p_j - \overline{N_e(N_e - 1)} \cdot \overline{p_t}^2 \\ B^2 &= -2\overline{p_t}\left(\overline{N_e^2 < p_t >_e} - \overline{N_e^2} \cdot \overline{p_t}\right) \end{aligned} \tag{16.11}$$

A^2 is effectively a weighted integral of net two-particle correlation, and B^2 is a measure of N-$<p_t>$ correlations.

Contributions to A^2 include but are not limited to quantum statistics, Coulomb effects, particle decays, instrumental effects and true dynamical fluctuations. We study this term in more detail by forming a two-particle space (p_1, p_2) for sibling pairs and mixed pairs. We plot the ratio of density distributions ρ_{sib} and ρ_{mixed} for sibling and mixed pairs, and also project these densities onto momentum difference q and mean momentum k by analogy with standard HBT analysis. The connection of ρ_{sib} and ρ_{mixed} to A^2 is given by

$$A^2 \approx \int p_1 p_2 \cdot \{\rho_{sib} - \rho_{mixed}\} dp_1 dp_2 \tag{16.12}$$

The density ratio ρ_{sib}/ρ_{mixed} is shown in Fig. 16.5 for two combinations of like-sign pairs, one of unlike-sign pairs and one for all pairs taken together. One observes a prominent ridge along the main diagonal corresponding to quantum statistics and a peak at low k in the unlike-signed pairs due to the Coulomb interaction. We can project these densities onto q for four k slices and take the ratio of projections as shown in Fig. 16.6.

One sees quite prominently the expected correlation peak at small q and k due to BE statistics, but there is also a very strong *anti*correlation at large q in the like-signed pair distributions and a strong correlation at large q in the unlike-signed pair distribution. These results are unexpected and increase the difficulty of interpreting the Φ_{p_t} analysis. Certainly a separate treatment of positive, negative and unlike-signed pairs

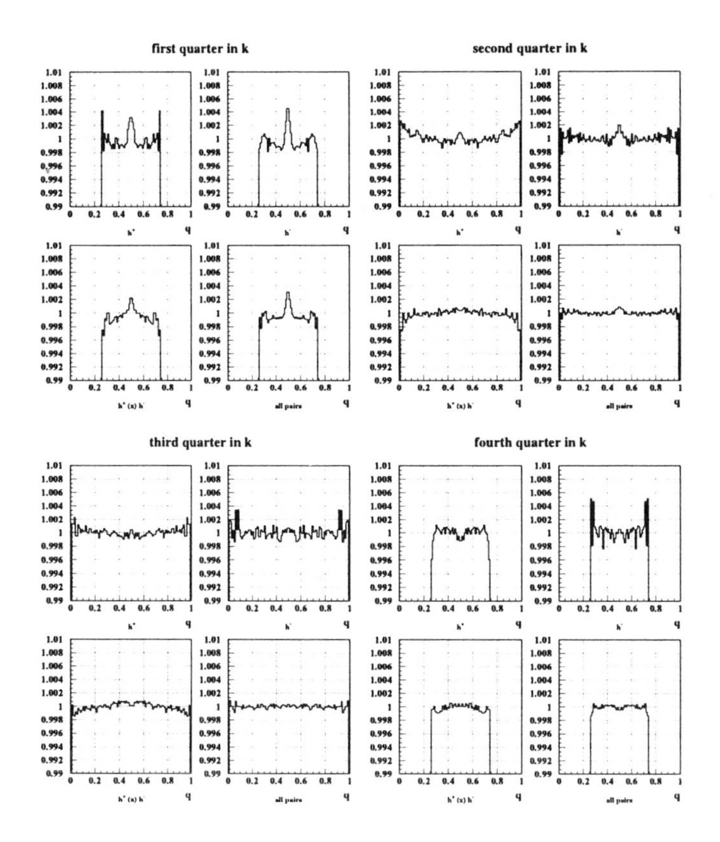

Figure 16.6 Projection of pair-density ratio on pair momentum difference q for four bins in mean pair momentum k.

is necessary in order to extract complete information. Different physics may affect each combination.

Contributions to B^2 come from correlations between event multiplicity and $<p_t>$. In N-N collisions these correlations are well-known and understood as a manifestation of energy conservation. The correlation is significantly different for negative and positive hadrons. The results of direct calculation of the linear correlation coefficient (lcc) between N and $<p_t>$ are summarized in Fig. 16.7. One sees that even for Pb there is a very significant nonzero lcc for total hadrons. The correlation is not completely removed by final-state rescattering but remains measurable.

For the NA49 Φ_{p_t} analysis it is found that A^2 and B^2 make contributions to Φ_{p_t} which are comparable in magnitude ($\simeq 20MeV/c$) and opposite in sign. This may be a case where different physical phenomena conspire to produce a small value for a given composite measure.

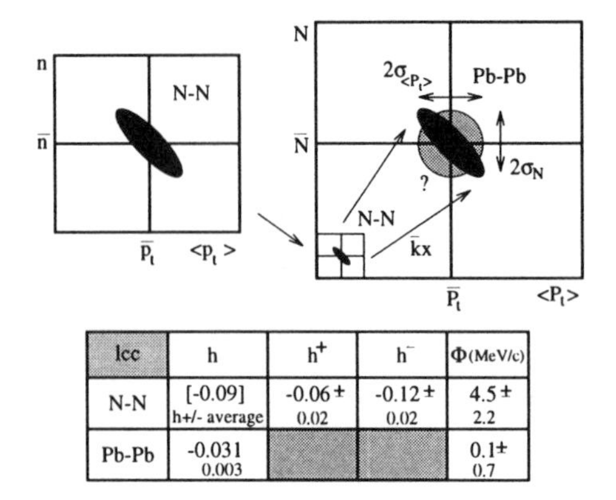

lcc	h	h+	h-	Φ(MeV/c)
N-N	[-0.09] h+/- average	-0.06± 0.02	-0.12± 0.02	4.5± 2.2
Pb-Pb	-0.031 0.003			0.1± 0.7

Figure 16.7 Sketch of N-$<p_t>$ linear correlation for N-N (top left), Pb-Pb (top right) and summary of measured values of the corresponding linear correlation coefficient (bottom).

5. CONCLUSIONS

Interpretation of event-by-event studies of relativistic heavy-ion collisions requires an approach to statistical and thermodynamical measures which includes limitations of finite system size and departures from classical statistics. The issue of critical-point phenomana on finite systems is interesting in its own right, and must be an important component of the study of the QCD phase boundary. The relationship between fluctuation variances and linear response coefficients in the neighborhood of the phase boundary is also important. Deviations from gaussian theory may in fact provide additional information about the free-energy surface and full QCD. Statistical analysis of global momentum-space variables requires differential analysis techniques which are capabable of separating several competing physical phenomena contributing to the correlation structure of the final state. The Φ_{p_t} measure represents an important step in the study of event-by-event fluctuations. Elaborations of this measure are now required for more detailed studies.

References

[1] M. Stephanov, K. Rajagopal, E. Shuryak, Phys. Rev. Lett. **81** (1998) 4816.

[2] L. E. Reichl, *A modern course in statistical physics* University of Texas Press, Austin, 1980.

[3] G. Roland, International Workshop XXV on *Gross Properties of Nuclei and Nuclear Excitations*, 1997, Hirschegg, Austria.

[4] M. Gaździcki, St. Mrówczyński, Z. Phys. **C26** (1992) 127.

[5] G. Roland for the NA49 collaboration, to be subm. to Phys.Lett.B.

[6] M. Gaździcki, A. Leonidov, G. Roland, Eur. Phys. J. **C6**, (1999) 365.

Chapter 17

TARGET N/Z EFFECTS ON PROJECTILE FRAGMENTATION

S. J. Yennello, R. Laforest, E. Ramakrishnan, D. J. Rowland, A. Ruangma, E. M. Winchester and E. Martin

Cyclotron Institute

Texas A&M University

College Station, TX, USA

Yennello@tbear.tamu.edu

Abstract Peripheral reactions of ^{28}Si with ^{112}Sn and ^{124}Sn at 30,40 and 50 MeV/-nucleon were used to elucidate the effect of the neutron content of the target on the process of projectile fragmentation. It is demonstrated that the fragments that result from these projectile fragmentation reactions can be divided into those which are the result of statistical emission of the quasi-projectile and those that are part of a direct component. The statistical part is independent of the target whereas the isotopic composition of fragments from the direct component are dependent on the neutron content of the target.

Keywords: Projectile Fragmentation, Isospin

1. INTRODUCTION

Projectile fragmentation has long been used to study the decay of excited nuclear systems[1][2][3][4]. It has traditionally been thought to be a two-step process consisting of excitation through a quasi-elastic collision followed by the breakup of the projectile[5][6][7]. In that scenario one could study the breakup of a hot nuclear system independent of the formation of the system. In fact much work has been done via projectile fragmentation to study whether the decay of excited nuclear matter is simultaneous or sequential[2] [3]. Recently, it was shown that projectile breakup can occur in close proximity with the target and that particles are emitted in a mixture of statistical decay (either sequential or prompt) and by direct emission [1].

In the two-step model the target nucleus has no or limited effect on the fate of the decaying excited projectile. However, since projectile breakup occurs in peripheral collisions with the target and the excitation energy is determined by friction and nucleon exchange between the target and the projectile, the use of two targets with different neutron content could help us to discriminate between what was emitted by the breakup of the projectile and what was directly emitted at the time of contact. Moreover a neutron-rich nucleus will have more neutrons at its surface thus showing a neutron skin. This neutron skin results from the different proton and neutron density distribution at the surface of a heavy nucleus and has been predicted by theoretical calculations that include a proper asymmetry potential [8][9][10]. The differences caused by differing neutron content of the target may be best observed by looking at isotopically resolved fragments.

Isotopic indentification of emitted fragments has been used by several groups to study the dynamics of intermediate mass fragment (IMF) and light charged particles (LCP) emission[11][12][13] [14][15][16][17]. These studies were done for mid-peripheral and central collisions. The goal of this study is to use the available isospin information to investigate the dynamics of fragment production and light charged particle emission in collisions where the dynamics was traditionally believed to have no effect.

2. EXPERIMENTAL SETUP

This experiment was done with a beam of ^{28}Si impinging on $\sim$1 mg/cm^2 of ^{112,124}Sn self supporting targets. The beam was delivered at 30 and 50 MeV/nucleon by the K500 superconducting cyclotron at the Cyclotron Institute of Texas A & M University. The detector setup for this experiment is composed of an arrangement of 68 Silicon - CsI(Tl) telescopes covering polar angles from 1.64° to 33.6° in the laboratory. Each element is composed of a 300μm surface barrier silicon detector followed by a 3cm CsI(Tl) crystal. The detectors are arranged in five concentric rings. The geometrical efficiency is more than 90% for each ring. A more detailed description of the detectors and electronics can be found in ref [18] . These detectors allow for isotopic identification of light charged particles and intermediate-mass fragments up to a charge of Z=5.

3. RESULTS

3.1 PROJECTILE RECONSTRUCTION

Fig. 17.1 shows the pseudo-momentum (sum of the charge multiplied by the parallel velocity) of all measured particles plotted as a function of total charge (sum of the charge) for each event. The well-detected quasi-projectile breakup events can be easily identified in the upper right.

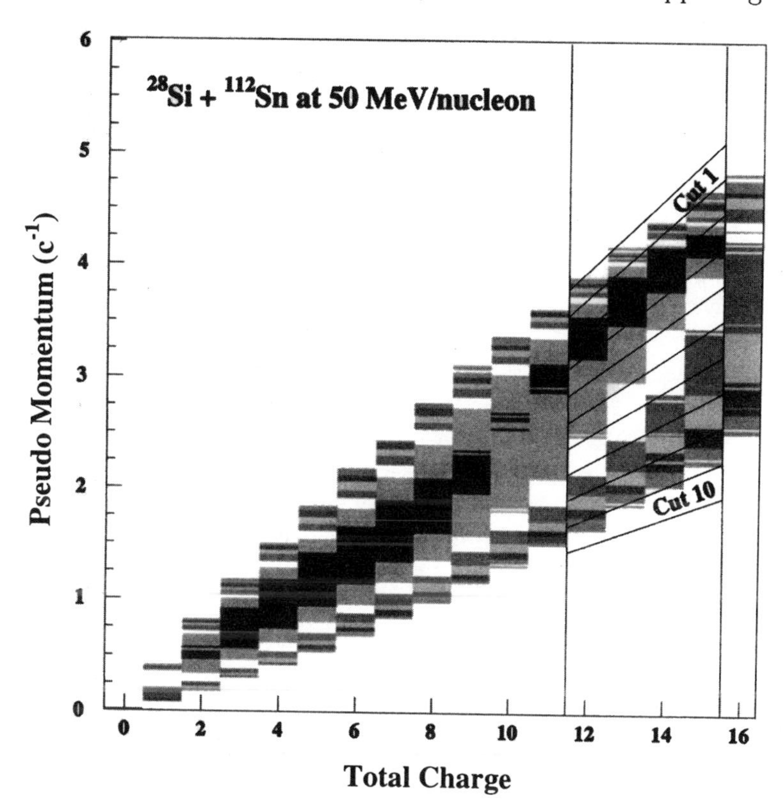

Figure 17.1 Pseudo-momentum plotted as a function of the total charge. The selection of quasi-projectile breakup events and the dissipation cuts are shown by the straight lines and are described in the text.

The analysis performed for this paper deals only with events with a total detected charge of at least 12 and no more than 15 thus keeping only the quasi-projectile breakup events. The forward momentum was used as a control variable for the violence of the collisions. Since some of the fragments and LCPs emitted from the statistical decay of the target could not be detected due to the detector geometry and energy threshold, the amount of forward momentum was less for more violent collisions. Ten cuts in forward momentum were made and are labeled from 1 to

10. These cuts are shown on Fig. 17.2 by the diagonal lines. The lines are diagonal since as more particles are detected the pseudo-momentum will be larger. Cut number 1 corresponds to the least violent collisions and 10 to the most violent. Although it is reasonable to believe that more violent collisions imply a smaller impact parameter, no attempts were made to relate them. Each cut corresponds to a dissipation of approximately 500GeV/c. One could have used the mass of the particle instead of the charge but this could lead to self-correlation in the ratios of yield that will be presented later in this paper.

3.2 CHARGE AND MULTIPLICITY DISTRIBUTIONS

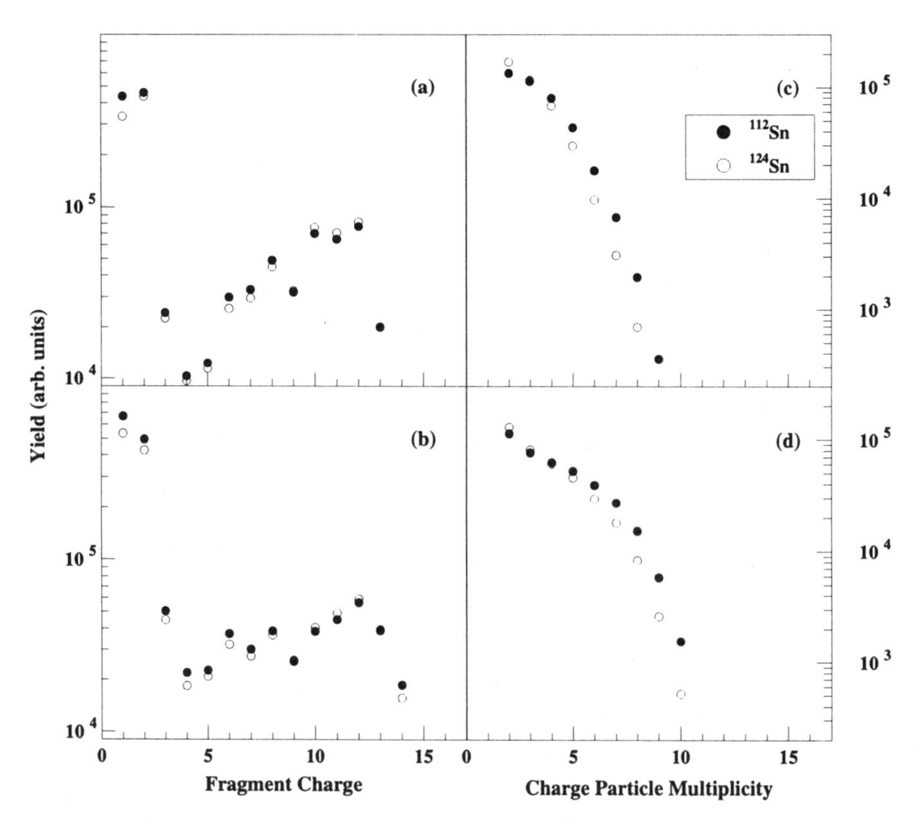

Figure 17.2 Charged particle distribution and multiplicity at 30 and 50 MeV/nucleon. Open circle correspond to data obtained with ^{124}Sn target, and close circle to ^{112}Sn target. The same notation will be used in the rest of the paper.

Fig. 17.2 shows the charge distribution and charged particle multiplicity distribution obtained for particles detected in the forward array for the quasi-projectile breakup events. One can see that the events are mostly composed of a projectile-like fragments (PLFs) accompanied by light charged particles. The distributions were normalized for the same number of events for the two targets at a given beam energy. The ^{112}Sn target produces slightly more Z=1 and Z=2 particles which results in slightly larger multiplicities. The 50 MeV/nucleon data reveal a smaller PLF and larger multiplicities due to the increase of excitation in the quasi-projectile. The larger proton yield with the ^{112}Sn target can be also interpreted by more pre-equilibrium or direct emission due to the peripheral nature of the collision. In a grazing collision with a more neutron-rich target like ^{124}Sn, this pre-equilibrium emission could appear in the neutron multiplicities which were not measured in this experiment. This would be consistent with a neutron skin at the surface of the target.

3.3 STATICTICAL VS. DIRECT EMISSION

An estimate of the light charged particles emitted by stastical decay of the excited quasi-projectile can be extracted for the total yield in the following manner. Particles emitted in the forward direction relative to the projectile like fragment are considered to be produced by the in-flight decay of the quasi-projectile. This forward contribution as well as its reflection around the projectile like fragment velocity is then subtracted from the total yield to extract the yield of the direct component. Fig. 3.3 shows the parallel velocity distribution of LCPs where the forward contribution (plus its reflection) is removed (see inset).

These spectra were created in the reference frame of the PLF for events containing a projectile-like fragment of charge six or heavier. In order to take into account the differential target Coulomb repulsion that results in velocity shifts[1], the reflection was performed around small offset relative to the PLF. The offsets were determined by the centroid of a Gaussian distribution fitted on the forward side of the distribution. These offsets were set to zero for deuterons and alpha particles since those have N/Z =1. The offset was set to a small positive value for protons and ^{3}He, and to a small negative number for ^{3}H. The necessity of considering different velocity shifts in the re-acceleration in the target Coulomb field is due to the short decay time of the quasi-projectile. The decay was estimated to occur within 15 fm from the target in a similar reaction[1]. The same method and offsets were applied for both targets. The ratios of detected particles, normalized to the same number of events for both targets, obtained by this method are summarized in table 17.1.

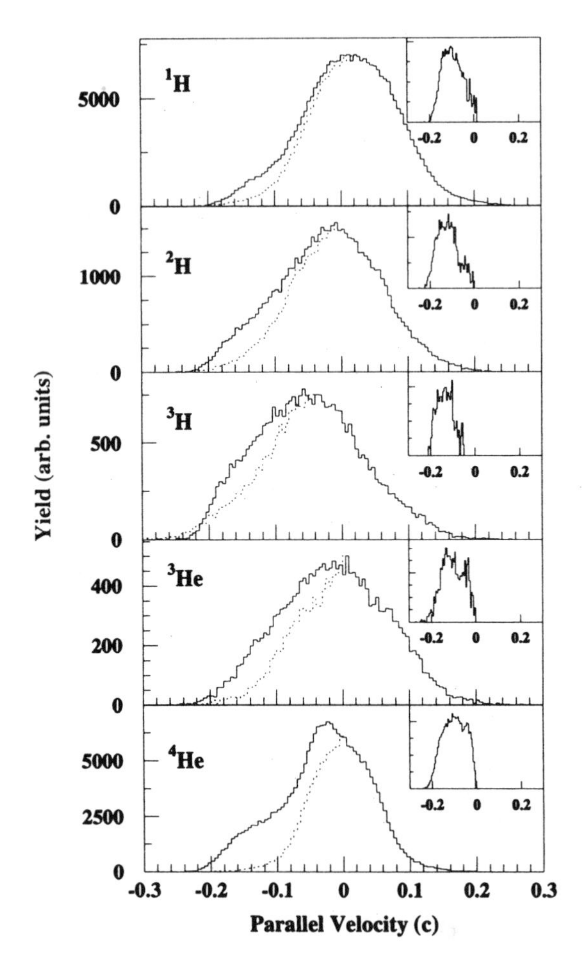

Figure 17.3 Parallel velocity spectra of light charged particles in the PLF reference frame from the reaction of ^{28}Si $+^{112}$Sn at 50 MeV/nucleon. The solid line represents the total velocity spectrum. The dotted line represents the projectile decay component. Inset shows the subtraction of the projectile decay component from the total spectrum, yielding the direct component.

All ratios are close to one for the quasi-projectile decay component meaning that the excited projectiles produced are fairly identical for both targets. However, the direct emission component differs substantially. A lot more protons and ^{3}He are directly emitted by a grazing collision with a neutron poor target than can be expected. Also, the number of ^{3}H directly emitted is significantly less for the ^{112}Sn target while the number of deuterons and alphas remains approximately the same. The statistical errors on those ratios are at most 5% (typically 1%). An evaluation of the systematic error was performed by changing the velocity offsets by a small amount and resulted in fluctuations of at most 20% in the evaluation of the ratios of the direct component only. The systematic errors for the quasi-projectile decay component evaluated by this method are very small.

Table 17.1 Yields of light charged particles from reconstructed events of ^{28}Si + ^{112}Sn at 50 MeV/nucleon divided by the yields from ^{28}Si + ^{124}Sn. The ratios were normalized for the same number of events for both targets. All LCPs were measured in the range of 1.64° to 33.6° in the laboratory.

Particle	Overall	Q-P Decay	Direct Emission
1H	1.36	1.24	2.58
2H	1.15	1.11	1.40
3H	0.96	1.01	0.53
3He	1.32	1.23	1.74
4He	1.12	1.10	1.18

4. CONCLUSIONS

This paper shows that projectile fragmentation depends on the N/Z of the target because fragment and light charge particle emission appears as a mixture of directly emitted light charged particles, neck rupture for the fragments and statistical decay of the excited quasi-projectile.

The multiplicity distribution of charged particles is larger for the reaction on the ^{112}Sn target. This is related to the shift in the charge distribution which has more smaller particles and fewer heavy particles with the ^{112}Sn target relative to the ^{124}Sn target. More neutron-rich light particles arise from the neck region as seen in the velocity distribution and angular distribution of the isotope ratios. This effect is enhanced for the more neutron-rich target. When the contribution due to statistical emission from the quasi-projectile is factored out the direct component reflects the relative neutron content of the target.

Acknowledgements

The authors wish to thank the cyclotron institute staff for the excellent beam quality. This work was supported in part by the NSF through grant # PHY-9457376, the Robert A. Welch Foundation through grant number# A-1266, and the Department of Energy through grant # DE-FG03-93ER40773

References

[1] R.J. Charity, L.G. Sobotka, N.J. Robertson, D.G. Sarantities, J. Dinius, C.K.Gelbke, T. Glasmacher, D.O. Handzy, W.C.Hsi, M.J. Huang, W.C. Lynch, C.P. Montoya, G.F. Peaslee, C. Schwarz and M.B. Tsang, Phys. Rev. C52 (1995) 3126.

[2] R. Laforest, D. Dore, J. Pouliot, R. Roy, C.St-Pierre, G. Auger, P. Bricault, S. Groult, E. Plagnol and D. Horn, Nucl. Phys. A568 (1994) 350.

[3] B. Harmon, J. Pouliot, J.A. Lopez, J.Suro, R. Knop, Y. Chan, D.E. Digregorio, R.G. Stockstead, Phys Lett B 235, 234 (1990)

[4] J. Pouliot, Y. Chan, D.E. DiGregorio, B.A. Harmon, R. Knop, C. Moisan, R. Roy and R.G. Stockstead, Phys. Rev. C43 (1991) 735.

[5] L.Beaulieu, R.Laforest, J.Pouliot, R.Roy, C.St-Pierre, G.C.Ball, E. Hagberg, D.Horn, R.B.Walker, Nucl.Phys. A580 (1994) 81.

[6] A.Badal'a, R.Barbera, A.Palmeri, G.S.Pappalardo, F.Riggi, G.Bizard, D. Durand and J.L.Laville, Phys. Rev. C45 (1992) 1730.

[7] R. Charity, J.Barreto, L.Sobotka, D.G.Sarantites, D.W.Stracener, A. Chbihi, H.G.Nicolis, R.Auble, C.Baktash, J.R.Beene, F. Bertrand, M. Halbert, D.C.Hensley, D.J.Horen, C.Ludemann, M.Thoennessen, R.Varner Phys. Rev C46 (1992) 1951.

[8] L.G. Sobotka, J.F. Dempsey, R.J. Charity, P. Danielewicz, Phys. Rev. C55 (1997) 2109.

[9] L.G. Sobotka, Phys. Rev. C50 (1994) R1272.

[10] Z.Z. Ren, W. Mittig, B.Q.Chen and Z.Y. Ma, Phys. Rev. C52(1995) 20.

[11] J.F. Dempsey, R.J. Charity, L.G. Sobotka, G.J. Kunde, S. Gaff, C.K. Gelbke, T. Glasmacher, M.J. Huang, R.C. Lemmon, W.G. Lynch, L. Manduci, L. Martin, M.B. Tsang, D.K. Agnihotri, B. Djerroud, W.U. Schröder, W. Skulski, J. Toke and W.A. Friedman, Phys. Rev. C54 (1996) 1710.

[12] E. Ramakrishnan, H. Johnson, F. Gimeno-Nogues, D.J. Rowland, R. Laforest, Y.-W. Lui, S. Ferro, S. Vasal and S.J. Yennello, Phys. Rev. C57 (1998) 1803.

[13] S.J. Yennello, B. Young, J. Yee, J.A. Winger, J.S. Winfield, G.D. Westfall, A. Vander Molen, B.M. Sherrill, J. Shea, E. Norbeck, D.J. Morrissey, T. Li, E. Gualtieri, D. Craig, W. Benenson, and D. Bazin, Phys. Lett. B321 (1994) 15.

[14] S.J. Yennello, E. Ramakrishnan, H.Johnson, G.Gimeno-Nogues, D.J. Rowland, R. Laforest, Y.-W. Lui and S. Ferro, Proc. of the VI

International Summer School on Heavy-Ion Physics, Dubna, Russia (1997).

[15] S.J. Yennello, R. Laforest, E. Martin, E. Ramakrishnan, D. J. Rowland, A. Ruangma and E. Winchester, Proc. of the XXXVI International Winter Physics Meeting, Bormio, Italy (1998)

[16] H. Johnston, T. White, J. Winger, D. ROwland, B. Hurst, F. Gimeno-Nogues, D. OKelly and S.J. Yennello, Phys. Lett. B 371 (1996) 186.

[17] H. Johnston et. al. Phys. Rev. C 56 (1997)1972.

[18] F. Gimeno-Nogues, D.J. Rowland, E. Ramakrishnan, S. Ferro, S. Vasal, R.A. Gutierrez, R. Olsen, Y.W. Lui, R. Laforest, H. Johnson and S.J. Yennello, Nucl. Inst. and Meth. In Phys. Res. A399 (1997) 94.

[19] J.P. Biersack and J.F. Ziegler, SRIM-96, Stopping and Range of Ion in Matter.

[20] D. Fox, D.R. Bowman, G.C. Ball, A. Galindo-Uribarri, E. Hagberg, D.Horn, L. Beaulieu and Y. Larochelle, Nucl. Instr. and Meth. A374 (1996) 63.

[21] D. Fox, private communication

[22] R. Tribble, R.H. Burch, C.A. Gagliardi, Nucl. Inst. Meth. A 285 (1989) 441 .

[23] R. Laforest, Progress in Reseach, TAMU(1997).

[24] W. Skulski, M. Fatyga, K. Kwiatkowski, H. Karwowski, L.W. Woo and V.E. Viola, Phys. Lett. B 218 (1989) 7.

[25] Y. Larochelle, C. St-Pierre, L. Beaulieu, N. Colonna, L. Gingras, G.C. Ball, D.R. Bowman, M. Colonna, G. D'Erasmo, E. Fiore, D. Fox, A. Galindo-Uribarri, E. Hagberg, D. Horn, R. Laforest, A. Pantaleo, R. Roy and G. Tabliente To be submitted

Chapter 18

NEW PHASES OF QCD; THE TRICRITICAL POINT; AND RHIC AS A "NUTCRACKER"

E.V.Shuryak

Department of Physics and Astronomy, State University of New York, Stony Brook, NY 11794-3800, USA

shuryak@dau.physics.sunysb.edu

Abstract Because too many interesting things are going on now, I have tried to squeeze three different subjects into one talk. The first is a brief summary of the color super-conductivity. During the last year we learned that instanton-induced forces can not only break chiral symmetry in the QCD vacuum, but also create correlated scalar diquarks and form new phases, some similar to the Higgs phase of the Standard model. The second issue I discuss is the remnant of the so called *tricritical* point, which in QCD with physical masses is the endpoint of the first order transition. I will argue that exchange of sigmas (which are massless at this point even with quark masses included) create interesting event-by-event fluctuations, which can be used to locate it. Finally I describe first results in *flow* calculations for non-central collisions at RHIC. It was found that it is extremely sensitive to Equation of State (EOS). Furthermore, the unusual "nutcracker" picture emerges for lattice-motivated EOS, which is formation of two *shells* which are physically separated *before* the freeze-out.

Keywords: color super-conductivity, chiral symmetry, the QCD tricritical point, RHIC, collective flow

1. QCD AT HIGH DENSITY

New phases of QCD at high density and the color super-conductivity issue are a part of a broader context, the studies of how the confining and chirally asymmetric QCD ground state is substituted by other phases as

Advances in Nuclear Dynamics, 5,
Edited by Bauer and Westfall, Kluwer Academic / Plenum Publishers, New York, 1999.

the temperature, the chemical potential, the number of flavors (or any combination of those) are increased. The key player in most of those effects (except confinement) happen to be instantons, see recent review [1]. In the QCD vacuum, for example, the quark condensate is simply the density of (almost) zero modes, originating from a superposition of zero modes associated with isolated instantons and anti-instantons.

At high temperature we expect to find the quark-gluon plasma (QGP) phase in which chiral symmetry is restored. So the density of (almost) zero modes goes to zero. This can only be realized if the instanton ensemble changes from a nearly random one to a correlated system with finite clusters, e.g. instanton-anti-instanton ($\bar{I}I$) molecules. The same is expected to happen (even at T=0) for sufficiently large number of flavors: in this case the expected next phase is the so called *conformal* phase.

The QCD at finite baryon density lay dormant since 70's, when basic applications of QCD like Debye screening were made. It was revived recently when it was realized that not only we expect the high density phase of QCD to be a color superconductor, as proposed in [5, 6, 7] with gaps in the MeV range, but that the instanton-induced effects lead to much larger gaps on the order of 100 MeV [2, 3].

In the next year it was realized that the phase structure of QCD at finite baryon density is actually very rich. In addition to the dominant order parameter, which is a scalar-isoscalar color anti-triplet ud diquark, many other condensates form. The overall picture can be characterized by some kind of "triality", both of three major phases under consideration, as well as of three competing attractive channels. These basic phases are: (i) the *hadronic* (H) phase, with (strongly) broken chiral symmetry (ii) the *color superconductor* (CSC) phase, with broken color symmetry ; and (iii) the *quark-gluon plasma* (QGP) phase, in which there are no condensates but the instanton ensemble is non-random. The three basic channels are the instanton-mediated attraction in (i) $\bar{q}q$ and (ii) qq channels (responsible for H and CSC phases) and the (quark-mediated) attraction between $\bar{I}I$, confining the topological charge in the QGP phase. The interrelation between these three attractive channels and phases is not straightforward: e.g. the $< \bar{q}q >$ may or may not be present in the CSC phase, and $\bar{I}I$ molecules have non-zero presence everywhere. However, this paper is still basically about a competition between these three attractive forces in different conditions.

The overview of the situation on the phase diagram is given by Fig.18.1, where one can see an approximate location of color super-conducting phases, as well as few schematic trajectories of excited matter, as it expands and cools in heavy ion collisions. One may see from those that unfortunately this new phase region corresponds to rather cool matter, and so it is *not* crossed by them. Therefore, color super-conductivity should only exists in compact stars. This created a challenge, known as the "pulsar cooling problem": a naked Fermi sphere is not allowed, because it generates too rapid cooling rate in contradiction to data.

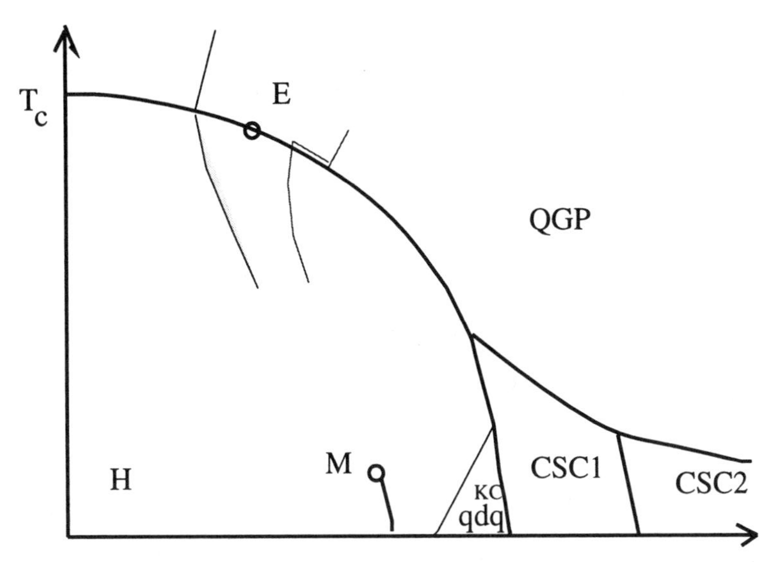

Figure 18.1 Schematic phase diagram of QCD phases as a function of temperature T and baryonic chemical potential μ, as we understand it today. The phases denoted by H and QGP are the usual hadronic phases (with nonzero $< \bar{q}q >$ and the quark-gluon plasma (no condensates). The color super-conducting phases CSC1 and CSC2 have various $< qq >$ condensates, the latter has broken chiral symmetry and tends asymptotically to color-flavor locking scenario. KC and QDQ are two possible phases, with Kaon condensate or quark-diquark gas. E is the endpoint of the 1-st order transition, M (from multi-fragmentation) is the endpoint of another 1-st order transition, between liquid and gas phases of nuclear matter. Two schematic trajectories corresponding to adiabatic expansion in heavy ion collisions are also indicated.

The scenario depends crucially on the number of quark flavors N_f. We start with discussion of (i) $N_f = 2$ massless quarks, u,d; (ii) then move on to $N_f = 3$ massless quarks, and finally to (iii) real QCD with non-zero quark masses.

In the $N_f = 2$ case the instanton-induced interaction is 4-fermion one. Its role in breaking chiral symmetry and making pion light and η' heavy is well documented, see e.g. [1]. One can Fierz transform it to diquark channels, which contain both color antisymmetric $\bar{3}$ and symmetric 6 terms. The scalar and the tensor are attractive:

$$\mathcal{L}_{diq} = \frac{g}{8N_c^2}\left\{-\frac{1}{N_c-1}\left[(\psi^T\mathcal{F}^T C\tau_2\lambda_A^a\mathcal{F}\psi)(\bar{\psi}\mathcal{F}^\dagger\tau_2\lambda_A^a C\mathcal{F}^*\bar{\psi}^T)\right.\right.$$
$$\left.+(\psi^T\mathcal{F}^T C\tau_2\lambda_A^a\gamma_5\mathcal{F}\psi)(\bar{\psi}\mathcal{F}^\dagger\tau_2\lambda_A^a\gamma_5 C\mathcal{F}^*\bar{\psi}^T)\right] \tag{18.1}$$
$$\left.+\frac{1}{2(N_c+1)}(\psi^T\mathcal{F}^T C\tau_2\lambda_S^a\sigma_{\mu\nu}\mathcal{F}\psi)(\bar{\psi}\mathcal{F}^\dagger\tau_2\lambda_S^a\sigma_{\mu\nu}C\mathcal{F}^*\bar{\psi}^T)\right\}$$

where τ_2 is the anti-symmetric Pauli matrix, $\lambda_{A,S}$ are the anti-symmetric (color $\bar{3}$) and symmetric (color **6**) color generators (normalized in an unconventional way, $tr(\lambda^a\lambda^b) = N_c\delta^{ab}$, in order to facilitate the comparison between mesons and diquarks). As discussed in ref.[2], in the case of two colors there is the so called Pauli-Gürsey symmetry which mixes quarks with anti-quarks. So diquarks (baryons of this theory) are degenerate with the corresponding mesons. It manifests itself in the Lagrangians given above: in this case the coupling constants in $\bar{q}q$ and qq channels are the same and the scalar diquarks, like pions. have the mass vanishing in the chiral limit.

Standard BCS-type mean field treatment leads to gap equation, from which one extract all properties of the color superconductor. Let me omit details and only mention the bottom line. The *chiral* symmetry is restored, while *color* SU(3) is broken to SU(2) by the colored condensate.

In the $N_f = 3$ case the situation becomes more interesting. (Since the critical chemical potential $\mu_c \sim 300 - 350 MeV$ is larger than the strange quark mass $m_s \simeq 140$ MeV, strange quarks definitely have to be included.) There are several qualitatively new features. First, since $N_f = N_c$, there are new order parameters in which the color and flavor orientation of the condensate is locked [8]. Second, the instanton induced interaction is a four-fermion vertex, so it does not directly lead to the BCS instability, unless there is also a $< \bar{q}q >$ condensate as well. So we need a superconductor where chiral symmetry is still *broken*. This is indeed what we have found [4], after a rather tedious calculation.

In the $N_f = 3$ case with variable strange quark mass the algebraic difficulties increase further. There are dozens of qq and $\bar{q}q$ condensates present, all competing for the resources. The largest, ud one, is still in the 100 MeV range, but the smallest are just few MeV, or comparable with light quark masses. Still, those small condensates are enough to solve the "pulsar cooling problem" (while without strangeness it remained unsolved).

We have found that two cases discussed above are in fact separated by a first order transition line, as a function of density or m_s. Partially this is caused by simple kinematic-al mismatch between $p_F(u,d)$ and $p_F(s)$ preventing their pairing, if m_s is large enough.

Finally let me mention that a transition region between nuclear matter and CSC (see Fig. 18.1) was claimed before by such exotic phase as Kaon condensation. In [4] we propose another (also exotic) quark-diquark (QDQ in Fig. 18.1) phase, in which nucleons dissociate into Fermi gas of constituent quarks plus Bose gas of constituent ud diquarks.

2. EVENT-BY-EVENT FLUCTUATIONS AND POSSIBLE SIGNATURES OF THE TRICRITICAL POINT

We now discuss the part of the phase diagram shown in Fig. 18.1 for densities below those for color super-conductivity. At high T and zero density it is believed to be second order if quark masses and strangeness is ignored, and a simple crossover otherwise. The discussion of the previous section (and many models e.g. the random matrix one [9]) suggest that it is likely to turn first order at some critical density. This means that there should be a tri-critical point in the phase diagram with $N_f = 2$ massless quarks, or the Ising-type endpoint E if quarks are not massless. The proposal to search for it experimentally was recently made by Stephanov, Rajagopal and myself [10]. A detailed paper about event-by-event fluctuations around this point [11] is the basis of this section.

The main idea is of course based on the existence of truly massless mode at this point, the sigma field, which is responsible for "critical opalescence" and large fluctuations. The search itself should be partially similar to "multi-fragmentation" phenomenon in low energy heavy ion collisions, which is also due to the endpoint M (see Fig. 18.1) of another first order transition. The "smoking gun" is supposed to be a non-monotonous behavior of observable as a function of such control parameters as collision energy and centrality. One can use pions as a "thermometer" to measure this fluctuations.

(Note that in many ways it is the opposite of the DCC idea: in that case the pion was the light fluctuating field, while its coupling to heavy and wide and strongly damping sigma field is the main obstacle.)

We have studied three ways in which sigmas can show up. First and the simplest is the "thermal contact" idea: at the critical point the sigmas specific heat becomes large, and this shows up in the pion fluctuations just due to energy conservation. The second is "dynamical exchange": pions can exchange the sigmas and this leads to long-range effects, over the whole correlation range. Both effects are in 10-20 percent range, after realistic account for correlation length is made. It is not large, but much larger than the accuracy of the measurements. The third effect is due to sigma decays into pions, which affect spectra at small p_t and (even more so) the multiplicity fluctuations.

Large acceptance detectors can study the event-by-event (ebe) fluctuations quite easily. The first data by NA49 detector at CERN on distributions of N, the charged pion multiplicity, and p_T (the mean transverse momentum of the charged pions in an event) for central PbPb collisions at 160 AGeV display beautiful Gaussians. Since any system in thermodynamic equilibrium exhibits Gaussian fluctuations, it is natural to ask how much of the observed fluctuations are thermodynamic in origin [13, 14]. We have answered this question quantitatively in this paper, considering fluctuations in pion number, mean p_t and their correlation. We model the matter at freeze-out as an ideal gas of pions and resonances in thermal equilibrium, and make quantitative estimates of the thermodynamic fluctuations in the resulting pions, many of which come from the decay of the resonances after freeze-out. The conclusion is that nearly all answers are reproduced by the resonance gas, with remaining part likely to be due to experimental corrections, due to two-particle track resolution and non-pion admixture. The good agreement between the non-critical thermodynamic fluctuations we analyze in Section 3 and NA49 data make it unlikely that central PbPb collisions at 160 AGeV freeze out near the critical point.

Estimates suggest that the critical point is located at a μ_f such that it will be found at an energy between 160 AGeV and AGS energies. This makes it a prime target for detailed study at the CERN SPS by comparing data taken at 40 AGeV, 160 AGeV, and in between. We are more confident in our ability to describe the properties of the critical point and thus *how* to find it than we are in our ability to predict where it is. If it is located at such a low μ that the maximum SPS energy is insufficient to reach it, it would then be in a regime accessible to study by the RHIC experiments.

3. RHIC AS A *NUTCRACKER*

This section is a brief account of unusual pattern of space-time evolution, found for non-central collisions at RHIC energies by D.Teaney and myself [15].

Let me begin with a pedagogic consideration of two opposite schematic models of high energy heavy ion collisions, leading to quite different conclusions about even such global thing as *duration* of the collision till freeze-out. This will set a stage for more elaborate considerations later, based on hydrodynamic approach.

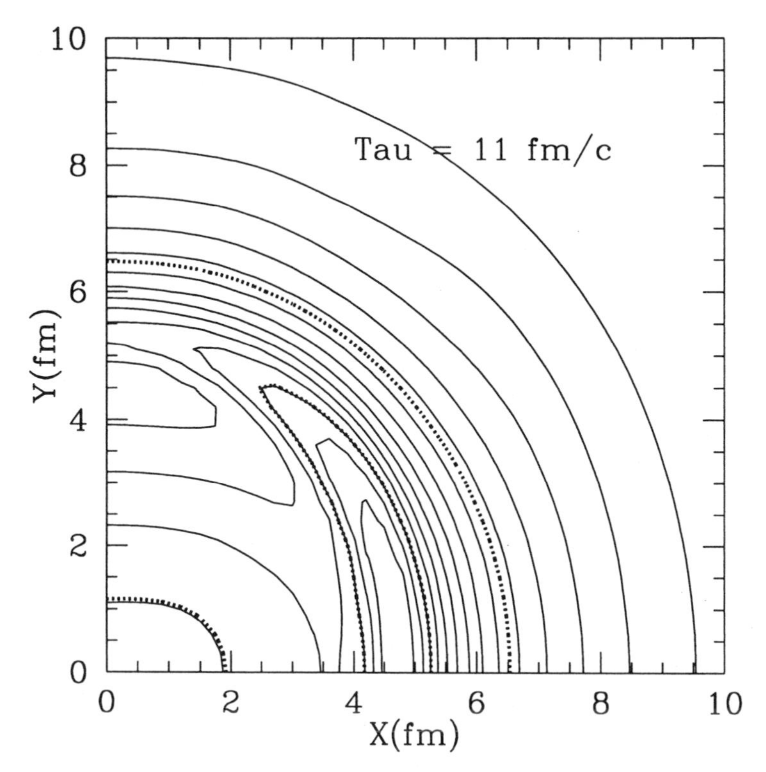

Figure 18.2 Typical matter distribution for AuAu collision at b=8 fm, at time t=11 fm/c, in the transverse plane x-y, resulting from hydro calculation with lattice-inspired EOS. Transverse expansion is assumed to be rapidity-independent, while the longitudinal expansion is Bjorken-like. The final multiplicity assumed is $dN_{ch}/dy = 850$. Lines show levels with fixed energy density, with step ??? Gev/fm^3. The dotted contours are for $T_f = 120MeV$ (the outer one), and T=140 MeV (two inner ones).

The "model A" is just a picture of longitudinal expansion without any transverse one, except maybe very late in the process. (One may therefore call it a "late acceleration model".) For rapidity-independent (Bjorken) expansion the dynamics is very simple: each volume element expands linearly in proper time τ. If the total entropy S is conserved, its density $s \sim 1/\tau$. The *initial* value of entropy density at RHIC s_i^{RHIC} is of course unknown, but it is believed to be several times that at SPS, say $s_i^{RHIC} = (2-4)s_i^{SPS}$. The *final* values should be roughly the same (At one hand, the larger the system the more should it cool down. On the other hand, at RHIC the fraction of baryons is expected to be significantly lower, and it should reduce re-scattering). Therefore, this model predicts total duration of expansion $\tau^{RHIC} = (2-4)\tau^{SPS} \sim 40 fm/c$.

The "model B" includes the transverse expansion, but in the opposite manner: the observed radial flow velocity at freeze-out $v_f \approx .4$ (the value we expect to see at RHIC) is now assumed to be there all the time. By contrast to model A, it assumes an "early acceleration". Including simple geometric expansion in the decrease of the entropy density $s \sim 1/(\tau(r_0 + v_f\tau)^2)$ one finds then much shorter duration of the collision $\tau^{RHIC} \sim 10 fm/c$ predicted by model B.

Which model is closer to reality depends on the real acceleration history, which is in turn determined by the interplay of the collision geometry, the energy and the EOS. Qualitatively speaking, the main message of the previous section is that while at AGS/SPS energy domain the collective flow appears late,like in model A, at RHIC/LHC it is expected to be generated early, as in the model B. The reason for that is specific behavior of the QCD EOS, which is soft in the "mixed phase" region of energy density, *very* soft in QGP at $T \approx T_c$, and then rapidly becoming hard at $T \approx 2 - 3T_c$.

One consequence[16] is that duration of the collision *grows* with energy in the AGS domain, but expected to *decrease* from SPS to RHIC. The maximal is expected to be when the initial conditions hit the "softest point" of the EOS, roughly at beam energy 30-40 GeV*A. Somewhat counter-intuitive, around the same energy one expects also a maximum value of the radial flow: longer acceleration time seem to win over softness! Especially interesting is the dynamics of the "elliptic" flow in the SPS-low RHIC energy domain. It is quite possible that its energy dependence would be sufficient to see the onset of QGP plasma.

Of course, the magnitude of the flow depends not only on hydro-EOS but also on kinetics of the freeze-out itself. We have already mentioned two factors which enter into consideration here: the absolute size of the system (hydro itself is scale-invariant!) and the baryon/meson ratio. Only careful systematic study of various systems at various energies will clarify the actual role of all these effects.

The magnitude of collective flow and its acceleration history can be understood as follows. We expect to have rapid change of pressure to energy density ration in QGP around the phase transition. Higher density QGP has $p/\epsilon \approx 1/3$, but at the transition there is the minimum of this ratio (the so called "softest point" [16]) where p/ϵ is small (0.1-0.05). So at AGS/SPS energies the expansion is slow and QGP just "burns inward". At some point, the outward expansion of the QGP and the inward burning may cancel each other, leading to near-stationary "burning log" picture [17]. At higher collision energies, the burning discontinuity is blown out and the situation returns to much simpler hydro picture typical for simple EOS $p = \epsilon/3$.

We have recently found that the lattice-inspired EOS leads to very unusual picture of the expansion, with quite characteristic inhomogeneous matter distribution (to be referred below as a *nutshells*). Stiff QGP at the center pushes against soft matter in the transition region: as a result some piling of matter occurs, in a shell-like structure. Furthermore, for non-central collisions the geometry drives expansion more to the direction of impact parameter (called x axis) rather than y, starting rather early. As a result, the two half-shells *separate* by freeze-out, and so (at least) two separate fireballs are actually produced. (We called this scenario a *nutcracker*.) Nothing like this happens for simple EOS, which always lead to matter distribution with a maximum at the center.

There is not much place here to display this interesting phenomenon. In Fig.18.2 we show a typical mater distribution. The time 11 fm/c is around (or slightly before) the freeze-out for most matter (it is not changed much anyway, the longitudinal expansion simply dilute it more). One can clearly see two shells in x direction and holes in y ones.

How can such phenomenon be seen experimentally?

(i) We have calculated several harmonics (in angle ϕ) of flow, v_n. We found quite observable deviations from a directed+elliptic (n=1 and 2 only) distributions seen before up to n=6.

(ii) The distribution of pions can be sufficiently accurate to see it, but with nucleons and, even better, heavier particles like deuteron-s we find much stronger signals for "nutcracker" scenario in production/flow patterns.

(iii) Another dramatic changes are found if one calculates correlators used by two-particle interferometry (HBT). Strong flow plus inhomogeneous distribution make *visible* HBT radii to be significantly *smaller* to what one might naively expect: we see only smaller "patches" of the picture in any given direction. (By the way, it significantly reduces conditions for momentum resolution of the detectors.). But for the same reason taking these patches all together, into a unified picture, is becoming more complicated.

Finally, let me emphasize it once again: the expected "nutcracker" pattern is supposed to be seen in typical non-central events. Because most of the RHIC detectors are able to detect the impact parameter plane in most events, there is no doubt that absolutely *any* phenomenon, from particle single-body distribution to $J/\psi, \Upsilon$ suppression or "jet-quenching" would be found strongly ϕ dependent, if it takes place. We will see exciting results on that, right from the first day of RHIC operation.

References

[1] T. Schäfer and E.V. Shuryak, Rev. Mod. Phys. 70, 323 (1998).

[2] R. Rapp, T. Schäfer, E. V. Shuryak and M. Velkovsky, Phys. Rev. Lett. **81** (1998) 53.

[3] M. Alford, K. Rajagopal and F. Wilczek, Phys. Lett. **B422** 247 (1998).

[4] R. Rapp, T. Schäfer, E. V. Shuryak and M. Velkovsky, High density QCD and Instantons. To be submitted to Phys.Rev.D.

[5] S. C. Frautschi, Asymptotic freedom and color superconductivity in dense quark matter, in: Proceedings of the Workshop on Hadronic Matter at Extreme Energy Density, N. Cabibbo, Editor, Erice, Italy (1978)

[6] F. Barrois, Nucl. Phys. B129, 390 (1977)

[7] D. Bailin and A. Love, Phys. Rep. 107, 325 (1984) .

[8] M. Alford, K. Rajagopal and F. Wilczek, hep-ph/9804403.

[9] M. A. Halasz, A. D. Jackson, R. E. Shrock, M. A. Stephanov and J. J. M. Verbaarschot, Phys. Rev. **D58** (1998) 096007.

[10] M. Stephanov, K. Rajagopal, E. Shuryak, Phys. Rev. Lett. **81** (1998) 4816

[11] M. Stephanov, K. Rajagopal, E. Shuryak,Event-by-Event Fluctuations in Heavy Ion Collisions and the QCD Critical Point, in preparation.

[12] H.Appelshauser et al (NA49 Collaboration), Event-by-event fluctuations..." In preparation.

[13] L. Stodolsky, Phys. Rev. Lett. **75** (1995) 1044.

[14] E. V. Shuryak, Phys. Lett. **B423** (1998) 9.

[15] D.Teaney and E. V. Shuryak, Unusual space-time evolution in non-central heavy ion collisiosn at RHIC energies, hep-ph 99????.

[16] C.M. Hung, E.V. Shuryak. Phys.Rev.Lett.75:4003-4006,1995; Phys.Rev.C57:1891-1906,1998.

[17] D.H.Rischke, M.Gyulassy Nucl.Phys.A597:701-726,1996

Chapter 19

HBT STUDIES WITH E917 AT THE AGS: A STATUS REPORT

Burt Holzman[1,6]
Physics Department
University of Illinois at Chicago and Argonne National Laboratory
Chicago, IL, USA
bholzm1@uic.edu

For the E917 collaboration:
B. B. Back[1], R. R. Betts[1,6], H. C. Britt[5], J. Chang[3], W. C. Chang[3], C. Y. Chi[4], Y. Y. Chu[2], J. Cumming[2], J. C. Dunlop[8], W. Eldredge[3], S. Y. Fung[3], R. Ganz[6,9], E. Garcia-Solis[7], A. Gillitzer[1,10], G. Heintzelman[8], W. Henning[1], D. J. Hofman[1], B. Holzman[1,6], J. H. Kang[12], E. J. Kim[12], S. Y. Kim[12], Y. Kwon[12], D. McLeod[6], A. Mignerey[7], M. Moulson[4], V. Nanal[1], C. A. Ogilvie[8], R. Pak[11], A. Ruangma[7], D. Russ[7], R. Seto[3], J. Stanskas[7], G. S. F. Stephans[8], H. Wang[3], F. Wolfs[11], A. H. Wuosmaa[1], H. Xiang[3], G. Xu[3], H. Yao[8], C. Zou[3]
[1] *Argonne National Laboratory, Argonne, IL*
[2] *Brookhaven National Laboratory, Upton, NY*
[3] *University of California at Riverside, Riverside, CA*
[4] *Columbia University, Nevis Laboratories, Irvington, NY*
[5] *Department of Energy, Division of Nuclear Physics, Germantown, MD*
[6] *University of Illinois at Chicago, Chicago, IL*
[7] *University of Maryland, College Park, MD*
[8] *Massachusetts Institute of Technology, Cambridge, MA*
[9] *Max Planck Institute für Physik, München, Germany*
[10] *Technische Universität München, Garching, Germany*
[11] *University of Rochester, Rochester, NY*
[12] *Yonsei University, Seoul, Korea*

Advances in Nuclear Dynamics, 5,
Edited by Bauer and Westfall, Kluwer Academic / Plenum Publishers, New York, 1999.

Abstract Two-particle correlations between pions in Au+Au collisions have been measured at beam kinetic energies of 6, 8, and 10.8 GeV/u at the Alternating Gradient Synchrotron (AGS) over a wide range of rapidities using a magnetic spectrometer. The data have been analyzed in the Hanbury-Brown and Twiss (HBT) framework to extract source parameters. The event-by-event orientation of the reaction plane has also been measured using a scintillator hodoscope at far forward rapidities, and beam vertexing detectors upstream of the target. A preliminary analysis of the dependence of the source parameters on the reaction plane is presented.

Keywords: AGS, HBT, reaction plane

1. INTRODUCTION

Identical particle correlations have previously been used to determine source sizes and lifetimes of the emission regions formed in heavy-ion collisions [1, 2, 3, 4]. In the case of non-central collisions, such measurements usually integrate over the reaction plane of the collision and therefore obscure physics which depends on the relative orientation of the two colliding nuclei.

In this report we discuss preliminary results from our investigations of the dependence of HBT source parameters on the reaction plane. This approach should eventually give us another perspective into the source dynamics. Together with other HBT dependencies of the source geometry and dynamics, we may thus be able to form a more complete picture of the nature and evolution of the emission region.

1.1 THE HBT CORRELATION FORMALISM

Quantum statistics require that the amplitudes for identical particles be added together before interpreting the square modulus as a probability. Specifically, identical bosons, such as pions, are required to have symmetric two-particle amplitudes, and the HBT correlation arises from the cross-term in the square of the symmetrized amplitude.

For a pair of identical bosons, emitted by an extended source $\rho(\mathbf{x})$ with four-momenta $\mathbf{p}_1$ and $\mathbf{p}_2$, detected at space-time coordinates $\mathbf{x}_1$ and $\mathbf{x}_2$, assuming that the bosonic wavefunctions can be described by plane waves, the symmetrized amplitude of this process is:

$$\Psi_{12}(\mathbf{x}_1, \mathbf{x}_2, \mathbf{p}_1, \mathbf{p}_2) = \frac{1}{\sqrt{2}}(e^{i(\mathbf{p}_1 \cdot \mathbf{x}_1 + \mathbf{p}_2 \cdot \mathbf{x}_2)} + e^{i(\mathbf{p}_1 \cdot \mathbf{x}_2 + \mathbf{p}_2 \cdot \mathbf{x}_1)}) \qquad (19.1)$$

Which leads to the probability of detecting a pair with relative four-momentum $\mathbf{q} = \mathbf{p}_1 - \mathbf{p}_2$:

$$C_2(\mathbf{q}) \equiv \frac{1}{N(\mathbf{q})}\int d^4x_1 d^4x_2\ \rho(\mathbf{x}_1)\rho(\mathbf{x}_2)|\Psi_{12}|^2 = \frac{1+|\tilde{\rho}(\mathbf{q})|^2}{N(\mathbf{q})} \tag{19.2}$$

where the two-particle correlation function $C_2(\mathbf{q})$ is simply related to the Fourier transform of the source density. The normalization $N(\mathbf{q})$ is discussed below in Section 1.4.

The above formulation is only strictly valid for completely incoherent sources. To account for partial coherence effects, as well as contamination from long-lived resonances, an empirical variable λ is added to the definition of the correlation function as a coherence scaling parameter:

$$C_2(\mathbf{q}) \equiv \frac{1+\lambda|\tilde{\rho}(\mathbf{q})|^2}{N(\mathbf{q})} \tag{19.3}$$

1.2 HBT PARAMETERIZATIONS

In practice, the source $\rho(\mathbf{r})$ has usually been assumed to be Gaussian in configuration-space. In this case, the momentum-space distribution is also Gaussian:

$$\rho(\mathbf{r}) \sim e^{\frac{-|\vec{r}|^2}{\mathbf{R}^2}} \Rightarrow \tilde{\rho}(\mathbf{q}) \sim e^{-|\vec{q}|\mathbf{R}^2} \tag{19.4}$$

For multi-dimensional HBT analyses, the relative momentum variable $\mathbf{q}$ can be expressed in terms of a variety of orthogonal components, each of which has model-dependent significance [5, 6].

In the preliminary analysis presented here, a simple 3-D Cartesian parameterization is chosen: (Q_x, Q_y, Q_z), with conjugate source parameters (R_x, R_y, R_z). R_z is taken along the beam axis; R_x is lies in the reaction plane; and R_y is orthogonal to both.

In this case, $C_2(\mathbf{q})$ has the following form:

$$C_2(\mathbf{q}) = \frac{1}{N(\mathbf{q})}[1+\lambda e^{-(q_xR_x)^2-(q_yR_y)^2-(q_zR_z)^2}] \tag{19.5}$$

1.3 REACTION PLANE DETERMINATION

The reaction plane is determined using the relative orientation of two axes: the direction of the incoming beam particle, $\hat{z}$, and the direction of the impact parameter $\hat{b}$. In our experiment, the beam axis is defined by a beam vertexing detector (BVER). The BVER detector consists of four planes of scintillating fibers each read out by a multi-anode photomultiplier tube. The fiber planes each consist of $\sim$ 150 200 x 200 μm^2

fibers, situated 5.84 m and 1.72 m upstream from the target [7]. The position of the projection of $\hat{z}$ onto the hodoscope can be determined with an accuracy of 1.5 mm at 11.4 m downstream from the target.

Charged projectile spectator fragments are detected in a hodoscope (HODO), and their charge centroid calculated for each event. HODO consists of two orthogonal planes of 38 plastic scintillator slats with 1 cm^2 cross-sections, centered on the beam line, and situated 11.4 m downstream from the target. The response of individual scintillators to deposited charge was calibrated on a run-by-run basis and this information was used to find the charge-weighted centroid $\vec{Q} = Q_x\hat{\imath} + Q_y\hat{\jmath}$ for each event, where

$$Q_x = \frac{\sum Q \cdot x}{\sum Q} \tag{19.6}$$

The direction of the impact parameter $\hat{b}$ is then $\hat{Q}$, defined with the origin at the projected beam position on HODO, and the reaction plane is defined as the plane spanned by $\hat{Q}$ and $\hat{z}$. $\hat{x}$ is then redefined to lie along $\hat{Q}$. Implicit in this definition of the impact parameter is the assumption that the direction of proton flow – the deflection of the spectator fragment – is along the reaction plane.

An estimate of the reaction plane resolution is determined by randomly dividing each event into two sub-events and looking at the $(\phi_1 - \phi_2)$ difference distribution for the two reaction planes calculated from each sub-event (see Fig. 19.1). The actual resolution for the reaction plane determined using the full event statistics is roughly half this value [8], and in this manner we obtain an estimate of $\delta\phi \approx 32°$.

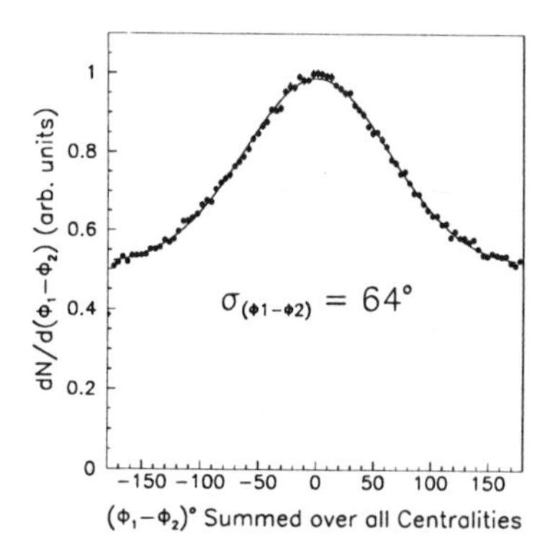

Figure 19.1 Distribution of the reaction plane angle difference for angles ϕ_1 and ϕ_2 determined by splitting each event into two sub-events. The angle resolution for the full event is roughly half this value ($\delta\phi \approx 32°$).

1.4 NORMALIZATION

The correlation function is, by definition, a normalized quantity. To find the proper normalization N(**q**), a background is generated which creates two-particle "events" out of single tracks from different events. $C_2(\mathbf{q})$ is then simply the ratio between real data and the event-mixed background.

1.5 CORRECTIONS TO THE CORRELATION FUNCTION

The shape of the measured two-particle correlation function mainly depends on three effects: the Coulomb repulsion between the two particles, the two-particle resolution of the detector, and the HBT correlation itself.

For the Coulomb effect, a correction f_C is numerically calculated for a simple extended source [9, 10] of size R_0. This correction is applied iteratively to the background until the value of R_{inv} obtained from the one-dimensional correlation function converges to that of R_0.

The two-particle resolution arises from the finite resolution of the tracking detectors in the spectrometer. Two close tracks are, at some point, indistinguishable from single tracks, and do not, therefore, appear in the measured two-particle correlation. A two-dimensional cut in relative coordinate space, f_{TPR}, is applied to both signal and background events. The separation between two tracks, projected onto the first plane in the spectrometer, is shown in figure 19.2, together with f_{TPR}.

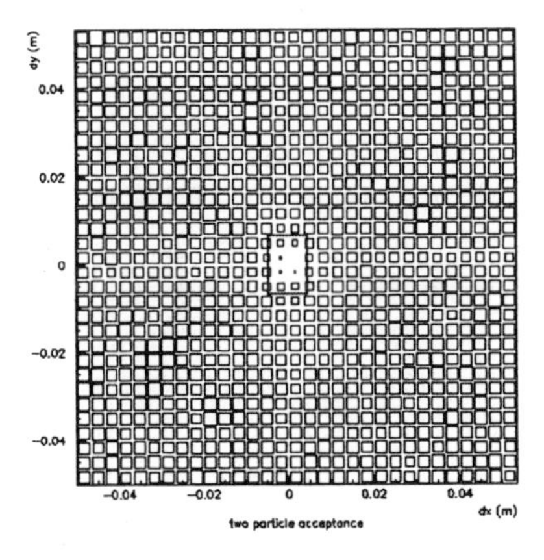

Figure 19.2 Separation between two tracks on T1. The f_{TPR} cut is the solid line.

The final correlation function is fitted from a spectrum in **q**, which is generated by dividing signal events by background events. Both signal and background have been corrected by $f_{\rm TPR}$; the background has additionally been corrected by f_C.

2. RESULTS

The reader is reminded that the following results and analysis are preliminary, and only contain a limited subset of the E917 data. Thus far, only about 10% of the data has passed through the HBT analysis. By December 1999, we expect at least another 50% will have been analyzed.

The measured correlation for identical pairs of pions is shown in Fig. 19.3, plotted as a function of Q_x and Q_y, where x and y are defined relative to the reaction plane of the event as discussed in Section 1.3. To improve the statistics, all of Q_z has been integrated over. In addition, since the radii for $\pi^+\pi^+$ and $\pi^-\pi^-$ pairs are similar [10], both datasets were combined in the present analysis.

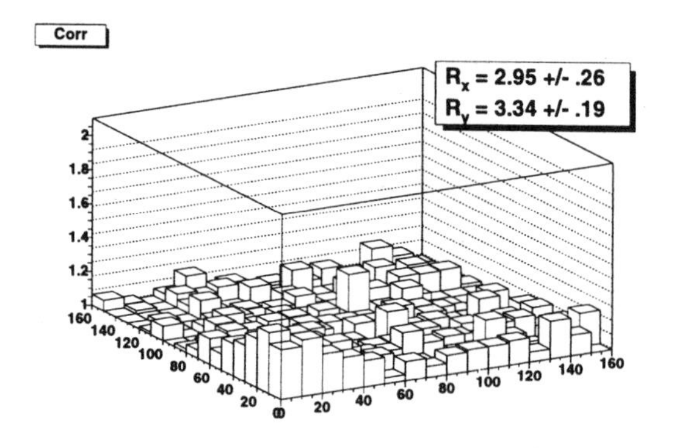

Figure 19.3 $|Q_y|$ vs. $|Q_x|$

Central slices are shown in Figures 19.4 and 19.5 along with the projections from the two-dimensional fit. These data have been fit with the function in Eq. 19.5, and values for R_x and R_y obtained. We find that $R_x = 2.95 \pm 0.26$ fm and $R_y = 3.34 \pm 0.19$ fm, the latter value being larger than the former. The error bars are solely from the fitting procedure and do not include systematic errors.

To establish a baseline for comparison, the data were also analyzed in the same fashion, but with $\hat{x}$ chosen along a random direction rather than along the reaction plane (Figs. 19.6, 19.7, 19.8). In this analysis, the values of R_x and R_y obtained are consistent with each other, indicating that the observed difference relative to the reaction plane is a real effect.

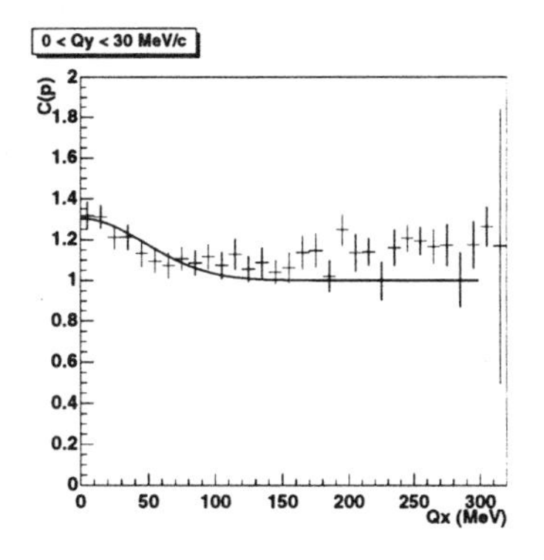

Figure 19.4 $|Q_x|$, $|Q_y| < 30$ MeV

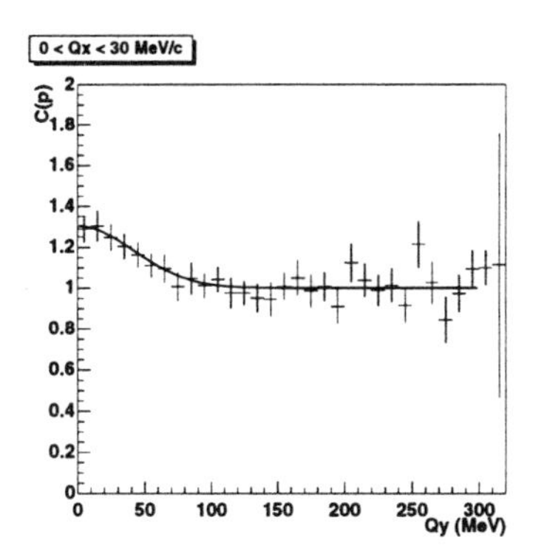

Figure 19.5 $|Q_y|$, $|Q_x| < 30$ MeV

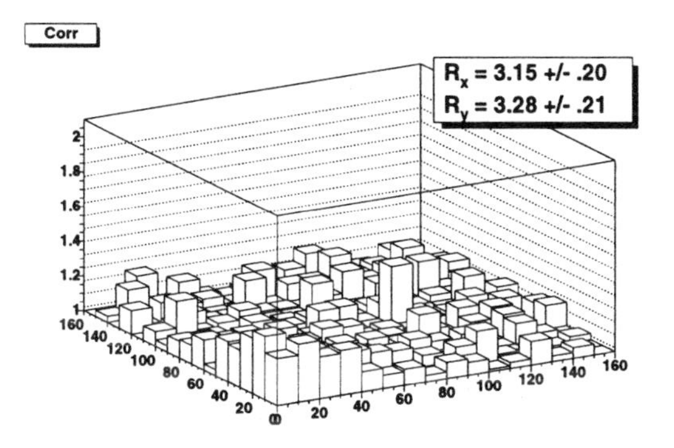

Figure 19.6 $|Q_y|$ vs. $|Q_x|$ – systematic check

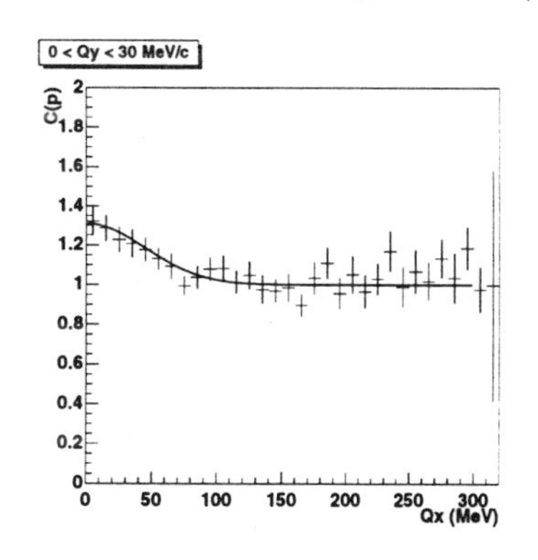

Figure 19.7 $|Q_x|$, $|Q_y| < 30$ MeV

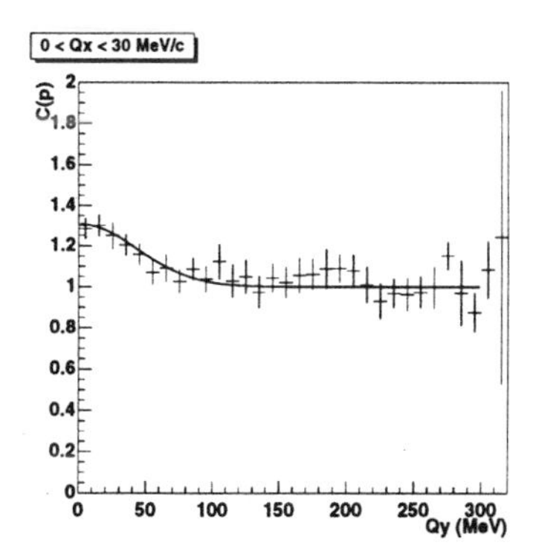

Figure 19.8 $|Q_y|$, $|Q_x| < 30$ MeV

3. CONCLUSIONS

In a naive geometrical model of the source, we would expect R_y to be larger than R_x. This is consistent with the simple overlap of two non-centrally colliding spheres for which the overlap region resembles an almond in shape, with $R_y > R_x$.

Other data and analyses [11, 12], however, indicate that the emission region is not a simple static source at these energies, with dN/dy distributions and flow studies indicating both longitudinal and transverse expansion. The shape of the source may also be obscured by particle absorption by spectator matter.

Future analyses with better statistics will allow the investigation of the source asymmetry with respect to the reaction plane and its dependence on collision centrality, pair m_T, and rapidity. These, taken together with dN/dy distributions and flow studies, may be able to further disentangle expansion and absorption effects from the source shape and size, and thus assemble a more complete picture of the collision dynamics. This work is in progress.

References

[1] Appelshäuer, H. *et al.* (1998) Nucl. Phys. **A638**, 91c.

[2] Bearden, I. G. *et al.* (1998) Phys. Rev. **C58**, 1656.

[3] Cianciolo, V. *et al.* (1995) Nucl. Phys. **A590**, 459c.

[4] Barrette, J. *et al.* (1997) Phys. Rev. Let. **78**, 2916.

[5] Boal, D. H., Gelbke, C., and B. K. Jennings (1990) Rev. Mod. Phys. **62**, 553.

[6] Wu, Y.-F., Heinz, U., Tomásik, B., and U. A. Wiedemann (1998) Eur. Phys. J. **C1**, 599.

[7] Back, B. B. *et al.* (1998) Nuclear Instruments and Methods **A412**, 191.

[8] Ahle, L. *et al.* (1998) Phys. Rev. **C57**, 1416

[9] Pratt, S. (1986) Phys. Rev. **D33**, 72.

[10] Baker, M. *et al.* (1996) Nucl. Phys. **A610**, 213c.

[11] Back, B. B. *et al.* (1999) Contribution to these proceedings.

[12] Ollitraut, J.-Y. (1998) Nucl. Phys. **A638**, 195c.

Chapter 20

LIGHT NUCLEI MEASUREMENTS FROM E864

N.K. George for the E864 Collaboration*

Physics Department

Yale University

New Haven, USA†

manhat@kiwi.physics.yale.edu

*Univ Bari-BNL-UCLA-UC Riverside- Iowa State- Univ. Mass-MIT-Penn State-Purdue-Vanderbilt-Wayne State-Yale

†Work supported in part by the DOE under contract number DE-FG02-91ER-40609

Abstract The yields of produced light nuclei within one unit of mid-rapidity for three different event centralities are presented. Sensitivity of these yields to source dynamics is explored, comparisons to theoretical models are made, and source radii are extracted.

Keywords: BNL, E864, light nuclei

1. INTRODUCTION

BNL experiment E864 was primarly designed to search for novel states of matter, such as strangelets, possibly produced in heavy ion collisions. This design naturally allows the measurement of known, but rarely produced baryons from such collisions. This proceedings focuses on the bound light nuclei measurements of E864 (up to A=6) [1],[2],[3] and their utility in exploring collision dynamics as well as improving the understanding of rare particle production in such collisions.

The proceedings are organized as follows. First, the apparatus and how the invariant yields are obtained is described. Then, general trends seen in the data are presented. This is followed by a discussion of their interpretation, radii extraction and comparisons to microscopic models.

2. EXPERIMENT

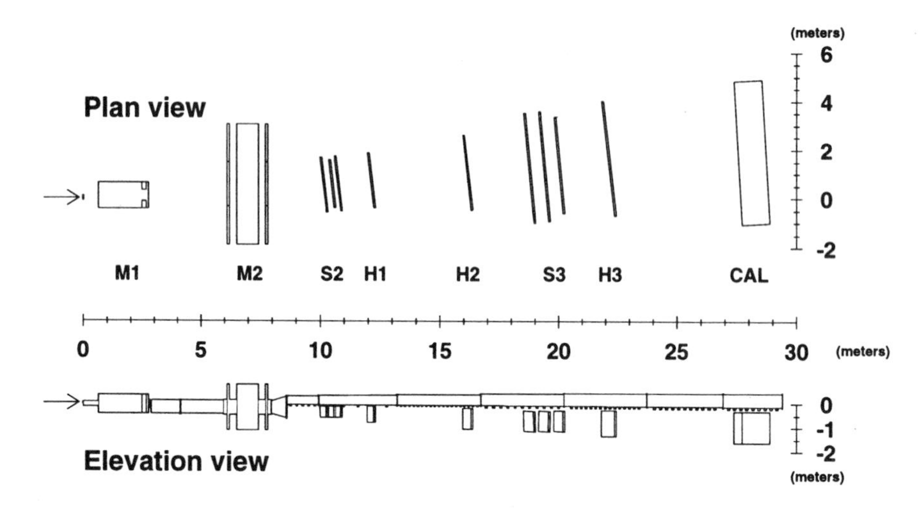

Figure 20.1 Plan and elevation view of the E864 spectrometer.

Plan and elevation views of the detectors comprising the experiment are shown in Figure 20.1. The 11.5A Gev/c Au beam enters from the right and is incident upon a Pb target. The charged reaction products, bent by two dipole analyzing magnets (M1 and M2), traverse a variety of tracking detectors. These consist of two straw tube stations (S2 and S3) which each measure a high resolution space point, and three hodoscope planes (H1, H2 and H3) which each measure a space-time point and charge. Reconstructed tracks are assigned a rigidity based upon their deflection in the field. The hodoscope's information allows a velocity and charge to be assigned to the track. From this, the track's mass and charge are determined and allow its unique particle identification. Figure 20.2 shows a typical mass plot for charge one tracks. Clear identication of proton, deuterons and tritons is possible.

It is also possible for E864 to measure neutral particles via a 13x58 tower hadronic calorimeter (labelled CAL in Figure 20.1) which is situated behind the tracking detectors. Each tower measures the energy deposited as well as the time of flight to that tower. Using the upstream detectors to remove charged depositions, neutral showers are identified and assigned a mass based on the energy and time information, from which unique particle identification is made.

The invariant yield of the light nuclei is calculated by counting the number of a particular species measured in a particular kinematic bin and correcting for acceptance, detector and identication cut inefficiencies.

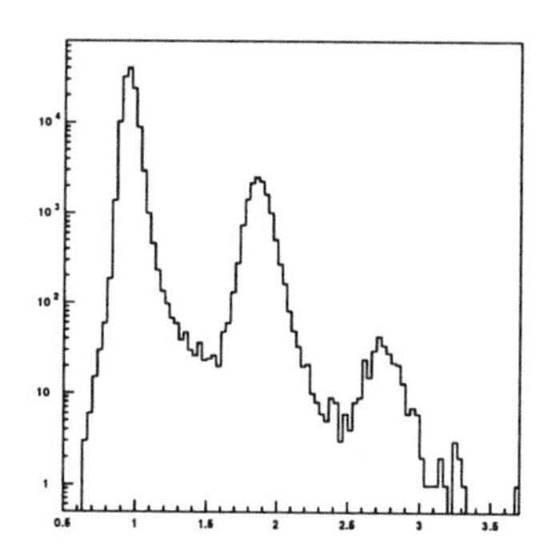

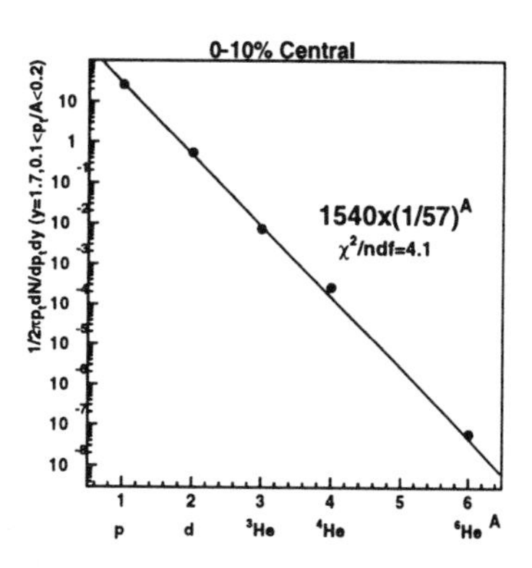

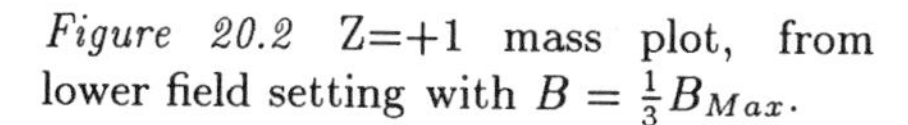

Figure 20.2 Z=+1 mass plot, from lower field setting with $B = \frac{1}{3}B_{Max}$.

Figure 20.3 Invariant yield as a function of mass number A. Result shown for the 10 percent most central collisions, at midrapidity,over the p_t range $0.1 \leq \frac{p_t}{A} \leq 0.2\text{GeV}/c$

An event centrality trigger was used to preselect events according to their centrality. The centrality of an event was determined via a charged particle multiplicity measurement. The multiplicitly detector is an annulus of scintillator surrounding the beam downstream of the target. It detects charged particles within polar angles $16.6 - 45$ degrees. The centralities presented are expressed as a percentage of the total geometric cross-section of Au+Pb collision (6240mb). The centrality data presented correspond to $0 - 10\%$,$10 - 38\%$ and $38 - 66\%$ of the total geometric cross-section of Au+Pb.

To measure very rarely produced objects, a Level II trigger was used. It is known as a late energy trigger (L.E.T.). It uses the fact that each calorimeter tower with energy and time measurements can make a rough mass measurement. Only events that produce high mass objects within the acceptance fire the trigger. This enables the measurement of the A=6 nuclei. Typical efficiencies of this trigger are $80 - 90\%$ for the A=6 nuclei, with a non-interesting event rejection factor around 60.

3. RESULTS

The yields of the light nuclei fall approximately exponentially as a function of the mass number, A, over the range $1 \leq A \leq 6$. This is shown in Figure 20.3. The data show that the yield falls by about a factor of 60 every time an additional nucleon is added to a cluster. A similar trend is observed over different centralities, although measurements are only made up through A=3.

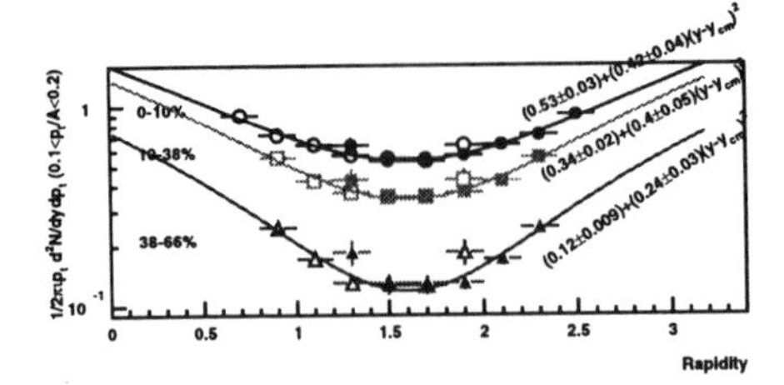

Figure 20.4 Deuteron invariant yield as a function of rapidity for three centralities. Each point represents yield averaged over the p_t range, $0.1 \leq \frac{\mathrm{p_t}}{\mathrm{A}} \leq 0.2\mathrm{GeV}/c$. Open symbols represent a reflection of the data points around $\mathrm{y_{cm}}$. The lines represent the resulting fits to a function of the form $\mathrm{A} + \mathrm{B}(\mathrm{y} - \mathrm{y_{cm}})^2$. The resulting fit parameters are also shown.

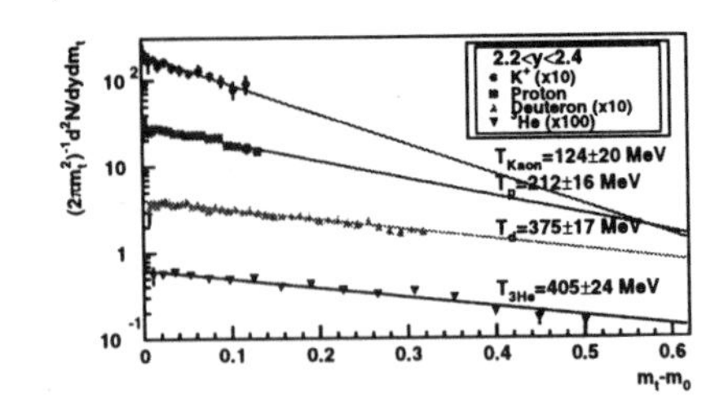

Figure 20.5 Invariant yield, divided by m_t, as a function of the transverse momentum less the rest mass of species K^+,proton,deuteron and ^{3}He. Measured over the rapidity interval $2.2 \leq y \leq 2.4$. The lines are the results of fits of the yields to a Boltzmann function of the form,$Am_T exp\left(\frac{-m_T}{T}\right)$. The resulting inverse slope parameters, T, are shown.

The E864 acceptance is such that the rapidity dependence of the yields can only be examined over a limited p_t range in each rapidity slice. To allow comparisons between different species, a p_t range that corresponds to $0.1 \leq \frac{\mathrm{p_t}}{\mathrm{A}} \leq 0.2\mathrm{GeV}/c$ per nucleon is chosen. Figure 20.4 shows the rapidity dependence of the deuterons for different centralities. A concave shape is observed. In fact, it is generally found that the relative concavity increases as the even t becomes less central and the mass of the cluster increases. The rapidity dependent yields are fit to a functional form $\mathrm{A} + \mathrm{B}(\mathrm{y} - \mathrm{y_{cm}})^2$. The ratio $\frac{B}{A}$ then gives the relative concavity.

The transverse momentum dependence of the yields can only be examined at the highest rapidity interval in E864. Figure 20.5 shows the invariant yield versus the transverse mass for K^+ to ^{3}He for the 10% most central Au+Pb collisions. The Boltzmann inverse slope parameters increase as the mass of the species increases. This behavior is consistent with the predicted effects of flow on the light nuclei spectra [4].

4. DISCUSSION

A common quantity associated with light nuclei production is the $\mathrm{B_A}$ parameter. It is defined in Equation 20.1

$$\mathrm{B_A} = \frac{\mathrm{E_A}\frac{d^3N_A}{dP_A^3}}{\left(\mathrm{E_n}\frac{d^3N_n}{dp_n^3}\right)^N\left(\mathrm{E_p}\frac{d^3N_p}{dp_p^3}\right)^Z} \tag{20.1}$$

where $(\mathrm{E_A}, \vec{\mathrm{P}}_\mathrm{A})$ is the four momentum of a nuclei of mass number A and $(\mathrm{E_{n/p}}, \vec{\mathrm{P}}_\mathrm{n/p})$ is the four momentum of a neutron/proton, where the yields are evaluated at the same velocities.

Using either a coalescence or thermal prescription, assuming uncorrelated position and momenta and instanteous emmission nucleon sources, various authors [5], [6], [7] have derived expressions for $\mathrm{B_A}$. These various predicitions have the common feature that the $\mathrm{B_A}$ parameter is independent of momentum, but dependent on the physical properties of the cluster and the volume over which it is formed.

The applicablility of such models is tested in Figure 20.6, where the B_2 parameter is plotted as a function of the transverse momentum and as a function of rapidity over the transverse momentum range $0.1 \leq \frac{p_t}{A} \leq 0.2\mathrm{GeV}/c$. To evalute this parameter the proton contribution from hyperonic decays to the E864 proton yield was removed. It was also assumed that the neutron yield equals the proton yield. Neutron yields measured by E864 [8] show this to be a reasonable approximation to within 20%. Although there appears to be a general trend for the B_2 parameter to increase as a function of transverse momentum, it is difficult to draw a definite conclusion given the error bars. However, there appears to be a definite rapidity dependence in the B_2 parameter. Similar results were found for the B_3 parameter using ^{3}He. Hence, the simple scaling law defined in Equation 20.1 is not observed to hold, at least as a function of rapidity.

Flow is known to exist in such collisions. Flow results in a correlation between position and momentum in the nucleon source. Hence one can understand the momentum dependence of the B_A parameter as being a manifestion of flow.

Additional evidence of the sensitivity of the light nuclei yields to flow is found from the inverse slope mass dependence. Mattellio et al [4] argue that the shape of the nuclei spectra will change depending on its mass due to the presence of flow. This is observed in the E864 data as shown in Figure 20.5. Polleri at al [9] also argue that the shape of the inverse slope's mass dependence is sensitive to both flow and freeze-out density profiles. They find if a uniform nucleon freeze-out density is assumed, a linear mass depencence is expected. If, however, the transverse freeze-out density is gaussian a weeker dependence is expected. Figure 20.7 shows the inverse slope mass dependence observed by experiment E864. For the top two centrality bins it is seen that the inverse slope parameter tends to flatten out as a function of mass. Thus it appears to not be linear, consistent with the idea that the spectra are sensitive to the effects of flow as well as the nucleon freeze-out density profile.

In an attempt to exploit this sensitivity, RQMDV2.3[10] with a coalescence afterburner [4] was used to determine the expected light nuclei yields. RQMD was run under two different dynamic conditions. One condition included the mean field potentials ("potentials mode") whereas the other did not("cascade mode"). Figure 20.8 shows the comparison of the calculations with the data. A disagreement is found both in the cascade and potential modes. One notes that the disagreement increases for larger masses. The cascade mode underpredicted the alpha yields by about a factor of 30. This large disagreement has made it difficult to draw any conclusion about the source dynamics at this time.

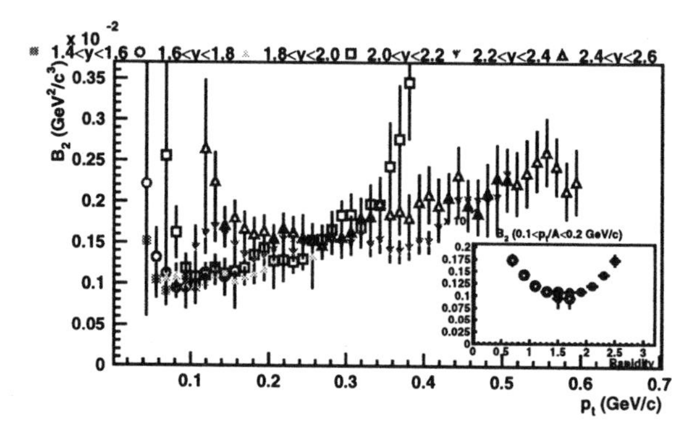

Figure 20.6 $B_2(p_t, y)$. $B_2(= \frac{d}{p^2})$ plotted as a function of the transverse momentum of the proton. Each rapidity slice is denoted by a different symbol. The insert shows the rapidity dependence of B_2 averaged over $0.1 \leq \frac{p_t}{A} \leq 0.2\text{GeV}/c$

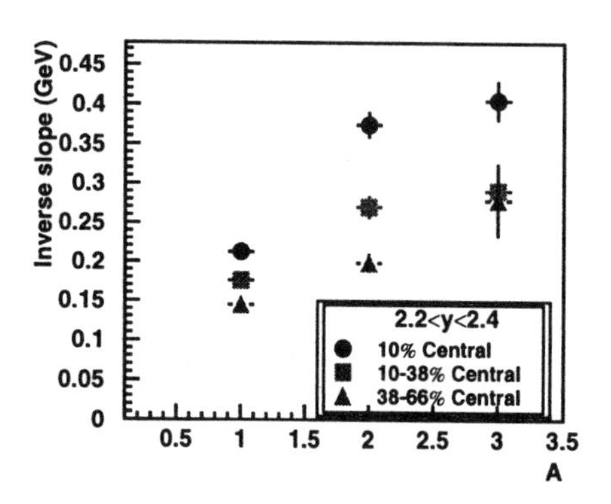

Figure 20.7 Boltzmann inverse slope parameter as a function of composites mass for three different centralities.

We use the fragment coalescence model [11] to extract quantitative source information. This model is better suited to flowing sources, because it evaluates the radius parameter in the rest frame of the compos-

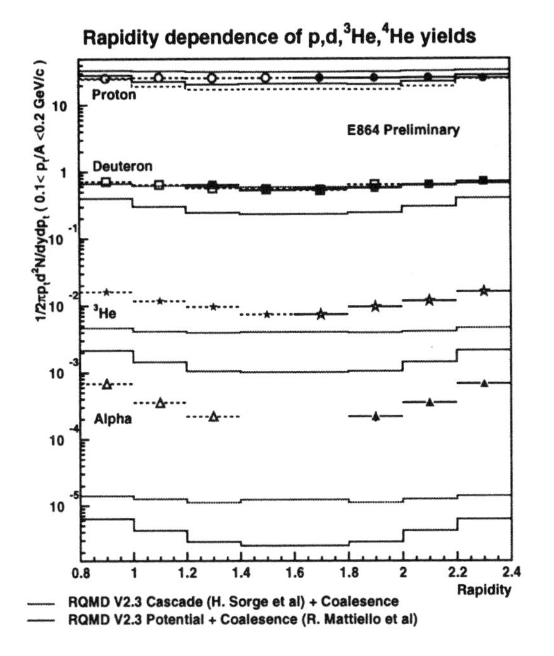

Figure 20.8 Comparison of data, with RQMDV2.3[10]+coalescence after-burner [4]. Data represented by solid symbols. Open symbols represent reflection around y_{cm}. For each species the upper solid lines represent predicted yields using RQMD in cascade mode. The lower solid lines represent predicted yields using RQMD in potentials mode.

ite. This model assumes nuclei are formed from the coalesence of two smaller nuclei, e.g. $d+p \rightarrow {}^3$He. It assumes the constituent clusters have a source distribution, $S(x,p) = \delta(t)\exp(-\frac{r^2}{2R^2})$. The final expression for the radius, in terms of the yields is given by Equation 20.2.

$$R^3 = \pi^{\frac{3}{2}} \frac{(2S_c+1)}{(2S_a+1)(2S_b+1)} \frac{m_c}{m_a m_b} \frac{E_a \frac{d^3N_a}{dp_a^3} E_b \frac{d^3N_b}{dp_b^3}}{E_c \frac{d^3N_c}{dp_c^3}} \tag{20.2}$$

where S and m are the spin and mass of the cluster and the invariant yields are evaluated at the same velocity. Figure 20.9 shows the resulting transverse mass dependence of the calculated radii. A radius between 4-3fm is extracted. The data is not necessarily consistent with the m_t scaling observed in two particle correlation studies.

5. CONCLUSION

The yields of light nuclei up to A=6 near mid-rapidity have been measured by E864. These yields appear to fall exponentially with A. The yields at low p_t show a concave rapidity dependence. The inverse

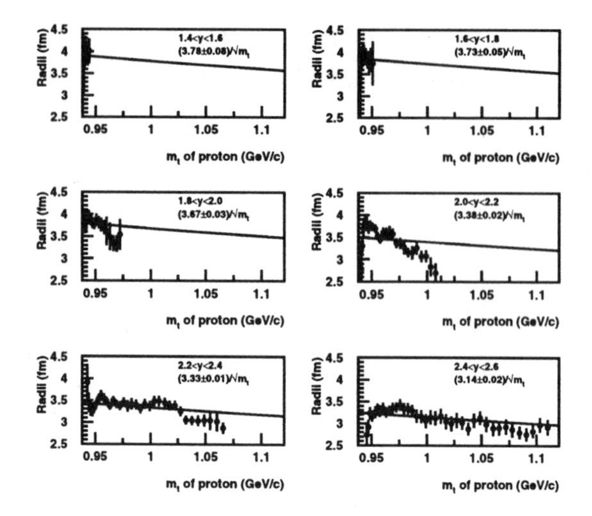

Figure 20.9 Radii extracted from fragment coalescence model[11] as a function of m_t of proton, for different rapidity slices. Radii extracted from deuteron to proton ratio. The lines are results of fits to $A/\sqrt{m_t}$

slope parameters show a mass dependence. Qualitatively, the light nuclei yields appear to be sensitive to the effects of flow. Quantitatively, increasing disagreement between theoretical predictions and data are seen as the mass increases. Therefore, the potential of the data has not yet been realized.

References

[1] J.K.Pope, PhD thesis, Yale University, 1998

[2] N.K. George, PhD thesis, Yale University, 1999

[3] Z. Xu, PhD thesis, Yale University, 1999

[4] R.Mattiello *et al.* (1997) Phys. Rev. C **55** 1443.

[5] A.Schwarzchild and C. Zupancic (1963) Phys. Rev. **129** 1963.

[6] R.Bond *et al.* (1977) Phys. Lett. B **71** 43.

[7] H.Sato *et al.* (1981) Phys. Lett. B **98** 153.

[8] L.E. Finch, PhD thesis, Yale University, 1999

[9] A.Polleri *et al.* (1997) nucl-th/9711011

[10] H.Sorge *et al.* (1989) Ann. Phy. (N.Y.) **192** 266.

[11] W.J. Llope *et al.* (1995) Phys. Rev. C **52** 2004.

Chapter 21

COLLECTIVE FLOW IN PB+PB COLLISIONS AT THE CERN-SPS

Shunji Nishimura, for the WA98 collaboration

M.M. Aggarwal,[1] A. Agnihotri,[2] Z. Ahammed,[3] A.L.S. Angelis,[4] V. Antonenko,[5] V. Arefiev,[6] V. Astakhov,[6] V. Avdeitchikov,[6] T.C. Awes,[7] P.V.K.S. Baba,[8] S.K. Badyal,[8] A. Baldine,[6] L. Barabach,[6] C. Barlag,[9] S. Bathe,[9] B. Batiounia,[6] T. Bernier,[10] K.B. Bhalla,[2] V.S. Bhatia,[1] C. Blume,[9] R. Bock,[11] E.-M. Bohne,[9] Z. Böröcz,[9] D. Bucher,[9] A. Buijs,[12] H. Büsching,[9] L. Carlen,[13] V. Chalyshev,[6] S. Chattopadhyay,[3] R. Cherbatchev,[5] T. Chujo,[14] A. Claussen,[9] A.C. Das,[3] M.P. Decowski,[18] V. Djordjadze,[6] P. Donni,[4] I. Doubovik,[5] S. Dutt,[8] M.R. Dutta Majumdar,[3] K.El Chenawi,[13] S. Eliseev,[15] K. Enosawa,[14] P. Foka,[4] S. Fokin,[5] V. Frolov,[6] M.S. Ganti,[3] S. Garpman,[13] O. Gavrishchuk,[6] F.J.M. Geurts,[12] T.K. Ghosh,[16] R. Glasow,[9] S. K.Gupta,[2] B. Guskov,[6] H. Å.Gustafsson,[13] H. H.Gutbrod,[10] R. Higuchi,[14] I. Hrivnacova,[15] M. Ippolitov,[5] H. Kalechofsky,[4] R. Kamermans,[12] K.-H. Kampert,[9] K. Karadjev,[5] K. Karpio,[17] S. Kato,[14] S. Kees,[9] H. Kim,[7] B. W. Kolb,[11] I. Kosarev,[6] I. Koutcheryaev,[5] T. Krümpel,[9] A. Kugler,[15] P. Kulinich,[18] M. Kurata,[14] K. Kurita,[14] N. Kuzmin,[6] I. Langbein,[11] A. Lebedev,[5] Y.Y. Lee,[11] H. Löhner,[16] L. Luquin,[10] D.P. Mahapatra,[19] V. Manko,[5] M. Martin,[4] A. Maximov,[6] R. Mehdiyev,[6] G. Mgebrichvili,[5] Y. Miake,[14] D. Mikhalev,[6] Md.F. Mir,[8] G.C. Mishra,[19] Y. Miyamoto,[14] D. Morrison,[20] D. S. Mukhopadhyay,[3] V. Myalkovski,[6] H. Naef,[4] B. K. Nandi,[19] S. K. Nayak,[10] T. K. Nayak,[3] S. Neumaier,[11] A. Nianine,[5] V. Nikitine,[6] S. Nikolaev,[6] P. Nilsson,[13] S. Nishimura,[14,b] P. Nomokonov,[6] J. Nystrand,[13] F.E. Obenshain,[20] A. Oskarsson,[13] I. Otterlund,[13] M. Pachr,[15] A. Parfenov,[6] S. Pavliouk,[6] T. Peitzmann,[9] V. Petracek,[15] F. Plasil,[7] W. Pinanaud,[10] M.L. Purschke,[11] B. Raeven,[12] J. Rak,[15] R. Raniwala,[2] S. Raniwala,[2] V.S. Ramamurthy,[19] N.K. Rao,[8] F. Retiere,[10] K. Reygers,[9] G. Roland,[18] L. Rosselet,[4] I. Roufanov,[6] C. Roy,[10] J.M. Rubio,[4] H. Sako,[14] S.S. Sambyal,[8] R. Santo,[9] S. Sato,[14] H. Schlagheck,[9] H.-R. Schmidt,[11] G. Shabratova,[6] T.H. Shah,[8] I. Sibiriak,[5] T. Siemiarczuk,[17] D. Silvermyr,[13] B.C. Sinha,[3] N. Slavine,[6] K. Söderström,[13] N. Solomey,[4] S.P. Sørensen,[20] P. Stankus,[7] G. Stefanek,[17] P. Steinberg,[18] E. Stenlund,[13] D. Stüken,[9] M. Sumbera,[15] T. Svensson,[13] M.D. Trivedi,[3] A. Tsvetkov,[5] L. Tykarski,[17] J. Urbahn,[11] E.C.v.d. Pijll,[12] N.v. Eijndhoven,[12] G.J.v. Nieuwenhuizen,[18] A. Vinogradov,[5] Y.P. Viyogi,[3] A. Vodopianov,[6] S. Vörös,[4] B. Wysłouch,[18] K. Yagi,[14] Y. Yokota,[14] G.R. Young[7]

[1] *Univ. of Panjab (India)* - [2] *Univ. of Rajasthan (India)* - [3] *VECC (India)* - [4] *Univ. of Geneva (Switzerland)* - [5] *Kurchatov (Russia)* - [6] *JINR (Russia)* - [7] *ORNL (USA)* - [8] *Univ. of Jammu (India)* - [9] *Univ. of Münster (Germany)* - [10] *SUBATECH (France)* - [11] *GSI (Germany)* - [12] *NIKHEF (Netherlands)* - [13] *Univ. of Lund (Sweden)* - [14] *Univ. of Tsukuba (Japan)* - [15] *NPI (Czech Rep.)* -

Advances in Nuclear Dynamics, 5,
Edited by Bauer and Westfall, Kluwer Academic / Plenum Publishers, New York, 1999.

[16] *KVI (Netherlands)* - [17] *INS (Poland)* - [18] *MIT (USA)* - [19] *Bhubaneswar (India)* -
[20] *Univ. of Tennessee (USA)* - [b] *CNS, Univ. of Tokyo (Japan)*

Abstract The preliminary results of anisotropic transverse flow will be reported in 158 A GeV Pb + Pb collisions.

The centrality dependence of the directed flow has been measured at the target rapidity region. The directed flow of the pions is opposite to that of the protons, where the magnitude of the directed flow of protons seem to be significantly smaller than observed at AGS energies and than RQMD. While, maximum directed flow is observed in more peripheral events. Near mid-rapidity region, the elliptic flow of $\pi^{\pm}$ mesons is studied. The shape of the two-pion correlation function is investigated as a function of the two-particle emission angle relative to the target proton flow. Our preliminary results show an indication of a dependence of the two-pion correlation function on the direction of emission relative to the target flow direction for semi-central collisions.

Keywords: Reaction plane, hadrons, directed flow, elliptic flow, HBT

1. INTRODUCTION

In the case of a QCD phase transition from ordinary nuclear matter to quark gluon plasma, it is expected that the equation of state (EOS) should exhibit a softening due to the increase number of degrees of freedom [1, 2, 3, 4, 5, 6]. According to the recent theoretical discussion, it is expected that the information about the EOS can be extracted from the study of the collective flow. Especially, studies of flow effects in terms of the azimuthal anisotropy of particle production have been discussed as an unique tool for extracting the information about pressure created at the early stage of the collisions.

At AGS and SPS energies, the observation of directed and elliptic transverse flow has recently been reported [7, 8, 9, 10, 11, 12], demonstrating its presence even at higher beam energies than the BEVALAC/SIS energy region [13, 14, 15].

In this article, we will report the recent results of the identified hadrons spectra measured by the CERN-WA98 experiment in 158AGeV Pb+Pb collisions, focusing on three topics. First of all, the systematic study of the directed flow for identified protons and π^+s will be reported at target rapidity region. Secondly, the dependence of elliptic transverse emission on the particle species near mid-rapidity region will be shown. Thirdly, the study on the shape of the two-pion correlation functions at mid-rapidity will be reported as a function of the azimuthal flow angle in the Plastic Ball. This study allows one to extract the information on the evolution of the hot and dense hadronic matter in the relativistic heavy-ion collisions [18] and possibly contains additional information of flow, which can not be accessed only by the single particle spectra.

2. ANALYSIS AND RESULTS

The data presented here were taken with a subset of the full WA98 experiment detector system using the 158 A·GeV ^{208}Pb beams of the CERN-SPS on a ^{208}Pb target of 213 μm thickness. In the WA98 setup the incident Pb beam is defined by a gas Cherenkov start counter with a timing resolution of 30 ps and a veto counter with a 3 mm diameter hole [19]. The centrality of the collision is determined by the total transverse energy, E_T, measured with the mid-rapidity calorimeter (MIRAC) which covers the pseudo-rapidity range of $3.5 < \eta < 5.5$.

One of the key detectors for the present analysis is the Plastic Ball which covers the pseudo-rapidity range of $-1.7 < \eta < 0.5$ with full azimuthal coverage. It identifies pions, protons, deuterons, and tritons with kinetic energies of 50 to 250 MeV by $\Delta E - E$ measurement [20].

The measurement of identified particles near mid-rapidity is performed using two tracking spectrometer arms with a large (1.6 m) aperture dipole magnet (GOLIATH). The particle identification is based on a measurement of momentum and time-of-flight. Detailed information of the experiment can be found elsewhere [12, 16].

2.1 REACTION-PLANE DETERMINATION

The reaction plane is determined from the azimuthal direction of the total transverse momentum vector of fragments (p, d, and t) detected by the Plastic Ball detector. The azimuthal angle of the reaction plane, Φ_0 is thus determined as

$$\Phi_0 = \tan^{-1}\left(\frac{\sum_{i=1}^{N} p_{Ti}\sin(\phi_i)}{\sum_{i=1}^{N} p_{Ti}\cos(\phi_i)}\right) \tag{21.1}$$

where the sum runs over all fragments. Here ϕ_i and p_{Ti} are the azimuthal angle in the laboratory coordinate and the transverse momentum of the i-th fragment, respectively. The multiplicity of protons in the Plastic Ball detector is around 8 in semi-central collisions. A minimum of three protons are required for this analysis. The observed Φ_0 distribution has a variation of less than 3% due to the detector biases such as dead channels and inefficiency. In the following flow analysis, we have corrected for this effect by weighting with the inverse of the yield.

In order to study how well the fragment flow direction is defined, we have performed the sub-event analysis [12]. As expected, the correlation observed for semi-central events is significantly larger than for very central events.

2.2 DIRECTED FLOW

Azimuthal anisotropies of the particle emission are evaluated by means of a Fourier expansion [21, 22]. The Fourier coefficients $v_n (n = 1, 2)$ are extracted from the azimuthal distribution of identified particles with respect to the reaction plane;

$$\frac{1}{N}\frac{dN}{d(\phi-\Phi_0)} = 1 + 2v_1'\cos(\phi-\Phi_0) + 2v_2'\cos(2(\phi-\Phi_0)), \qquad (21.2)$$

where ϕ is the measured azimuthal angle with respect to Φ_0. The Fourier coefficient v_1' quantifies the directed flow, whereas v_2' quantifies the elliptic flow. Using the accurate procedure and interpolation formula of Ref. [22] one obtains $\langle\cos(\Phi_0 - \Phi_r)\rangle = 0.377 \pm 0.018$ for the semi-central $(100 < E_T < 200$ GeV) event selection.

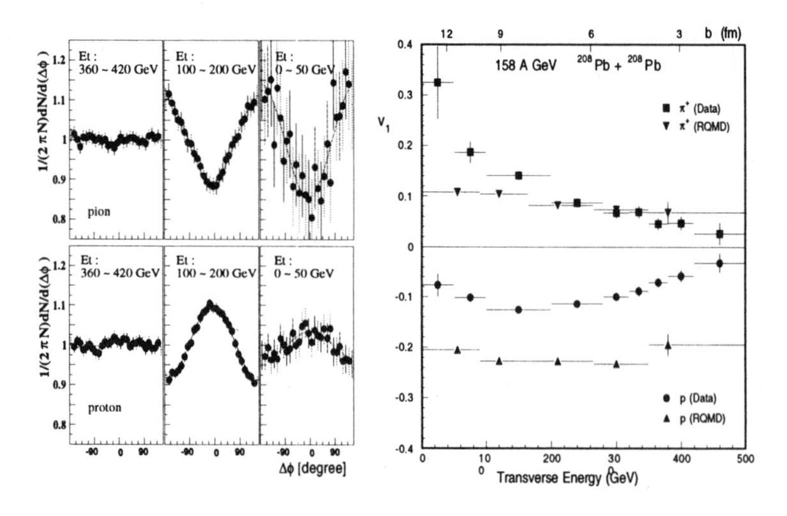

Figure 21.1 left: Azimuthal angle distributions with respect to the Φ_0 for identified pions and protons in various centrality regions. right: The centrality dependence of the directed flow coefficient v_1 for protons (circles) and π^+ (squares). Triangles are results from RQMD model calculations. The data have been corrected for the event-plane resolution. The vertical bars indicate the uncertainty of the fit and resolution correction. The horizontal bars indicate the E_T bin intervals (or impact parameter intervals for RQMD).

The dependence of the v_1 fit parameter on centrality, as determined by the measured transverse energy (E_T), is shown in Fig. 21.1. For convenience an impact parameter scale is also shown. The E_T scale has been converted to an impact parameter scale assuming a monotonic relationship between the two quantities, and equating $d\sigma/dE_T$ with $d\sigma/db$. As seen in Figure 21.1, the strength of the directed flow of protons increases with centrality and reaches a maximum value for semi-central collisions with $b \approx 8$ fm. It is interesting to note that the strongest flow effect appears at larger impact parameters than observed at lower incident energy for similar systems (where $b \approx 4$ fm) [8, 17]. For comparison, RQMD 2.3 [6] model predictions are shown subjected to the same analysis after applying the Plastic Ball detector acceptance, but using the true reaction plane. RQMD predicts a significantly stronger correlation for protons than observed.

Also shown in Fig. 21.1 is the strength of the directed flow of π^+, identified in the Plastic Ball. A clear anti-correlation, or anti-flow [23], is observed between the fragment and π^+ flow directions. This behavior has been observed at incident energies from 1 A GeV to SPS energies and has been explained as resulting from preferential absorption of the pions emitted in the target spectator direction [8, 23, 24, 25]. The absorption results in an oppositely directed apparent π^+ flow. The strength of the anti-correlation increases for the most peripheral events, indicating the increasing role of absorption.

2.3 ELLIPTIC FLOW

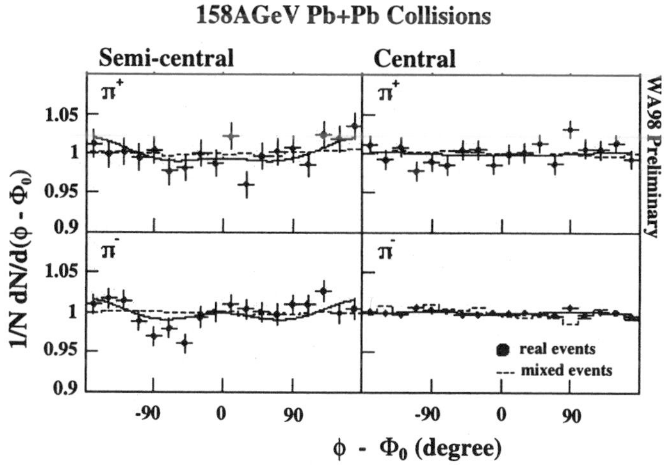

Figure 21.2 The azimuthal distributions of π^+ and π^- mesons with respect to the reaction plane for semi-central and central 158 A·GeV ^{208}Pb + ^{208}Pb collisions. The solid curves show the fits using Eq.(1.2). Solid circles show the results for real events. Dashed histograms show the mixed event results.

Particle	y	p_T (GeV/c)	v_1	v_2
π^+	$2.4 \sim 3.6$	$0.0 \sim 1.0$	-0.015 ± 0.009	0.044 ± 0.037
π^-	$2.0 \sim 3.2$	$0.05 \sim 1.0$	-0.015 ± 0.006	0.054 ± 0.028

Table 21.1 Results of the fit to the azimuthal distributions of π^+ and π^- mesons with respect to the reaction plane for semi-central 158 A·GeV ^{208}Pb + ^{208}Pb collisions. The values are integrated over the indicated y and p_T ranges. The v_1 and v_2 values are corrected for the experimental resolution of the reaction plane determination. The statistical fit errors are given.

Fig. 21.2 shows the azimuthal distributions of π^+ and π^- mesons with respect to the reaction plane for semi-central ($50 < E_T < 250$ GeV in MIRAC) and central ($320 < E_T < 500$ GeV) Pb+Pb collisions. In semi-central collisions, a $\cos(2(\phi - \Phi_0))$ component, namely elliptic flow, could be visible for $\pi^\pm$, while in central collisions, the distributions have less structure as expected from the azimuthal symmetry of the collision. For $\pi^\pm$ mesons in semi-central collisions, the azimuthal distributions exhibit maxima at $\phi - \Phi_0 = 0^o$ and $\phi - \Phi_0 = \pm 180^o$, which indicates an enhanced emission in the reaction plane. The azimuthal distributions for the mixed-events are flat which indicates that the observed anisotropies are not due to detector effects.

The values have been corrected for the reaction plane resolution as described above. At SPS energy, results from NA49 [11] have shown that protons and pions exhibit in-plane emission near mid-rapidity. Our $\pi^\pm$ data agree with the NA49 results within errors. The RQMD calculation agrees with the measured results for $\pi^\pm$.

2.4 TWO-PARTICLE CORRELATION

The correlation function as a function of the relative momentum of pion pairs on the transverse plane Q_t is studied near mid-rapidity region with an increased data sample [12]. The data is analyzed in the "Longitudinal Center-of-Mass System" (LCMS). Correction for the finite momentum resolution has not been applied in this analysis yet. The source parameters are extracted by fitting the corrected correlation function $C_2^{corr}(q)$ with a Gaussian distribution formula, where Q_l is required to be below 20MeV/c. Figure 21.3 shows the centrality dependence of the correlation function for identified pions. The preliminary results show larger transverse source size parameter R_t in central collisions compare to the semi-central collisions. The tendency of this results is consistent with the source distribution based on a simple geometrical model.

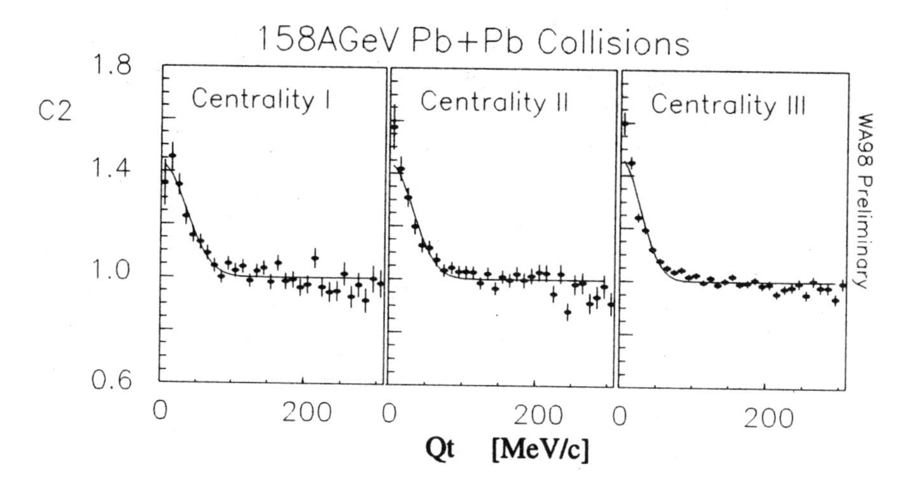

Figure 21.3 The centrality dependence of the two-pion correlation function in Pb+Pb collisions. Only statistical error is presented. The extracted source parameter R_t increases with the centrality.

Centrality	E_t	λ	R_t
Centrality I	50-290 GeV	0.43±0.03	4.0±0.3
Centrality II	290-340 GeV	0.43±0.04	4.4±0.4
Centrality III	340-500 GeV	0.46±0.02	5.0±0.2

Table 21.2 Centrality dependence of the pion source parameters R_t and λ.

The azimuthal angle dependence of the shape of the two-particle correlation function is performed with respect to the direction of the proton flow measured at target rapidity. The transverse momentum in the x component (px) is enhanced by limiting the azimuthal acceptance ϕ^{par} of ARM2 from -60° to 60° ($| px | > | py |$). In this analysis, the azimuthal flow angle Φ_0 is categorized into three types of events, $in - plane$ events, and $out - of - plane$ events to have less biases and higher statistics of pairs.

Fig. 21.4 (left) shows the experimental two-pion correlation functions as a function of Q_t for the event types, $in - plane$ and $out - of - plane$. Statistical errors are presented in this figure. The data correspond to the semi-central part of the reaction cross section, ($50GeV < Et < 290GeV$). Event mixing is carefully applied by taking the same event samples in each event type. One of the interesting results is that the shape of the correlation function might depend on the direction of the emitted particle relative to the reaction plane. The fitted results show

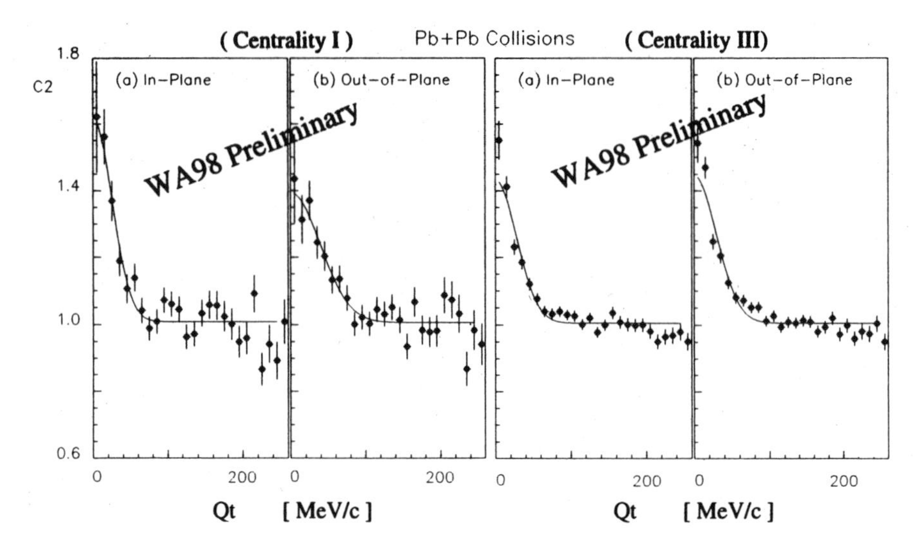

Figure 21.4 Two-pion correlation functions for semi-central (left: $50 < E_t < 290GeV$) and central (right: $340GeV < E_t$) Pb+Pb collisions. For semi-central collisions, relatively larger source size parameter is obtained for the $in - plane$ events compare to the $out - of - plane$ events. While, no significant difference is observed between two types of events in central collisions.

Centrality	E_t	React. Plane	λ	R_t
Semi-Cent.	(a) 50-290 GeV	In-Plane	0.61±0.09	5.6±0.6
	(b) 50-290 GeV	Out-of-Plane	0.39±0.05	3.6±0.4
Central	(a) 340-500 GeV	In-Plane	0.42±0.03	5.1±0.3
	(b) 340-500 GeV	Out-of-Plane	0.44±0.03	4.7±0.3

Table 21.3 Reaction plane dependence of the two-pion correlation function. The source parameters R_t and λ are extracted by fitting the Gaussian source distribution.

larger source size parameter R_t and λ parameter for the $in - plane$ events and relatively smaller R_t and λ for the $out - of - plane$ events in semi-central collisions. Fig. 21.4(right) shows the two-pion correlation functions for the most central collisions ($340GeV < Et$). The results show no significant dependence of correlation shape as a function of the azimuthal flow angles within statistical errors. Based on geometrical model, this results support that the differences in semi-central collisions is not caused by the detector biases. The difference of the correlation function could be caused by the geometrical propagation of freeze-out and flow in the reaction plane. On the other hand, it should be remembered that there is a positive correlation between the extracted parameters λ and source size R_t, so that possibility of the statistical fluctuation should be also taken into account in the analysis.

3. SUMMARY

In summary, the directed flow of protons and π^+ has been studied in 158 A GeV ^{208}Pb + ^{208}Pb collisions. Our preliminary results shows that the directed flow seem to be largest for impact parameter $\approx$ 8 fm, which is considerably more peripheral than observed at lower incident energies. The π^+ directed flow is in the direction opposite to the protons, similar to observations at 11 A GeV energy [8]. The magnitude of the proton directed flow seems to be much less than cascade mode RQMD model predictions. We have measured the elliptic emission patterns of $\pi^\pm$ mesons near mid-rapidity. In semi-central collisions, $\pi^\pm$ mesons seems to be emitted in the reaction plane. The RQMD cascade calculation reproduces the v_2 values for $\pi^\pm$. The shape of the two pion correlation function has been investigated relative to the event plane. Preliminary results show a possible dependence of the correlation shape with respect to the reaction plane in semi-central Pb+Pb collisions.

Acknowledgments

This work is supported by the Grant-in-Aid for Scientific Research (Specially Promoted Research & International Scientific Research) of the Ministry of Education, Science and Culture, JSPS Research Fellowships for Young Scientists and also by the University of Tsukuba Special Research Projects.

References

[1] H. Stöcker and W. Greiner, Phys. Rep. **137**, (1986) 277.

[2] J.-Y. Ollitrault, Phys. Rev. D **46**, 229 (1992); J.-Y. Ollitrault, Phys. Rev. D **48**, 1132 (1993); J.-Y. Ollitrault, Nucl. Phys. **A638**, 195c (1998).

[3] C.M. Hung and E.V. Shuryak, Phys. Rev. Lett. **75**, 4003 (1995).

[4] N.S. Amelin et al., Phys. Rev. Lett. **67**, 1523 (1991).

[5] D.H. Rischke, Nucl. Phys. **A610**, 88c (1996); D.H. Rischke and M. Gyulassy, Nucl. Phys. **A597**, 701 (1996).

[6] H. Sorge, Phys. Rev. **C52**, 3291 (1995); H. Sorge, Phys. Rev. Lett. **78**, 2309 (1997).

[7] E802 Collaboration, L. Ahle, *et al.*, Phys. Rev. **C57**, 1416 (1998).

[8] E877 Collaboration, J. Barrette, *et al.*, Phys. Rev. Lett. **73**, 2532 (1994); J. Barrette, *et al.*, Phys. Rev. **C55**, 1420 (1997); J. Barrette,

et al., Phys. Rev. **C56**, 3254 (1997); S.A. Voloshin, *et al.*, Nucl. Phys. **A638**, 455c (1998).

[9] E895 Collaboration, H. Liu *et al.*, Nucl. Phys. **A638**, 451c (1998).

[10] NA45 Collaboration, F. Ceretto, *et al.*, Nucl. Phys. **A638**, 467c (1998).

[11] NA49 Collaboration, T. Wienold, *et al.*, Nucl. Phys. **A610**, 76c (1996); H. Appelshäuser, *et al.*, Phys. Rev. Lett. **80**, 4136 (1998); A.M. Poskanzer, *et al.*, Nucl. Phys. **A638**, 463c (1998).

[12] WA98 Collaboration, M.M. Aggarwal, *et al.*, preprint nucl-ex/9807004,submitted to Phys. Rev. Lett.; S. Nishimura, *et al.*, in proceedings of Jaipur, 158p. ICPA-QGP(1997); M. Kurata, *et al.*, in proceedings of Jaipur, 549p. ICPA-QGP(1997). S. Nishimura, *et al.*, Nucl. Phys. **A638** 459c (1998).

[13] Plastic Ball Collaboration, H.Å. Gustafsson et al., Phys. Lett. **B142**, 141 (1984); H.Å. Gustafsson, *et al.*, Phys. Rev. Lett. **52**, 1590 (1984); H.H. Gutbrod, *et al.*, Phys. Lett. **B216**, 267 (1989); H.H. Gutbrod, *et al.*, Rep. Prog. Phys. **52**, 1267 (1989); H.Å. Gustafsson, *et al.*, Phys. Rev. **C42**, 640 (1990).

[14] KaoS Collaboration, D. Brill, *et al.*, Phys. Rev. Lett. **71**, 336 (1993); D. Brill, *et al.*, Z. Phys. **A355**, 61 (1996); D. Brill, *et al.*, Z. Phys. **A357**, 207 (1996).

[15] FOPI Collaboration, J.L. Ritman, *et al.*, Z. Phys. **A352**, 355 (1995); N. Bastid, *et al.*, Nucl. Phys. **A622**, 573 (1997).

[16] WA98 Collaboration, Proposal for a large acceptance hadron and photon spectrometer, 1991, Preprint CERN/SPSLC 91-17, SP-SLC/P260.

[17] W. Reisdorf and H.G. Ritter, Ann. Rev. Nucl. Part. Sci. **47**, 663 (1997).

[18] Goldhaber et al., Phys. Rev. **120**, 300 (1960).

[19] T. Chujo, et al., Nucl. Inst. and Methods, **A383** 409 (1996).

[20] A. Baden, et al., Nucl. Inst. **203**, 185 (1982).

[21] S.A. Voloshin and Y. Zhang, Z. Phys. **C70**, 665 (1996).

[22] A.M. Poskanzer and S. A.Voloshin, Phys. Rev. **C58**, 1671 (1998).

[23] A. Jahns, *et al.*, Phys. Rev. Lett. **72**, 3463 (1994).

[24] A. Kugler, *et al.*, Acta Phys. Pol. **25**, 691 (1994).

[25] WA80 Collaboration, T.C. Awes, *et al.*, Phys. Lett. B **381**, 29 (1996).

Chapter 22

STRANGENESS PRODUCTION IN AU+AU COLLISIONS AT THE AGS: RECENT RESULTS FROM E917

Wen-Chen Chang

Department of Physics, University of California, Riverside

for the E917 collaboration

B.B. Back[1], R.R. Betts[1,6], H.C. Britt[5], J. Chang[3], W.C. Chang[3], C.Y. Chi[4], Y.Y. Chu[2], J.B. Cumming[2], J.C. Dunlop[8], W. Eldredge[3], S.Y. Fung[3], R. Ganz[6,9], E. Garcia[7], A. Gillitzer[1,10], G. Heintzelman[8], W.F. Henning[1], D.J. Hofman[1], B. Holzman[1,6], J.H. Kang[12], E.J. Kim[12], S.Y. Kim[12], Y. Kwon[12], D. McLeod[6], A. Mignerey[7], V. Nanal[1], C. Ogilvie[8], R. Pak[11], A. Ruangma[7], D. Russ[7], R. Seto[3], P.J. Stanskas[7], G.S.F. Stephans[8], H. Wang[3], F.L.H. Wolfs[11], A.H. Wuosmaa[1], H. Xiang[3], G.H. Xu[3], H. Yao[8], C.M. Zou[3]

[1] Argonne National Laboratory, Argonne, IL 60439
[2] Brookhaven National Laboratory, Chemistry Department, Upton, NY 11973
[3] University of California Riverside, Riverside, CA 92521
[4] Columbia University, Nevis Laboratories, Irvington, NY 10533
[5] Department of Energy, Division of Nuclear Physics, Germantown, MD 20874
[6] University of Illinois at Chicago, Chicago, IL 60607
[7] University of Maryland, College Park, MD 20742
[8] Massachusetts Institute of Technology, Cambridge, MA 02139
[9] Max Planck Institut für Physik, D-80805 München, Germany
[10] Technische Universität München, D-85748 Garching, Germany
[11] University of Rochester, Rochester, NY 14627
[12] Yonsei University, Seoul 120-749, South Korea

Abstract Strangeness production in Au+Au collisions has been measured via the yields of K^+ , K^- at 6, 8 AGeV and of $\overline{\Lambda}$ at 10.8 AGeV beam kinetic energy in experiment E917. By varying the collision centrality and beam energy, a systematic search for indications of new phenomena and in-medium effects under high baryon density is undertaken.

Keywords: AGS, Strangeness, kaons, $\overline{\Lambda}$, $\overline{p}$, excitation function.

1. INTRODUCTION

The study of strangeness production in relativistic heavy-ion collisions has been of continuing interest as strangeness is predicted to be enhanced by the formation of a quark gluon plasma (QGP)[1]. At the same time, many particle properties such as effective mass, production threshold, and absorption cross section may be sensitive to a high baryon density. Their study might reveal the influence of a many-body mean-field potential and provide a signal of chiral symmetry restoration.

Experiment E917 at the AGS measured Au+Au collisions at beam kinetic energies of 6, 8 and 10.8 AGeV in the winter of 1996/97. The Henry Higgins spectrometer, used previously in experiments E802, E859 and E866 [2], and an upgraded data acquisition system enabled the experiment to take 280×10^6 kaon-pair/$\overline{p}$ triggered events. The quality of this data set allows for a detailed study of short lived vector mesons, baryons, anti-baryons and the systematics of two particle correlations of pion, kaon and proton pairs. E917 is unique among the AGS experiments in its ability to measure a wide variety of strangeness-carrying particles including K^+, K^-, Λ, $\overline{\Lambda}$ and ϕ-mesons. The study of the excitation function of kaon production may help identify a possible phase transition, and the study of ϕ-mesons may provide a direct probe of any in-medium effect. More details on the experimental setup and trigger condition can be found in Ref. [3, 4].

This article presents a systematic study of the spectra and yield of K^+ and K^- in Au+Au collisions at beam energies of 6 and 8 AGeV combined with published E866 data [2] at 10.8 AGeV. A previously reported discrepancy in the measured $\overline{p}$ yields between AGS experiments E878 and E864 has been hypothesized to arise from an abundance of $\overline{\Lambda}$ production and different acceptances for the $\overline{p}$ daughter from $\overline{\Lambda}$ decay in the two experiments. Experiment E917 is able to make the first direct measurement of $\overline{\Lambda}$ yields, for which preliminary results are presented.

2. MEASUREMENT OF KAONS

There is significant theoretical interest in the study of kaon properties in dense nuclear matter. Qualitatively, the models suggest that K^+ mesons experience a weak repulsive potential inside the nuclear medium resulting in a slight increase in the effective mass with baryon density, whereas a strong attractive potential for K^- mesons leads to a significantly reduced effective mass in the high baryon density environment [5]. Because the ΛK^+ production channel is expected to be larger than K^-K^+ pair production at AGS energies, this scenario results in a larger K^-/K^+ yield ratio for central events near mid-rapidity, where high baryon density is expected and the K^-K^+ channel is enhanced.

The rapidity distributions of kaons were obtained from exponential fits to the transverse mass, m_t, spectra at each rapidity bin. From these fits we obtain the integrated production probability, dN/dy, per unit of rapidity and the inverse slope, T_{inv}, of these spectra. We emphasize that T_{inv} should not be interpreted as the *temperature* of the emitting source as it is well known that collective effects, such as radial expansion, can mimic high source temperatures.

The rapidity distributions, dN/dy, for K^+ and K^- emission for 0-5% central collisions are shown in Fig. 22.1 for beam energies of 6, 8 and 10.8 AGeV . The rapidity distributions are observed to be peaked at mid-rapidity and the yields increase with beam energy without substantial change in the shape of the rapidity distributions. We also find that the rapidity distributions are essentially independent of centrality.

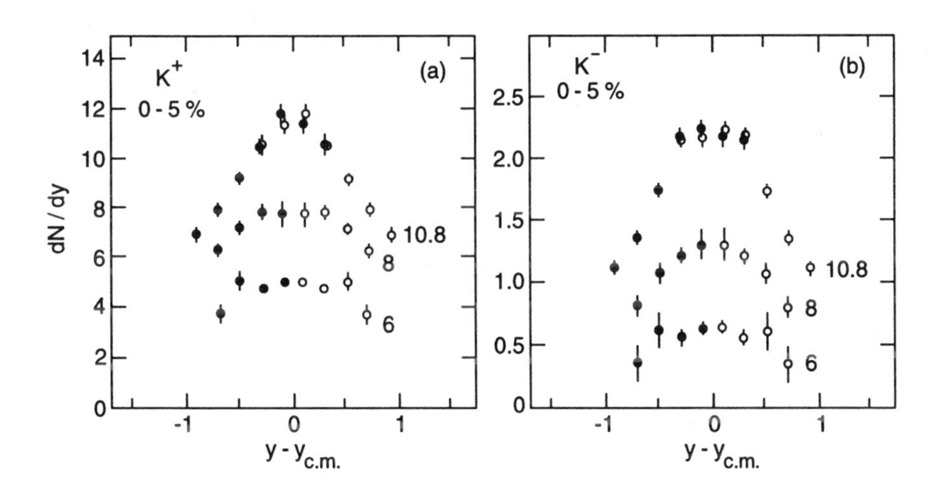

Figure 22.1 Rapidity distributions for K^+(panel a) and K^-(panel b) for 0-5% central events are shown for beam energies of 6, 8 (E917 preliminary) and 10.8 AGeV (E866 [2]).

Since the effects of the nuclear medium have opposite sign for K^+ and K^-, the ratio of K^-/K^+ production might be a very sensitive probe for studying such effects. It was studied as a function of global parameters for the collision, such as centrality and rapidity. In Fig. 22.2(a), the K^-/K^+ ratio is shown as a function of rapidity for the 5% most central events at 6, 8, and 10.8 [2] AGeV. We observe that the ratio increases with beam energy over the rapidity range of this study and that the rapidity distribution for K^- is narrower than for K^+ , an observation that has also been made in studies of Ni+Ni collisions at SIS energies [6]. The fact that the production of K^- relative to K^+ is more abundant around mid-rapidity might be expected, because:

- the available energy for producing particles is peaked around mid-rapidity.
- the baryon density is largest around mid-rapidity and the in-medium effect enhances the production of K^- relative to that of K^+.

It is, however, difficult to disentangle the relative importance of these two effects [7].

The measurements of the K^-/K^+-ratio at mid-rapidity is shown in Fig. 22.2(b) as a function of center-of-mass energy from SIS through AGS to SPS energies. The observed increase in the ratio with beam energy may be expected on the basis of the higher production threshold for K^-. This makes the production cross section of K^- increase faster than that of K^+ above the production threshold [7]. At SPS energy, the increase in the ratio is not as steep as that in the lower energy. This is probably caused by a near saturation in the population of the available phase space for both K^+ and K^-.

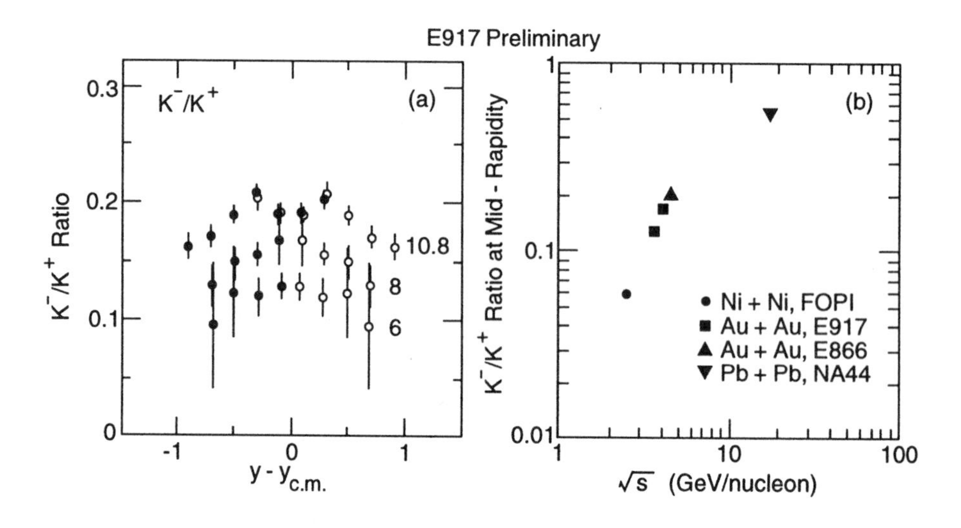

Figure 22.2 The rapidity distribution of K^-/K^+ ratio for the 0-5% central events at various beam energies (panel a) and the K^-/K^+ ratio at mid-rapidity at SIS, AGS and SPS [6, 8] (panel b).

We have also studied the centrality dependence of K^-/K^+ over a wide range of rapidities. The rapidity distribution of this ratio exhibits a very weak dependence on centrality at all three energies, similar to the observations at SIS energies [6, 9]. This weak dependence seemingly contradicts the naive expectation based on the in-medium effect. Thus, one might expect that the K^-/K^+ ratio at mid-rapidity should increase significantly towards central collisions, and that this increase should be most pronounced in the low-energy region close to the production threshold where the effect of the reduction (increase) of K^- (K^+) mass from the in-medium effect is expected to be strongest. Li and Brown have proposed [7] that a suppression of K^- production through the hyperon-feeding channel compensates for the increase in the K^-/K^+-ratio from the in-medium effect in the central collisions. It is, however, difficult to verify this hypothesis experimentally.

The inverse slope parameter, T_{inv}, derived from the fits to the m_t-spectra is found to peak at mid-rapidity for both K^+ and K^-. There is also a slight increase in T_{inv} with beam energy, but the rapidity dependence is virtually unchanged. The difference in T_{inv} for K^+ and K^- transverse mass spectra is found to be small, although the value of T_{inv} for K^+ is about 50 MeV larger than that of K^- at the 6 AGeV beam energy.

3. MEASUREMENTS OF $\overline{\Lambda}$ AND $\overline{P}$

AGS experiment E859 has measured a large $\overline{\Lambda}/\overline{p}$ ratio of $2.9\pm0.9\pm0.5$ for Si+Au at 13.7 AGeV [10]. This ratio is unexpectedly large relative to thermal model calculations or with reference to the results for NN collisions at AGS energies [11]. In addition, experiments E864 and E878 at the AGS have measured $\overline{p}$ production in the mid-rapidity region and zero p_t for Au+Pb at 10.6 AGeV [12] and Au+Au at 10.8 AGeV [13] respectively. There is a significant discrepancy between the two experiments in the reported anti-proton production probability for central events (about a factor of 3.5), but a good agreement for the most peripheral events. Since these two experiments have different acceptances for detecting the $\overline{p}$ from $\overline{\Lambda}$ and $\overline{\Sigma}$ decay, a large production of $\overline{\Lambda}$ might reconcile the results for the two experiments. If this discrepancy is attributed entirely to this effect, a $\overline{\Lambda}/\overline{p}$ ratio of 3.5 (most probable value) or larger than 2.3 (98% confidence level) is required.

Experiment E917 measured $\overline{p}$ in the rapidity range $1.0 < y < 1.4$ and $\overline{\Lambda}$ were reconstructed from $\overline{p}\pi^+$ pairs. The signal of $\overline{\Lambda}$ is clearly seen in the invariant mass distribution shown in the Fig. 22.3. The transverse mass spectra of $\overline{p}$ and $\overline{\Lambda}$ in the rapidity range $1.0 < y < 1.4$ for the central 0-23% events are shown in Fig. 22.4. The efficiency of detecting $\overline{p}$ from $\overline{\Lambda}$ decay is close to unity in our experiment. Assuming that the decay of $\overline{\Lambda}$ is the only source of hyperon feed-down into $\overline{p}$, the yield of $\overline{p}$

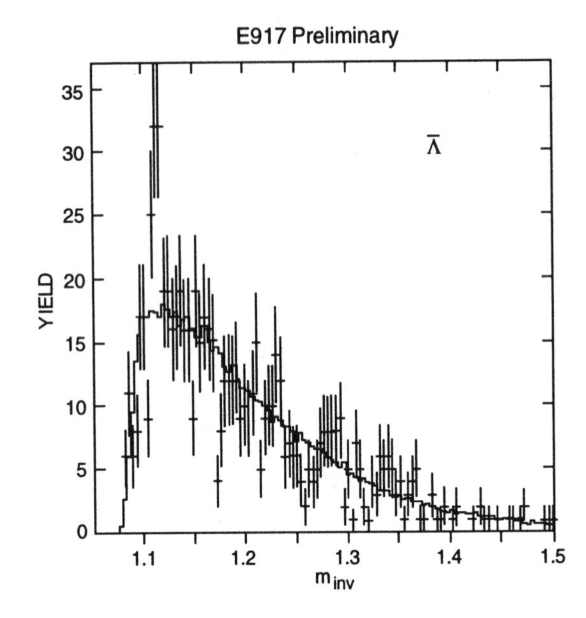

Figure 22.3 The invariant mass distribution of $\overline{\Lambda}$ reconstructed from the pair of $\overline{p}$ and π^+. The line is the fitted background from mixed events.

Table 22.1 The measurement of $\overline{p}$ and $\overline{\Lambda}$ by E917 in the rapidity interval $y = 1.0-1.4$.

	Minimum bias		Central events 0-23%	
Particle	dN/dy ($\times 10^{-3}$)	T_{inv} (MeV)	dN/dy ($\times 10^{-3}$)	T_{inv} (MeV)
$\overline{\Lambda}^a$	$4.3^{+1.8}_{-1.2}$	243^{+112}_{-59}	$12.9^{+5.5}_{-3.7}$	243^{+110}_{-60}
$\overline{p}_{measured}$	7.31 ± 0.17	179 ± 8	15.0 ± 0.6	196 ± 11
$\overline{p}_{direct}$	$4.6^{+0.7}_{-1.2}$		$6.8^{+2.3}_{-3.6}$	
Ratio $\overline{\Lambda}/\overline{p}_{direct}$	$0.9^{+0.9}_{-0.3}$		$1.9^{+3.8}_{-0.9}$	

[a] $dN(\overline{\Lambda})/dy = dN(\overline{\Lambda} \to \overline{p}\pi^+)/dy\ /\ 0.64$.

directly produced is $dN(\overline{p}_{\rm direct})/dy = dN(\overline{p}_{\rm measured})/dy - 0.64 dN(\overline{\Lambda})/dy$, thereby correcting for the 64% branching ratio of the $\overline{\Lambda}$ decay into the $\overline{p}\pi^+$ channel. The rapidity yield, dN/dy, for $1.0 < y < 1.4$ and the inverse slope parameter, T_{inv}, obtained from a fit to the m_t-spectra with an exponential function, are listed in Table 22.1. Details on this analysis are available in Ref. [14].

The E917 measurement of the $\overline{\Lambda}/\overline{p}$-ratio is greater than unity for the 0-23% central collisions and consistent with the E859 measurement in

Si+Au system. The ratio derived from the difference between the E864 and E878 $\overline{p}$ measurements lies within the upper bound of E917 results. It should be noted, however, that there exist several differences in the experimental measurements presented here and those of E864, E878, and E859, as listed in Table 22.2. Most important are the ranges in m_t, rapidity, and centrality measured by the different experiments. More data will be analyzed in the future to compare the results from the other experiments under similar centrality and rapidity cuts.

Table 22.2 The difference in experimental conditions for measuring $\overline{\Lambda}/\overline{p}$-ratio.

Exp.	Collision	E_{beam} (GeV/nucleon)	Centrality (%)	Rapidity	$m_t - m_0$ (MeV/c^2)
E859	Si+Au	13.7	0-15	1.0 – 1.4	> 250
E864	Au+Pb	10.6	0-10	1.6 – 2.0	= 0
E878	Au+Au	10.8	0-10	1.4 – 2.4	= 0
E917	Au+Au	10.8	0-23	1.0 – 1.4	> 250

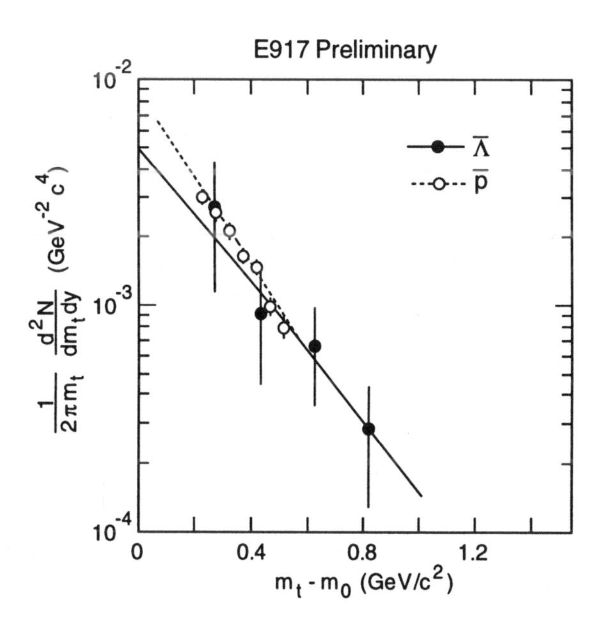

Figure 22.4 Transverse mass (m_t) spectra of $\overline{p}$ (open circles) and $\overline{\Lambda}$ (solid circles) for the 0-23% central events.

4. SUMMARY AND OUTLOOK

A complete measurement of kaon production at 6 and 8 AGeV has been presented. No dramatic change is evident in the excitation function of kaons from 6 to 10.8 AGeV. A straightforward expectation based on the scenario of many-body in-medium effect on kaons cannot explain the observation of a weak centrality dependence of K^-/K^+ ratios with changing beam energy.

The $\overline{\Lambda}/\overline{p}$ ratio was measured to be greater than unity for 0-23% central collisions. More data need to be analyzed to enable a detailed comparison with the other results.

The results presented in this talk are all very preliminary in nature. For this reason, we have not presented any comparison with, or analysis in terms of, theoretical models. These will be presented in future publications.

Acknowledgments

This work was supported by the Department of Energy (USA), the National Science Foundation (USA), and KOSEF (Korea).

References

[1] Koch, P., Müller, B., and Rafelski, J. (1986) Phys. Rep. **142**, 167.

[2] Ahle, L. *et al.*, (1998) Phys. Rev. **C58**, 3523.

[3] Back, B. (1999) these proceedings.

[4] Dunlop, J. C. (1999) Ph.D. thesis, MIT.

[5] Weise, W. (1996) Nucl. Phys. **A610**, 35c.

[6] Hong, B. (1998) in *Proceedings of APCTP Workshop on Astro-Hadron Physics*, edited by G. E. Brown, World Scientific, Singapore.

[7] Li, G. Q. and Brown, G. E. (1998) Phys. Rev. **C58**, 1698.

[8] Bearden, I. G. *et al.* (1998) Nucl. Phys. **A638**, 419.

[9] Barth, R. *et al.* (1997) Phys. Rev. Lett. **78**, 4027.

[10] Stephans, G. S. and Wu, Y. (1997) J. Phys. G **G23**, 1895.

[11] Wang, G. J. *et al.* (1998) Los Alamos Preprint Archive nucl-th/9807036 and nucl-th/9806006.

[12] Armstrong, T. A. *et al.*, (1997) Phys. Rev. Lett. **79**, 3351; Los Alamos Preprint Archive nucl-ex/9811002.

[13] Beavis, D. *et al.* (1995) Phys. Rev. Lett. **75**, 3633; (1997) Phys. Rev. **C56**, 1521.

[14] Heintzelman, G. (1999) Ph.D. thesis, MIT.

Chapter 23

STRANGE BARYON AND H-DIBARYON MEASUREMENTS IN AU-AU COLLISIONS IN AGS EXPERIMENT E896

R. Bellwied for the E896 Collaboration
Physics Department
Wayne State University
Detroit, USA
bellwied@physics.wayne.edu

Abstract The primary goal of E896 at the AGS was to perform a definitive search of the H-Dibaryon in central Au-Au collisions. In addition E896 has measured strange meson and baryon distributions at mid-rapidity. The different components of the experiment will be presented, with special emphasis on the first successful tracking detector based on Silicon Drift Detector technology. The physics motivation for the H-Dibaryon search as well as the measurement of strange baryon abundances will be detailed. Finally, preliminary results of the ongoing analysis will be shown.

Keywords: AGS, Quark Gluon Plasma, Strangeness Production, H-Dibaryon

1. INTRODUCTION

Experiment E896 at the AGS has two main goals, namely to perform a definitive search for the H-Dibaryon and to measure the production of strange mesons and baryons in central Au-Au collisions at 11.6 GeV/c. The experiment took data during a four week period in April '98 at the AGS at BNL.

The detector consists of two main, and largely independent, components. A Silicon Drift Detector Array (SDDA) is mounted very close to the target in a strong magnetic field (B = 6T). A Distributed Drift Chamber (DDC) is mounted in a moderate dipole field (B = 1.5T). The second magnet is located in such a way that the sweeping magnet removes most charged particles and the beam itself from the active area of the DDC, so that only neutral primary particles will reach the detector. The DDC is backed up by a time of flight array (TOF) and a neutron detector (MUFFINS) outside of the magnetic field. Two beam vectoring low pressure drift chambers were installed upstream of the target. A forward multiplicity array was used to determine the centrality of the interaction for trigger purposes. A schematic setup of the complete experiment was shown in the Proceedings of the 1997 Nuclear Dynamics Workshop [26].

The goal is to measure the charged decay particles of the H-Dibaryon and of neutral strange particles (Λ, $\overline{\Lambda}$, K_s^0) in the DDC. The distance of the chamber from the target has been optimized for a specific range of H-Dibaryon lifetimes, based on detailed theoretical predictions [1] and prior measurements of Hypernuclei candidates [2]. The rapidity coverage of the DDC ranges from 2.0 to 3.2.

The SDDA is located very close to the target and its primary Physics goal is the detailed measurement of strangeness production at mid-rapidity. Although the charged Kaon distributions in central Au-Au collisions at the AGS have been measured to great accuracy by E866 [3], no systematic measurements of strange baryons and neutral Kaon distributions have been performed yet. Presently two experiments, E917 [4] and E896 are analyzing Λ, $\overline{\Lambda}$, and Ξ^- distributions. These data will complement the extensive CERN program that recently led to spectacular results for the strange baryon enhancement factors [5],[6]. The SDDA covers about one unit of rapidity around mid-rapidity (y = 0.8-2.0). The SDDA consists of 15 layers of Silicon Drift Detectors [7]. The wafers and the electronics modules are prototypes of the production components for the Silicon Vertex Tracker for the STAR-RHIC experiment [8].

2. PHYSICS MOTIVATION

2.1 H-DIBARYON SEARCH

The H-Dibaryon was postulated as a possible metastable state by Jaffe [9] on the basis of detailed calculations in the framework of the MIT Bag Model. Since then, many different Quark models postulated the existence of a bound six-quark state, mostly due to attractive potentials in the color force interaction. At this point the existence of a stable six-quark state and even a weakly decaying six quark state, with Δs=2, can be excluded on the basis of detailed calculations and many experiments. Ashery gave a nice overview presentation at HADRON 97 [10] summarizing the state of H-Dibaryon search results. Presently about 20 experiments have been concluded without yielding any positive results. Based on these results, which were mostly obtained by Kaon induced experiments, and in particular a measurement reported by E836 at the AGS [11], it is safe to conclude that the H-Dibaryon, if existent as a non-resonant state, has to have a mass somewhere between 2200 MeV and 2230 MeV (the $\Lambda\Lambda$ threshold). In 1995 the E810 collaboration reported the discovery of potential H_0 candidates in this mass range in central Si+Pb collisions at the AGS [12]. The strongest branching ratio is believed to be to the Σ^-p channel. The two other possible channels are the $\Lambda p\pi$ channel and the Λn channel. A resonant state could theoretically also decay into the Ξ^-p channel, which can be analyzed as part of the Ξ reconstruction effort. E896 has capabilities to detect all of those channels, with the main emphasis on the Σ^-p channel. The reconstruction of this particular decay is further complicated by the subsequent decay of the Σ into a neutron and a π^-, and thus relies on the detection of a 'kink' at the position of the Σ^- to $n\pi^-$ decay.

2.2 STRANGENESS MEASUREMENTS

Recently several model interpretations of the CERN measurements of strangeness production in the light (S+S) and heavy (Pb+Pb) ion systems indicated that the enhancement factors can not be easily interpreted by the simple condition that the system is a very dense hadron gas. If chemical and thermal equilibrium in a hadron gas is assumed [13], the non-strange meson and baryon ratios are well described, but most of the strange and multi-strange baryon ratios can not be fully reproduced. Thus, various statistical models that include a QGP phase transition have been suggested [14],[15]. The latest theory presented by Rafelski at this conference [16], assumes thermal equilibration and explicit chemical non-equilibration to interpret the data. Here, the strangeness ratios are based on sudden hadronization of a QGP phase and no subsequent inelastic scattering to change the particle abundances or ratios. All final state interactions of the heavy baryons are elastic and cause only changes in the kinetic parameters, such as the transverse momentum. Thus chemical and thermal freezeout are decoupled. After hadronization the fireball expands with a well defined expansion velocity. Folding this velocity into the kinetic particle spectra leads to a common thermal freezeout temperature of about 130 MeV, whereas the particle abundance and ratio analysis yields a chemical freezeout temperature close to the critical temperature of about 180 MeV [17]. If the CERN measurements indeed indicate a phase transition, it will be intriguing to show whether these transition signatures in the strange baryon channels are reproduced at the lower AGS energies. On the other hand, many event generators based on hadronic cascade models or string models can describe some of the CERN data, by using specific reaction mechanisms, such as color ropes [18] or quark droplets [19] in a dense hadronic matter system. These models do not require a global phase transition, although both reaction mechanisms could be interpreted as requiring a local transition into the quark condensate. The main ratios that show unusual behavior at the SPS and that have now been measured with the SDDA are the multi-strange to strange baryon ratios (e.g. Ξ^-/Λ) and the anti-baryon ratios (e.g. $\overline{\Lambda}/\overline{p}$).

Fig. 23.1 shows compilations of the most spectacular results from SPS and AGS. Fig. 23.1a shows the strangeness enhancement factors measured in experiment WA97 [6]. pA-yields extrapolated to AA-yields are compared to yields measured in Au-Au collisions. The extrapolation is based on a linear scaling of particle yields with the number of participant nucleons. Fig. 23.1b shows a compilation of NA35/NA49 $\overline{\Lambda}/\overline{p}$ ratios as a function of the participant nucleon number [5]. The additional data points are deduced from two AGS measurements, one a direct ratio measurement by E866 in the Si+Pb system [3], and one an indirect measurement in the Au-Au system which is based on comparing the Anti-Proton yield in two experiments with different Anti-Lambda acceptance (E878 and E864). The comparative method that was applied to this analysis is described in detail in [20]. It is apparent that there seems to be very little energy dependence in the $\overline{\Lambda}$ enhancement. Our group tried to describe those large anti-particle ratios with two different models, a cascade model and a thermal model [21]. Both descriptions failed (see the open point in Fig. 23.1b), which seems to indicate that these ratios require a new reaction mechanism. The E896 data will allow to replace the indirect $\overline{\Lambda}/\overline{p}$ measurement of E864 by a statistically significant direct measurement.

3. RESULTS

Both the DDC and SDDA performed well throughout the four week AGS beam time. The SDDA is the first actual tracking detector based on Silicon Drift Detectors. These detectors have been used before in Heavy-Ion experiments as single plane multiplicity counters [22],[23], but with the advent of the SDDA it was proven that Silicon Drift Detectors are a mature technology that can be used to obtain high precision tracking results. The 15 layer detector contained 7,200 channels, less then 1% of these were inactive during the beam time. The hit position resolution measured during the beam time is about 20 μm in x (the anode direction) and 30 μm in y (the drift direction). The resolution might

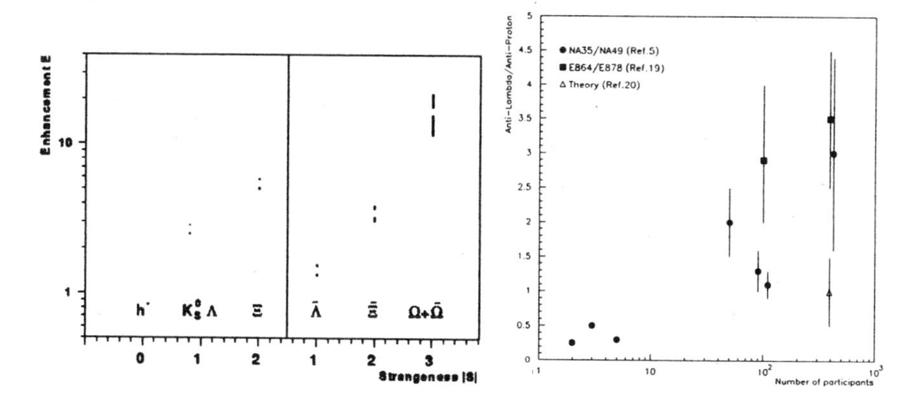

Figure 23.1 a.) Strangeness enhancement factors based on WA97 measurements, b.) Compilation of $\overline{\Lambda}/\overline{p}$ ratios at the SPS and AGS

further improve after the calibrations have been completed, as presently the resolution is slightly worse than bench test measurements [24]. The main calibrations of the SDDA are the alignment of the detectors and the drift velocity. A multi dimensional alignment code was developed for the STAR-SVT [25] and is presently applied to the SDDA. The drift velocity in a Silicon detector depends mostly on environmental effects, such as the magnetic field, the operating temperature, the operating voltage. There are also bulk defects which lead to drift non-linearities. These were determined before the beam time during bench tests. The dependence on field, temperature and voltage is calibrated by injecting a well defined charge into the detector at well defined time intervals during the run. Each detector contains eight charge injection lines equally spaced across the active area. The presently achievable position resolution leads to an average momentum resolution of about 1.7% and a mass resolution of about 7.3 MeV (= 0.65%) for the Λ reconstruction. This resolution is comparable to the mass resolution in the DDC.

The analysis of data is presently in progress in both detector subsystems. The DDC analysis focuses on the H-Dibaryon reconstruction and uses the Λ reconstruction mostly for calibration purposes. The large amount of unsuppressed data (1.5 TByte = 100 Million Events) was filtered with the condition of a large rigidity proton leaving the chamber. This filter is very efficient for H-candidates and reduced the data volume by an order of magnitude.

The SDDA is a slow detector recording data at a 1 Hz acquisition rate, but it has a large number of pixels (61,400 pixels/detector) and thus the unsuppressed data volume (1.3 TByte = 650,000 events) is comparable to that of the DDC. In the case of the SDDA the focus of the analysis is to reconstruct decay particles of strange baryons, which are present in every event. Therefore each event had to be recorded. The main data reduction mechanism here is the elimination of empty pixels (zero-suppression). In this way, the SDDA data volume was reduced by a factor 14 to 90 GByte.

Table 1 and 2 show the expected number of strange particles in the SDDA and DDC, based on the actual number of recorded central Au-Au events. Realistic reconstruction efficiencies, based on detailed simulations including detector response and noise levels, were taken into account. The initial particle multiplicities are based on RQMD [18] and a simulation by Carl Dover [1] for the strange baryons and the H-Dibaryon, respectively.

Table 23.1 Strange Particle Yields in the SDDA

SDDA yields/event	Λ	$\overline{\Lambda}$	Ξ	$\overline{p}$	H
generated in RQMD	15	0.045	0.4	0.015	0.1
geom. acceptance	3.5	0.01	0.04	0.011	0.003
reconstructed	0.3	6×10^{-4}	7×10^{-4}	4×10^{-4}	2×10^{-4}
num. part. in SDDA	2×10^{5}	400	450	250	100

Table 23.2 Strange Particle Yield in the DDC

DDC yields/event	Λ	H
generated in RQMD	15	0.1
geometrical acceptance	0.07	8×10^{-6}
reconstructed	0.0056	2.5×10^{-6}
number of particles in DDC	450,000	200

3.1 PRELIMINARY DDC RESULTS

Both detector systems developed independent tracking software. Typical online tracking results in the DDC were shown at a previous Nuclear

Dynamics conference [26]. Fig. 23.2a shows a preliminary Armenteros plot based on the complete filtered DDC data set. The filter favors the Λ reconstruction over the $\overline{\Lambda}$ reconstruction by requiring a stiff proton rather than a $\overline{p}$. This leads to the asymmetry in the plot. The resulting Λ mass peak is shown in Fig. 23.3.

The H-Dibaryon analysis also includes a cut on the Armenteros plot. Fig. 23.2b shows a simulation of the H-Dibaryon Armenteros distribution compared to the Λ distribution. A potential background in the H-Dibaryon sample is the K_s^0, but the stiff positive track requirement cleans up the sample considerably. The present status of the analysis is that we are in the process of eliminating background contributions in a sample of around 200 H-candidates.

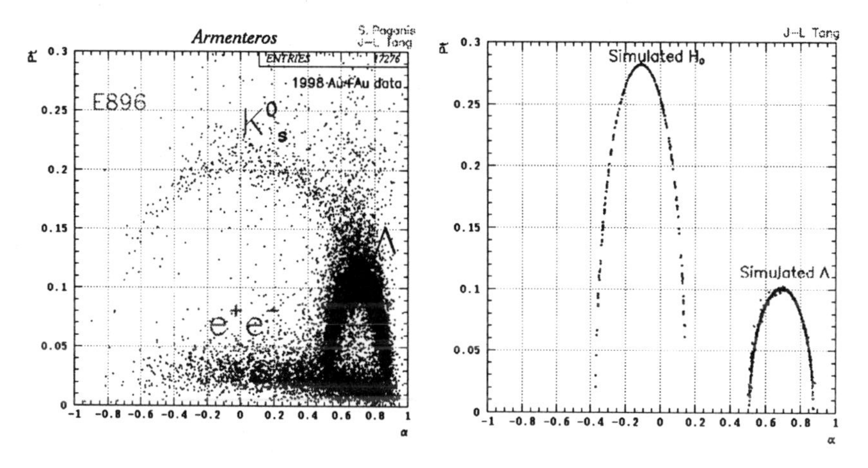

Figure 23.2 a.) Armenteros plot based on the complete filtered DDC data set; b.) Simulation of the H-Dibaryon signal in an Armenteros plot

3.2 PRELIMINARY SDDA RESULTS

Fig. 23.4 shows the reconstructed tracks of a typical central Au-Au event in the Silicon Drift Detector Array. The strong magnetic field leads to large deflections of the tracks. The field was mapped very carefully before the beam time, allowing us to use a Runge-Kutta algorithm based on the field map to correct for field asymmetries and to extrapolate the tracks accurately to the primary vertex. The primary vertex resolution is presently about 1 mm, we expect to improve on this number by about a factor three based on better calibrations.

Fig. 23.5 shows the preliminary Λ mass reconstruction result based on a very small event sample of 300 events (0.05% of the complete data set). This plot simply demonstrates that the tracking and v0 finding

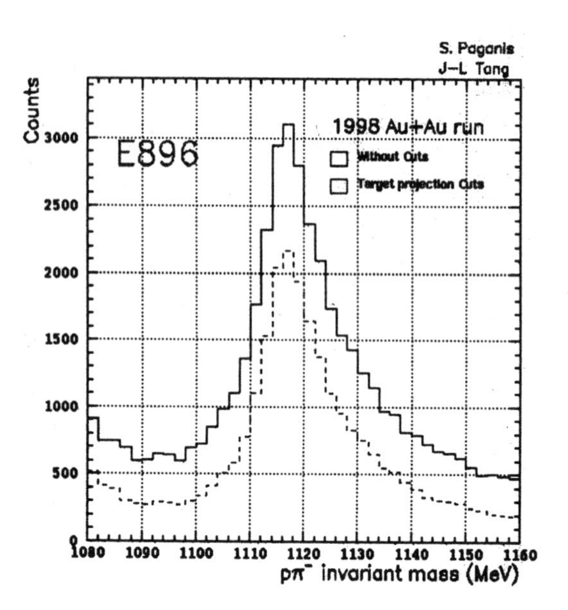

Figure 23.3 Λ mass reconstruction based on the measurements in the DDC

algorithms work. Any improvements in the calibrations will lead to further improvements in the reconstruction efficiency.

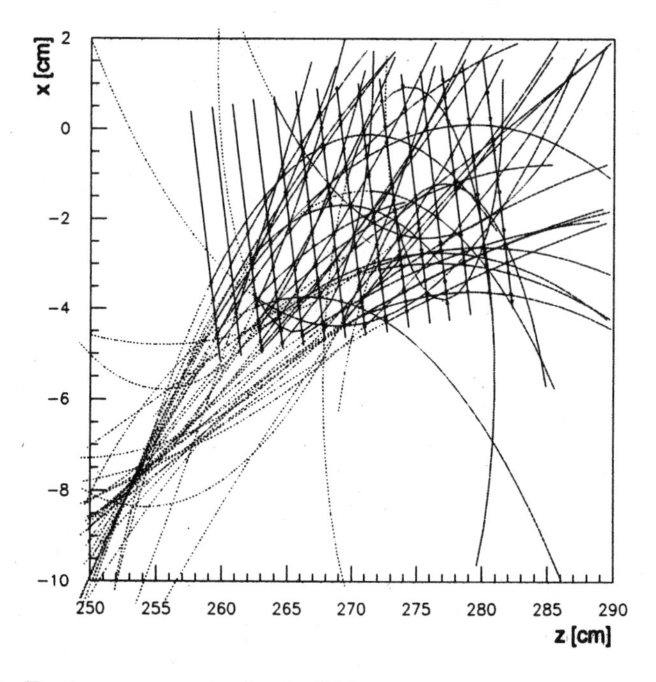

Figure 23.4 Track reconstruction in the SDDA for a typical central Au-Au event

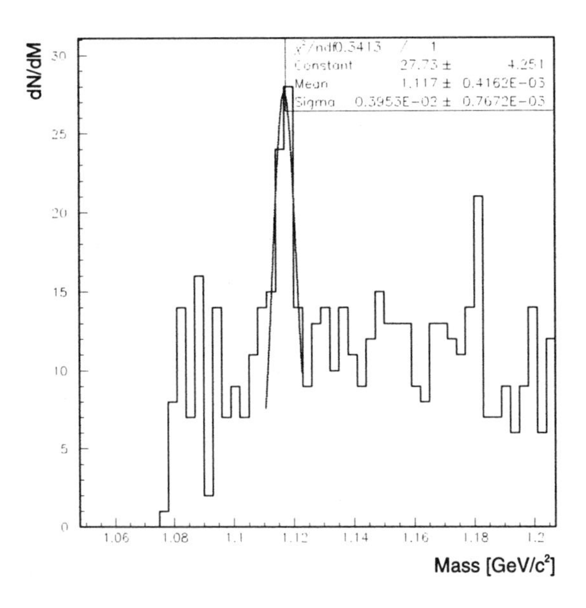

Figure 23.5 Preliminary Λ mass reconstruction based on 0.05% of the SDDA data set

4. SUMMARY

E896 has recorded a large number of strange baryon and possible H-Dibaryon candidates during its four week beam time in 1998. The present level of the analysis shows that these measurements will lead to conclusive results on strangeness production and strangeness enhancement at mid rapidity in the heaviest system at the AGS. These data will have to be compared to measurements from CERN which hint at an onset of new phenomena at higher energies.

Acknowledgments

Many thanks to the whole E896 collaboration, in particular Helen Caines, Salvo Costa, Gaspare LoCurto, Sadek Nehmeh, Stathes Paganis, Jeff Sheen, and Hank Crawford for their invaluable input into the manuscript.

References

[1] C.B. Dover et al., Phys. Rev. **C40** (1989) 115

[2] S. Aoki et al., Prog. Theor. Phys. **85** (1991) 1287

[3] G. Stephans and Y. Wu, J. Phys. **G23** (1998)

[4] R. Ganz for the E917 Collaboration, J. Phys. **G25** (1999) 247

[5] G. Odyniec, Nucl. Phys. **A638** (1998) 135c

[6] R. Lietawa for the WA97 Collaboration, J. Phys. **G25** (1999) 181

[7] E. Gatti and P. Rehak, Nucl. Instr. and Meth. **A225** (1984) 608

[8] J. Harris for the STAR Collaboration, Nucl. Phys. **A566** (1994) 277c

[9] R.L. Jaffe, Phys. Rev. Lett. **38** (1977) 195

[10] D. Ashery, AIP Conf. Proc. **432** (1998) 293

[11] R.W. Stotzer et al., Phys. Rev. Lett. **78** (1997) 3646

[12] S. Ahmad et al., Nucl. Phys. **A590** (1995) 477c

[13] P. Braun-Munzinger et al., Phys. Lett. **B365** (1996) 1

[14] U. Heinz, J. Phys. **G25** (1999) 263

[15] J. Letessier and J. Rafelski, Phys. Rev. **C59** (1999) 947.

[16] J. Rafelski, contribution to these proceedings, hep-ph/9902365 (1999)

[17] R. Stock, to be published and hep-ph/9901415 (1999)

[18] H. Sorge, Nucl. Phys. **A630** (1998) 522 and Phys. Rev. **C52** (1995) 3291

[19] K. Werner and J. Aichelin, Phys. Rev. **C52** (1995) 1584

[20] T.A. Armstrong et al. (E864 Collaboration) to be published and nucl-ex/9811002

[21] G.J. Wang et al., to be published and nucl-th/9807036

[22] W.Chen et al., Nucl. Instr. and Meth. **A326** (1993) 273

[23] M Aggarwal et al. (WA98 Collaboration), Nucl. Phys. **A638** (1998) 147c

[24] R. Bellwied et al., Nucl. Inst. and Meth. **A377** (1996) 1234.

[25] O. Barannikova et al., Internal STAR notes 356 and 364 (1998)

[26] M. Kaplan for the E896 Collaboration, Advances in Nuclear Dynamics 3, Plenum Press (1997) 205

Chapter 24

RECENT RESULTS FROM THE CERES/NA45 EXPERIMENT AND UPGRADE WITH A RADIAL-DRIFT TPC

ShinIchi Esumi for the CERES/NA45 collaboration
Physikalisches Institut, Universität Heidelberg
Philosophenweg 12, D-69120 Heidelberg, Germany
esumi@ceres1.physi.uni-heidelberg.de

Abstract The CERES/NA45 experiment measures low mass electron pair production in Pb-Au collisions at 158AGeV at the CERN SPS. In this paper, we present the electron pair enhancement in the mass range $0.2 < m_{ee} < 0.7$ GeV/c^2, it increases more than linear with charged multiplicity and is most pronounced at low pair $p_T < 0.5$ GeV/c. Transverse mass and rapidity distributions for charged hadrons $h^+ - h^-$ and azimuthal event anisotropies (v_1, v_2) of charged particles are shown as a function of centrality. Finally, we present the status of the upgrade with a radial-drift TPC.

Keywords: QGP, CERES, RICH, TPC, dilepton

1. INTRODUCTION

CERES is an experiment dedicated to the measurement of low-mass electron pair production in heavy-ion collisions at SPS energy. The production of dileptons is expected to convey information from the early stage of the collisions where quark-gluon-plasma (QGP) formation and restoration of chiral symmetry might happen. The CERES spectrometer is optimized to measure electron pairs from 50 MeV/c^2 up to 2 GeV/c^2 at mid-rapidity $2.1 < \eta < 2.65$ with 2π azimuthal coverage. The collision point (vertex) is determined by two silicon drift chambers (SDC) located just behind the target. Two ring imaging Cherenkov detectors (RICH) with methane gas radiator ($\gamma_{th} \sim 32$, which makes it blind to most of the hadrons) are used to identify electrons. A pad chamber (PC) behind the two RICH detectors, in conjunction with all other detectors, is used to track the particles. In between the two RICH detectors, there is a superconducting double solenoidal magnet to provide an azimuthal deflection for momentum / charge measurement of the tracks.

2. ELECTRON PAIR PRODUCTION

CERES has measured electron pair production in p-A collisions at the SPS and found that it is well explained by the decay of hadrons whose abundancies have been measured [1]. On the other hand, in sulphur and lead induced heavy-ion collisions, production of electron pairs has been found to be enhanced relative to the expected contributions from hadrons decays [2, 3].

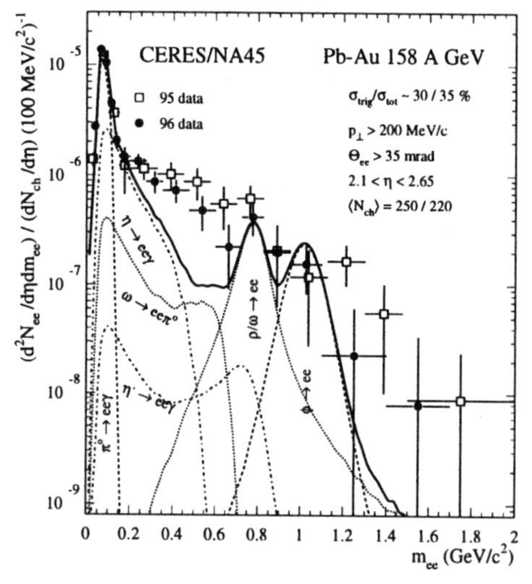

Figure 24.1 Electron pair mass spectra normalized to charged particle multiplicity. Two different symbols represent the data sets from the different years for the same system 158AGeV Pb-Au. The expected electron pair yield from hadron decays at freeze out is shown as the solid line, the different source contributions are shown as the dashed lines.

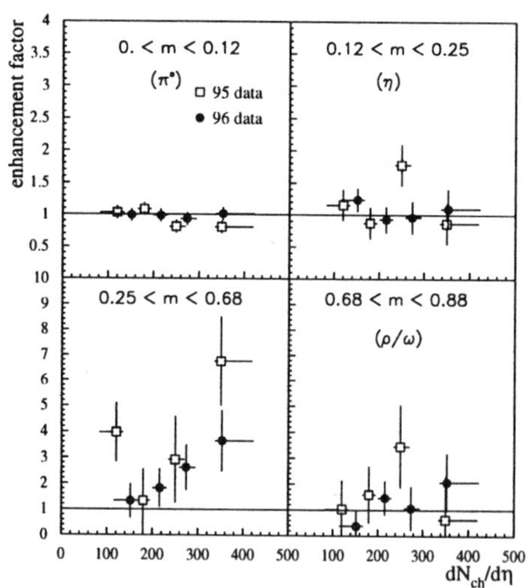

Figure 24.2 Centrality dependence of the enhancement factor relative to the expected yield ($N^{e^+e^-}_{measured}$ / $N^{e^+e^-}_{hadronic\ sources}$) for 4 different mass windows. Linear increase (quadratic increase with charged multiplicity, since $N^{e^+e^-}_{hadronic\ sources}$ is proportional to the charged multiplicity) is observed only within the mass window $0.25 < m_{ee} < 0.68$ GeV/c².

After background rejection cuts, an inclusive e^+e^- mass spectrum for 158AGeV Pb-Au collisions is obtained by subtracting the geometric mean of the like sign pair spectrum from the unlike sign pair spectrum. The result is shown in Fig. 24.1 for the data sets from the 1995 and

1996 runs. The invariant mass spectrum is in good agreement with the expectations for hadron decays below 200 MeV/c^2 where the expected sources are dominanated by π^0 and η Dalitz decays. For masses above 200 MeV/c^2, an enhancement of electron pair production is clearly seen in the mass region up to the ρ, ω peak compared to the expected contribution from hadronic sources as shown in Fig. 24.1. The enhancement factor, integrating over the mass range $0.25 < m_{ee} < 0.7$ GeV/c^2, is $2.6 \pm 0.5(stat.) \pm 0.5(syst.)$ for the '96 data and $3.9 \pm 0.9(stat.) \pm 0.9(syst.)$ for the '95 data [4, 5].

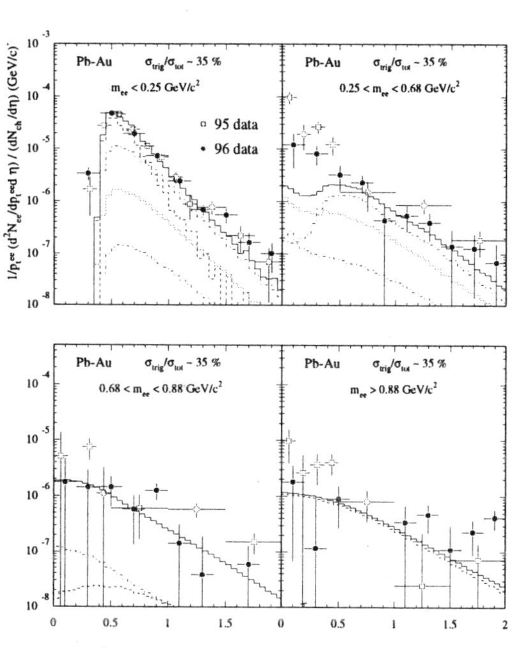

Figure 24.3 Inclusive e$^+$e$^-$ pair p_T spectra in GeV/c for 4 different mass windows, which show the observed enhancement is within the mass window $0.25 < m_{ee} < 0.68$ GeV/c^2 and mostly from the low pair p_T region < 0.5 GeV/c.

The effect has been studied as a function of centrality as shown in Fig. 24.2. If hadron decays are the only source of dielectron pairs, as is the case in p-p and p-A collisions, the data should scale linearly with charged multiplicity, as indicated by the horizontal line in Fig. 24.2. The data clearly indicate a stronger than linear dependence as a function of charged multiplicity within the mass window $0.25 < m_{ee} < 0.68$ GeV/c^2. Inclusive e$^+$e$^-$ pair p_T spectra are shown in Fig. 24.3 for 4 different mass windows. Our single particle p_T cut ($p_T > 200$ MeV/c) strongly affects the pair p_T distribution for masses below 400 MeV/c^2 at low pair p_T. The hadronic decay generator curves shown are folded with such kinematical cuts and the momentum resolution. The measured enhancement within the mass window $0.25 < m_{ee} < 0.68$ GeV/c^2 is most pronounced in the low pair p_T region at $p_T < 0.5$ GeV/c.

3. CHARGED HADRON DISTRIBUTIONS

The two silicon drift chambers just behind target and before the magnetic field and pad chamber after the magnetic field are used to allow charged particle tracking. By subtracting the negatively charged particle

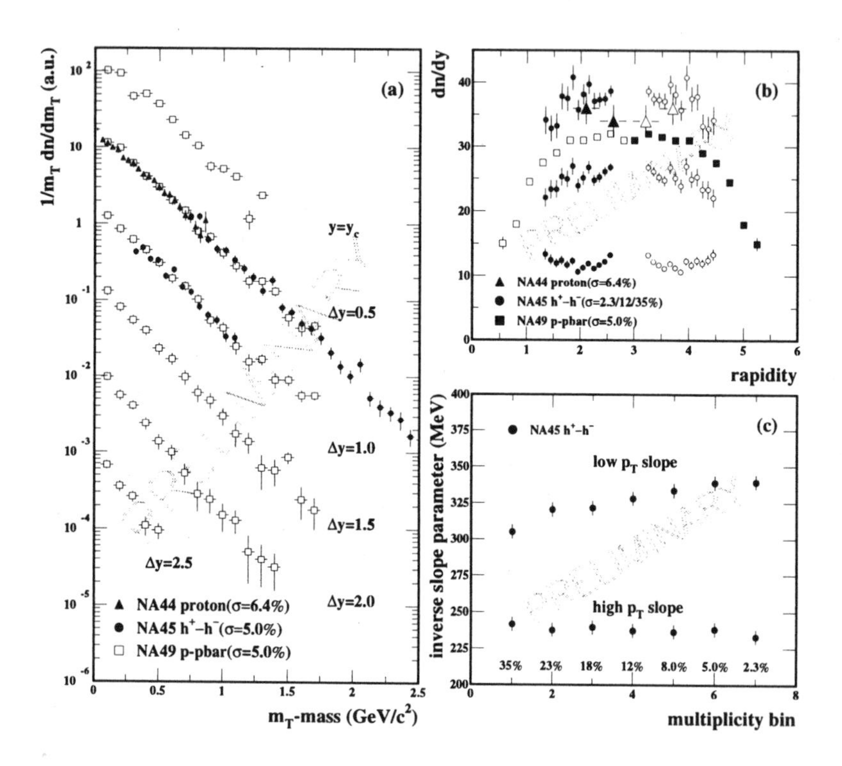

Figure 24.4 Part(a) shows the $(h^+ - h^-)$ transverse mass distribution at 158AGeV Pb-Au collisions (circle) compared with identified proton transverse mass distributions in 158AGeV Pb-Pb collisions from NA44 (triangles)[8] and $(p - pbar)$ transverse mass distribution from NA49 (squares)[9]. Part(b) shows the rapidity distribution of $(h^+ - h^-)$ for different centralities compared to similar results from two other experiments. Part(c) shows the fitted inverse slope parameters as a function of the centrality of the collision. The fitting is done with two exponential functions. The upper(lower) data points represent the lower(higher) $m_T - m_p$ region < 0.7GeV/c($>$ 0.7GeV/c).

distribution from the positively charged particle distribution and using the proton mass, transverse mass and rapidity distribution of $(h^+ - h^-)$ (related to the net proton distributions) are studied as a function of the centrality in 158AGeV Pb-Au collisions shown in Fig. 24.4. The first results were shown in [6, 7]. Within the angular (θ) acceptance of the CERES spectrometer which corresponds for protons to the rapidity acceptance $1.0 < y_p < 2.6$ with a varying transverse momentum coverage $m_T - m_p < 2$ GeV/c^2, the shape of the transverse mass and rapidity distributions of $(h^+ - h^-)$ are investigated here with the main emphasis to study the stopping power in 158AGeV Pb-Au collisions. The change of rapidity distribution from mid-central to central collisions in terms of yield and shape shows an increase of stopping power as a function of centrality as seen in Fig. 24.4(b). The data shows a better fit with two exponential functions than with single exponential function. The shape of transverse mass distribution at lower p_T region $m_T - m_p < 0.7$ GeV/c^2 is flattening with increasing centrality, while the inverse slope parameter in the higher p_T region $m_T - m_p > 0.7$ GeV/c^2 is unchanged with increasing centrality seen in Fig. 24.4(c).

Following the method outlined in [10, 11, 12, 13] a Fourier analysis of the azimuthal charged particle distribution is done with respect to the reaction plane. The directed (v_1) and elliptic (v_2) flow coefficients are measured as a function of pseudorapidity and as a function of centrality as shown in Fig. 24.5. We have chosen a negative sign of v_1 in the forward hemisphere, since the dominant charged particles are pions which are measured to be negative with respect to the proton flow in the other experiments at the same energy [14, 15]. Both v_1 and v_2 are decreasing from $3-5\%$ at mid-central (30%) down to about 1% at central collisions (5%).

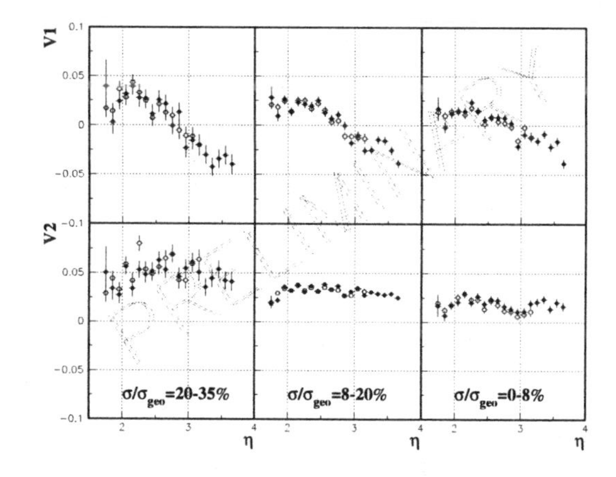

Figure 24.5 Flow parameters v_1 and v_2 as a function of pseudorapidity for different centrality. The reaction plane is determined by two independent subsamples of hits in SDC1. The solid symbols are coefficients calculated for hits in SDC1 and the open symbols are for SDC2. Autocorrelations are avoided by using the reaction plane from another sub-sample. Correction for the reaction plane resolution are applied.

4. UPGRADE WITH RADIAL-DRIFT TPC

The CERES spectrometer has been upgraded by the addition of radial-drift time projection chamber (TPC) with an additional magnetic field behind the RICH detectors. The new experimental setup is shown in Fig. 24.6. The TPC has a cylindrical geometry as all the other detectors in the experiment and covers the same angular acceptance in θ. Electrons produced by inonization due to the passage of charged particles in the TPC are drifted in a radial electric field defined by an inner cylindrical electrode and 16 read-out chambers at out side. The read-out chambers are multiwire proportional chambers with cathode pad read-out. There are 20 pad-rows in beam (z) direction with 16×48 pads in azimuthal direction, each sampling the radial (drift) direction in 256 time bins. A magnetic field is provided by two large coils surrounding the TPC with currents running in opposite direction. The measurement of track curving through the magnetic field will provide momentum of the track. The TPC in conjunction with the two SDC's will significantly improve the dielectron mass resolution less than 2% at the ω mass.

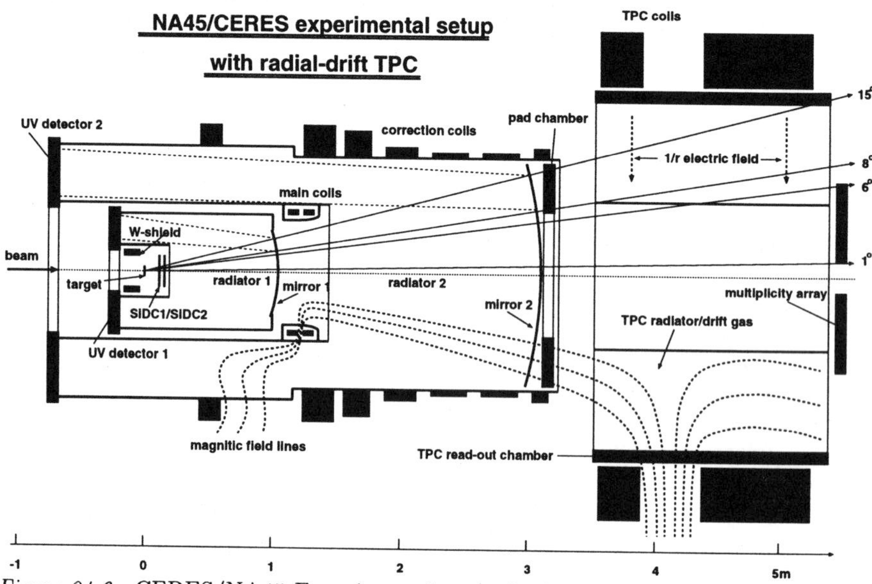

Figure 24.6 CERES/NA45 Experimental setup for '98 and '99 run with radial-drift TPC behind the RICH spectrometer

The goal is to achieve a momentum resolution which allows the experiment to be sensitive to the width of ω and ϕ mesons. The dielectron mass spectrum expected from hadronic decays is shown in Fig. 24.7. Here unchanged natural widths are used for the ρ, ω, ϕ mesons. The TPC was operational in an engineering run in November 1998 and a

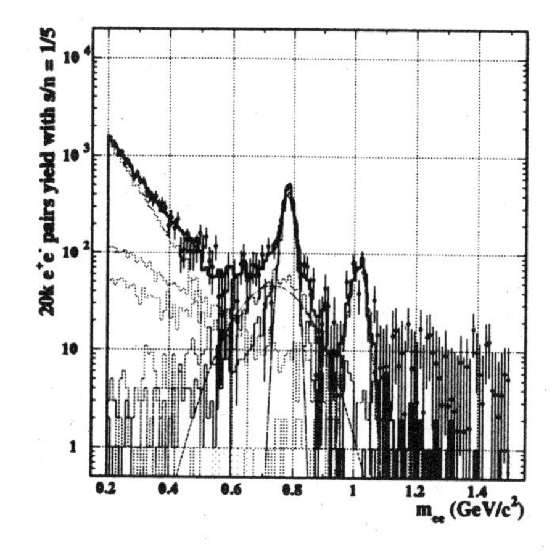

Figure 24.7 Expected electron pair mass spectrum from upgraded CERES/NA45 spectrometer with the radial-drift TPC is calculated by assuming a statistics of 20k pairs with signal to noise ratio at 1/5 in the masses above 200 MeV/c^2.

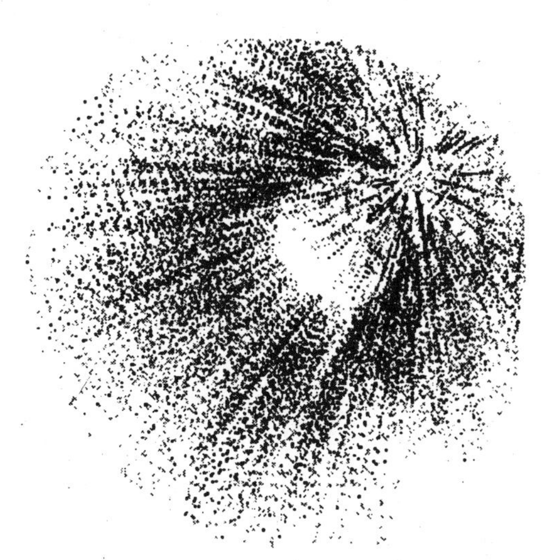

Figure 24.8 158AGeV central Pb-Au event display of the TPC looking from the downstream end of the TPC; all hits are displayed and track segments in the TPC are clearly recognizable.

typical event display from one central Pb-Au collision is shown in Fig. 24.8.

5. SUMMARY

Electron pair production in 158AGeV Pb-Au collisions is enhanced about quadratically in the mass region $0.2 < m_{ee} < 0.7$ GeV/c^2 and most pronounced in the low pair p_T region < 0.5 GeV/c. CERES has also measured hadronic variables: charged hadron distributions and azimuthal event anisotropies. Our plan is to study the correlation between those different variables including measured leptons. A Radial-drift TPC has been build in order to improve the mass resolution and first test runs with laser/muon/p-Pb/Pb-Au triggers have been performed during 1998. Analysis of those runs is now in progress and a low energy heavy ion run at 40AGeV Pb-Au is planned for 1999 to investigate dilepton production at higher baryon and lower energy density.

Acknowledgments

The authors wish to thank for the good performance of the CERN PS and SPS accelerators. We are grateful for financial support by the German BMBF, the U.S. DoE, the MINERVA Foundation, the German-Israeli Foundation for Scientific Research and Development, and the Israeli Science Foundation.

References

[1] G.Agakichiev et al., CERES collaboration, Eur. Phys. J. **C4**, (1998) 431.

[2] G.Agakichiev et al., CERES collaboration, Phys. Rev. Lett. **75**, (1995) 1727.

[3] G.Agakichiev et al., CERES collaboration, Phys. Lett. **B422**, (1998) 405.

[4] B.Lenkeit, Doctoral Thesis, Heidelberg (1998).

[5] B.Lenkeit et al., CERES collaboration, Proc. International Nuclear Physics Conference, Paris, August 1998, Nucl. Phys. **A** in print.

[6] F.Ceretto, Doctoral Thesis, Heidelberg (1998).

[7] F.Ceretto et al., CERES collaboration, Nucl. Phys. **A638**, (1998) 467c.

[8] I.G. Bearden et al., NA44 Collaboration, Phys. Lett. **B388**, (1996) 431.

[9] P.G.Jones et al., NA49 Collaboration, Nucl. Phys. **A610**, (1996) 188c.

[10] J.Barrette et al., E877 Collaboration, Phys. Rev.Lett. **73**, (1994) 2532.

[11] J.Barrette et al., E877 Collaboration, Phys. Rev. **C55**, (1997) 1420.

[12] J.Barrette et al., E877 Collaboration, Phys. Rev. **C56**, (1997) 3254.

[13] A.Poskanzer and S.Voloshin, Phys. Rev. **C58**, (1998) 1671.

[14] S.Nishimura et al., WA98 Collaboration, Nucl. Phys. **A638**, (1998) 459c.

[15] A.Poskanzer et al., NA49 Collaboration, Nucl. Phys. **A638**, (1998) 463c.

Chapter 25

THE MYTH OF LARGE P_T HADRON SPECTRA IN HIGH-ENERGY HEAVY-ION COLLISIONS

Xin-Nian Wang
Nuclear Science Division, MS 70A-3307
Lawrence Berkeley National Laboratory
Berkeley, CA 94720, USA
xnwang@lbl.gov

Abstract

Hadron spectra in high-energy pp, pA and AA collisions are analyzed in a pQCD parton model, including initial k_T and its broadening via initial multiple scattering. Within this model the hadron spectra in both pp and pA or AA collisions can be described very well for $p_T > 2$ GeV/c. The modification of the spectra in pA and AA as compared to that in pp collisions is consistent with multiple parton scattering scenario. Such an analysis will shed new light on the physics one can extract from the single inclusive hadron spectra, *e.g.*, freeze-out temperature and flow velocity in a thermal fire-ball model, and parton energy loss in dense matter.

Keywords: pQCD, RHIC, spectra

1. INTRODUCTION

The transverse momentum spectra are among the basic measurements one can make in high-energy heavy-ion collisions. Yet, there is a wealth of physics one can extract from such spectra. For example, within a thermal fire-ball model, one can extract the freeze-out temperature, collective radial flow velocity and chemical potential [1]. On the another hand, one can also use a dynamics model such as perturbative QCD parton model to study the spectra and the effects of multiple parton scattering and radiative parton energy loss in a dense medium. Apparently, these two approaches are valid only in different region of transverse momentum. While pQCD parton can only be used to calculate hadron spectra in large p_T region, thermal fire-ball model might only make sense in low p_T region if it can ever be applied. Therefore,it is quite important to investigate how well a pQCD parton model can describe hadron spectra in pp collisions and their modification in pA and AA collisions. In particular, the impact-parameter or A dependence of the spectra may be unique to distinguish the parton model from other thermal fire-ball or hydrodynamic models. One can then at least make a quantitative conclusion about the validity of different models at different p_T range. The values of temperature and flow velocity extracted from fire-ball model analysis, for example, will have then be looked at with caution and knowledge of limitations.

In addition, large-p_T partons or jets are good probes of the dense matter formed in ultra-relativistic heavy-ion collisions [2, 3]. If a dense partonic matter is formed during the initial stage of a heavy-ion collision with a large volume and a long life time, the produced large p_T parton will interact with this dense medium and, according to many recent theoretical studies [4, 5, 6], will lose its energy via induced radiation. The energy loss is shown to depend on the parton density of the medium. Therefore, the study of parton energy loss can shed light on the properties of the dense matter in the early stage of heavy-ion collisions. It is also a crucial test whether there is any thermalization going on in the initial stage of heavy-ion collisions. Since the large p_T single-inclusive particle spectra in nuclear collisions are a direct consequence of jet fragmentation as in hadron-hadron and hadron-nucleus collisions, they are also shown [7] to be sensitive to parton energy loss or modification of jet fragmentation functions inside a dense medium.

In this talk, I would like present a systematic study of high p_T hadron spectra in pp, pA and AB collisions. For the study of the parton energy loss, the reliability of the determination of a small effective parton energy loss in high-energy heavy-ion collisions crucially depends on the precision to which we understand the spectra without parton energy loss. We will include both the intrinsic k_T in $p+p$ collisions and k_T broadening due to multiple initial-state scattering in $p+A$ collisions, which are necessary to describe large p_T particle production at energies below $\sqrt{s} < 50$ GeV. We will then compare the calculations with the experimental data for $S+S$, $S+Au$ and $Pb+Pb$ collisions at the CERN SPS with different centrality cuts and verify the scaling behavior characteristic of hard processes. Using the same parton model we will then predict the high p_T spectra in pA collisions at the RHIC energy and then calculate same spectra in AA collisions with and without effects of parton energy loss.

2. HADRON SPECTRA IN *PP* COLLISIONS

We shall start with hadron spectra in *pp* collisions. In the parton model one would expect that partons inside a hadron carry an average intrinsic transverse momentum of about a few hundred MeV, reflecting the hadron size via the uncertainty principle. This is consistent with the fact that hadrons produced in parton fragmentation in e^+e^- annihilation also have an average transverse momentum of a few hundred MeV. However, experimental data of Dell-Yan dilepton pairs production [8] indicate an average initial parton transverse momentum (before the hard process that produces the dilepton pair) of about a few GeV much larger than expected of the intrinsic value. This is because higher order pQCD processes like initial state radiation or $2 \to 3$ subprocesses with additional radiated gluons can also give rise to an additional initial transverse momentum for the colliding partons. Such complications make it difficult to differentiate what is true intrinsic and what is QCD generated initial transverse momentum. In this paper, we introduce a simple scheme of including the intrinsic transverse momentum in the lowest-order (LO) perturbative calculation of the inclusive particle spectra in high-energy *pp* collisions.

The inclusive particle production cross section in *pp* collisions will then be given by [9]

$$\frac{d\sigma_{pp}^h}{dyd^2p_T} = K\sum_{abcd}\int dx_a dx_b d^2k_{aT} d^2k_{bT} g_p(k_{aT},Q^2) g_p(k_{bT},Q^2) f_{a/p}(x_a,Q^2) f_{b/p}(x_b,Q^2) \frac{D_{h/c}^0(z_c,Q^2)}{\pi z_c}\frac{d\sigma}{d\hat{t}}(ab\to cd) \quad (25.1)$$

where $D_{h/c}^0(z_c,Q^2)$ is the fragmentation function of parton c into hadron h as parameterized in [10] from e^+e^- data, z_c is the momentum fraction of a parton jet carried by a produced hadron. We choose the momentum scale as the transverse momentum of the produced parton jet $Q = p_T/z_c$. We also use a factor $K \approx 2$ (unless otherwise specified) to account for higher order QCD corrections to the jet production cross section [11].

We assume the initial k_T distribution $g_N(k_T)$ to have a Gaussian form,

$$g_N(k_T,Q^2) = \frac{1}{\pi\langle k_T^2\rangle_N} e^{-k_T^2/\langle k_T^2\rangle_N}. \quad (25.2)$$

Since the initial k_T includes both the intrinsic and QCD radiation-generated transverse momentum, the variance in the Gaussian distribution $\langle k_T^2\rangle_N$ should also depend on the momentum scale Q of the hard processes. We choose the following form in our scheme according to the

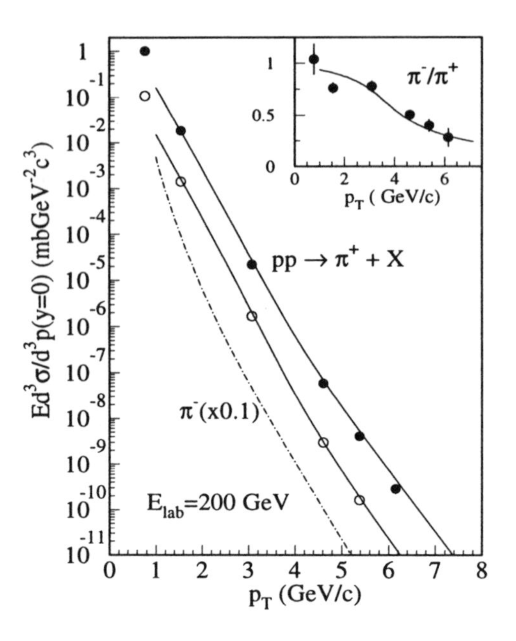

Figure 25.1 Single-inclusive pion spectra in $p+p$ collisions at $E_{\rm lab} = 200$ GeV. The solid lines are pQCD parton model calculations with intrinsic k_T and the dot-dashed line is without. Experimental data are from Ref. [13]. The inserted figure shows the corresponding π^-/π^+ ratio.

next-to-leading order (NLO) analysis of Dell-Yan (DY) processes [12],

$$\langle k_T^2\rangle_N(Q^2) = 1(\mathrm{GeV}^2) + 0.2\alpha_s(Q^2)Q^2. \qquad (25.3)$$

The parameters are chosen to reproduce the experimental data in our following discussions.

Shown in Figs. 25.1 are our calculated spectra for charged pions as compared to the experimental data [13, 14] for $p+p$ collisions at $E_{\rm lab}$ =200 GeV. As one can see from the figures that pQCD calculations with the initial k_T smearing (solid lines) fit the experimental data very well. However, without the initial k_T smearing (dot-dashed lines for π^-) the calculations significantly underestimate the experimental data, as much as a factor of 20 at $E_{\rm lab} = 200$ GeV. This is because the QCD spectra are very steep at low energies and even a small amount of initial k_T could make a big increase to the final spectra. As the energy increases, the QCD spectra become flatter and small amount of initial k_T does not change the spectra much. Such parton model calculations fit very well at energies from $\sqrt{s} = 20 - 1800$ GeV for pp and $p\bar{p}$ collisions [15].

In the inserted boxes in Figs. 25.1, we also plot π^-/π^+ ratio as a function of p_T. At higher p_T, particle production is more dominated by the leading hadrons from valence quark scattering. Since there are more up-quarks than down-quarks in $p+p$ system, one should expect the ratio to become smaller than 1 and decrease with p_T. The pQCD calculations describe this isospin dependence of the spectra very well.

3. *PA* AND *AA* COLLISIONS

We assume that the inclusive differential cross section for large p_T particle production is still given by a single hard parton-parton scattering. However, due to multiple parton scattering prior to the hard processes, we consider the initial transverse momentum k_T of the beam partons is broadened. Assuming that each scattering provide a k_T kick which also has a Gaussian distribution, we can in effect just change the width of the initial k_T distribution. Then the single inclusive particle cross section in minimum-biased $p+A$ collisions is,

$$\frac{d\sigma^h_{pA}}{dyd^2p_T} = K\sum_{abcd}\int d^2bt_A(b)dx_adx_bd^2k_{aT}d^2k_{bT}g_A(k_{aT},Q^2,b)g_p(k_{bT},Q^2)$$
$$f_{a/p}(x_a,Q^2)f_{b/A}(x_b,Q^2,b)\frac{D^0_{h/c}(z_c,Q^2)}{\pi z_c}\frac{d\sigma}{d\hat{t}}(ab\to cd), \quad (25.4)$$

where $t_A(b)$ is the nuclear thickness function normalized to $\int d^2bt_A(b) = A$. The parton distribution per nucleon inside the nucleus (with atomic mass number A and charge number Z) at an impact parameter b,

$$f_{a/A}(x,Q^2,b) = S_{a/A}(x,b)\left[\frac{Z}{A}f_{a/p}(x,Q^2) + (1-\frac{Z}{A})f_{a/n}(x,Q^2)\right], \quad (25.5)$$

is assumed to be factorizable into the parton distribution in a nucleon $f_{a/N}(x,Q^2)$ and the nuclear modification factor $S_{a/A}(x,b)$ which we take the parameterization used in HIJING [16] for now. The initial parton transverse momentum distribution inside a projectile nucleon going through the target nucleon at an impact parameter b is then,

$$g_A(k_T,Q^2) = \frac{1}{\pi\langle k_T^2\rangle_A}e^{-k_T^2/\langle k_T^2\rangle_A}, \quad (25.6)$$

with a broadened variance

$$\langle k_T^2\rangle_A(Q^2) = \langle k_T^2\rangle_N(Q^2) + \delta^2(Q^2)(\nu_A(b)-1). \quad (25.7)$$

The broadening is assumed to be proportional to the number of scattering $\nu_A(b)$ the projectile suffers inside the nucleus.

We will use the following k_T broadening per nucleon-nucleon collision,

$$\delta^2(Q^2) = 0.225 \frac{\ln^2(Q/\mathrm{GeV})}{1 + \ln(Q/\mathrm{GeV})} \quad \mathrm{GeV}^2/c^2. \tag{25.8}$$

The p_T dependence of the broadening is to consider the fact that the distribution of soft k_T kick for each scattering does not necessarily have a Gaussian form.

Shown in Fig. 25.2 are our calculated ratios of charged pion spectra in $p+W$ over that of $p+Be$ each normalized by the atomic number of the target nucleus. If there was no nuclear dependence due to multiple scattering, the ratios would be approximately 1. As shown in the figure, our model can roughly describe the general feature of the nuclear dependence of the spectra at large p_T due to multiple parton scattering. The ratios should become smaller than 1 at very small p_T because of the absorptive processes. But here our perturbative calculation will eventually breaks down because of the small momentum scale. At larger p_T, the spectra are enhanced because of multiple parton scattering. As p_T increases further, the ratios decrease again and saturate at about 1. The decrease follows the form of $1/p_T^2$ consistent with the general features of high twist processes. Since the transverse momentum broadening due to multiple parton scattering is finite, its effect will eventually become smaller and disappear. Therefore, the p_T location of the maximum enhancement can give us the scale of average transverse momentum broadening.

Similarly one can also incorporate the initial k_T broadening due to multiple parton scattering in $A + A$ collisions,

$$\begin{aligned} \frac{d\sigma^h_{AB}}{dyd^2p_T} = & K \sum_{abcd} \int d^2bd^2rt_A(r)t_B(|\mathbf{b}-\mathbf{r}|) \int dx_a dx_b d^2k_{aT} d^2k_{bT} \\ & g_A(k_{aT}, Q^2, r) g_B(k_{bT}, Q^2, |\mathbf{b}-\mathbf{r}|) f_{a/A}(x_a, Q^2, r) \\ & f_{b/B}(x_b, Q^2, |\mathbf{b}-\mathbf{r}|) \frac{D_{h/c}(z_c, Q^2, \Delta L)}{\pi z_c} \frac{d\sigma}{d\hat{t}}(ab \to cd), \end{aligned} \tag{25.9}$$

where $D_{h/c}(z_c, Q^2, \Delta L)$ is the modified effective fragmentation function [18, 19] for produced parton c which has to travel an average distance ΔL inside a dense medium. The modified fragmentation function with depend on dE/dx and mean-free-path of parton interaction in medium.

Shown in Fig. 25.3 are the calculated inclusive spectra for produced π_0 in $S + S$, $S + Au$ and $Pb + Pb$ collisions, both minimum-biased and central events. The pQCD parton model calculations with the k_T broadening due to initial multiple scattering (solid lines) agree with the

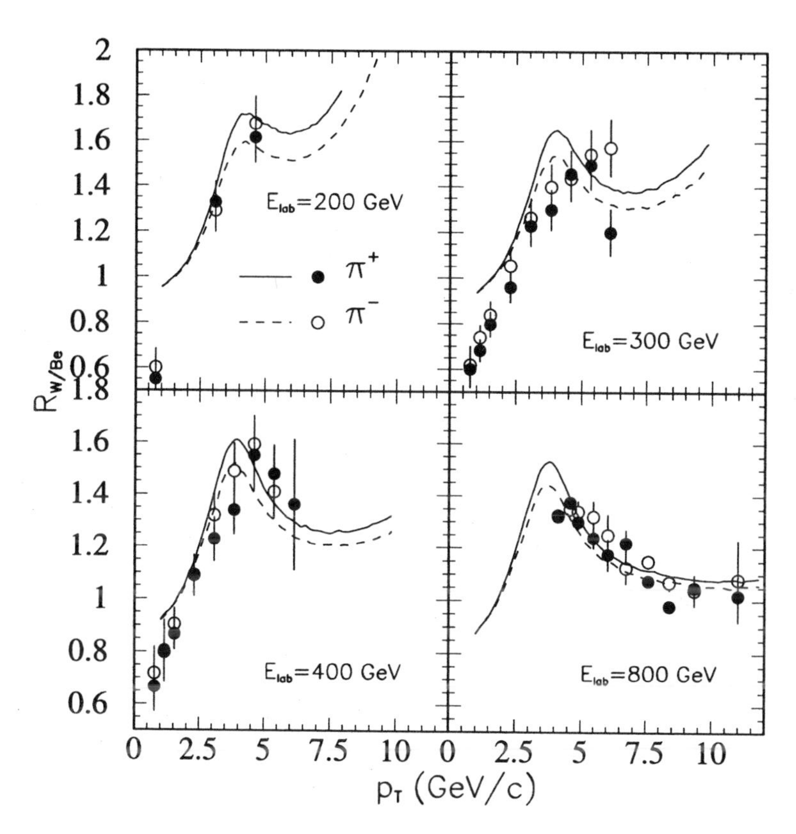

Figure 25.2 Ratios of charged pions spectra in $p+W$ over $p+Be$ each normalized by the atomic number of the target nucleus. The lines are the parton model calculation with k_T broadening due to multiple parton scattering. Experimental data are from Refs. [13, 17].

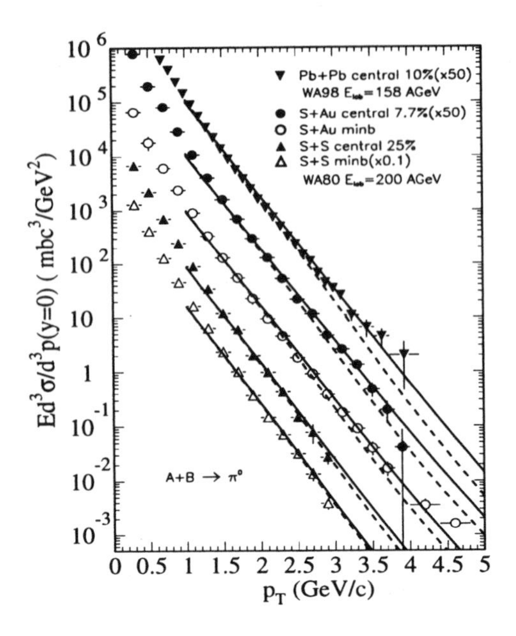

Figure 25.3 Single-inclusive spectra of π^0 in $S+S$, $S+Au$ and $Pb+Pb$ collisions (both minimum-biased and central events) at the CERN SPS energies. The solid lines are pQCD parton model calculations with k_T broadening due to initial multiple parton scattering and the dashed lines are without. Experimental data are from Refs. [20, 21].

experimental data (WA80 and WA98) [20, 21] well very at p_T above 1 GeV/c. No parton energy loss has been assumed in the calculations. The dashed lines are the spectra in pp collisions at the same energy multiplied by the nuclear geometrical factor as given in Eq. (25.10). The difference between the solid and dashed lines is simply caused by effects of k_T broadening and nuclear modification of parton distributions inside nuclei. These effects are similar as in $p+A$ collisions and are more important in collisions of heavier nuclei. Without these nuclear effects the high p_T hadron spectra in $A+B$ collisions are exactly proportional to the average number of binary collisions as shown by the dashed lines. This is a common characteristic of hard processes in $p+A$ and $A+B$ collisions.

4. *A* SCALING OF HADRON SPECTRA

According to the pQCD parton model, the hadron spectra at large p_T should scale with the number of binary nucleon-nucleon collisions if no nuclear effect is included. So if one defines a ratio,

$$R_{AB}(p_T) \equiv \frac{d\sigma^h_{AB}/dyd^2p_T}{\langle N_{\rm binary}\rangle d\sigma^h_{pp}/dyd^2p_T} \tag{25.10}$$

between spectra in AB and NN collisions normalized by the averaged number of binary collisions $\langle N_{\rm binary}\rangle$ which one can easily calculate from Wood-Saxon nuclear distribution, the ratio will be approximately one for spectra from hard parton collisions. Once initial multiple scattering is included, the spectra will be enhanced (Cronin effect) so that ratio will be above one.

Because of absorptive processes, low p_T particle production, which can be considered as coherent over the dimension of nuclear size, has much weaker A-dependence. In the wounded-nucleon model, soft particle production cross section is proportional to the average number of wounded nucleons which is much smaller than the number of binary collisions. So the ratio as defined in Eq. (25.10) will be smaller than one at low p_T and larger than one at large p_T. Such a general feature has been found to be almost universal in both pA and AB collisions. One interesting feature from this analysis is that the transition between soft coherent interaction to hard parton scattering happen roughly around p_T=1.5 GeV. This is also the place where hadron spectra in pp collisions start to deviate from a pure exponential form. One can the expect that for spectra above this value of p_T the underlying mechanism of hadron production become dominated by hard processes.

At even higher p_T, the effect of multiple scattering becomes less important, so the ratio R_{AB} will approach to 1 again (higher twist effect should be suppressed by $1/p_T^2$), as shown by the data for pW collisions at $E_{\rm lab} = 800$ GeV in Fig. 25.2. However, at SPS energy, such a feature cannot be be fully revealed because of the kinetic limit. One will then only see the initial increase of the ratio due to the transition from soft to hard processes. Such a change of spectra from pp to pA and AA collisions in a limited kinetic range looks very similar to the effect of collective flow in a hydrodynamic model. Therefore one should take caution about the values of temperature and flow velocity extracted from such a fire-ball analysis of the spectra, especially if one has to reply on the shape of the spectra in the p_T region around 1 GeV.

5. CONCLUSIONS

In the parton model calculations presented so far, effects of parton energy loss have not been considered yet. If there is parton energy loss and the radiated gluons become incoherent from the leading parton, the resultant leading hadron spectra at large p_T from the parton fragmentation should be suppressed as compared to $p+p$ and $p+A$ collisions. As we have shown, however, the parton model calculations without parton energy loss fit the experimental data very well. Including the energy loss effect, we found that the large p_T spectra in $Pb+Pb$ collisions at the CERN SPS energy can put very stringent limits on the interaction of energetic partons with dense medium and the induced energy loss. Within the parton model, one can exclude from the observed hadron spectra a parton energy loss larger than $dE_q/dx = 0.01$ GeV/fm and a mean free path shorter than $\lambda_q = 7$ fm. This is much smaller than the most conservative estimate of parton energy loss in a dense medium [4, 5, 6].

There are several implications one can draw from this analysis. Most of the recent theoretical estimates of parton energy loss are based on a scenario of a static and infinitely large dense parton gas. If the system produced in a central $Pb+Pb$ collision only exists for a period of time shorter than the interaction mean free path of the propagating parton, one then should not expect to see any significant parton energy loss. Using the measured transverse energy production $dE_T/d\eta \approx 405$ GeV [22] and a Bjorken scaling picture, one can indeed estimate [23] that the life time of the dense system in central $Pb+Pb$ collisions is only about 2 – 3 fm/c before the density drops below a critical value of $\epsilon_c \approx 1$ GeV/fm^3. Even if we assume that a dense partonic system is formed in central $Pb+Pb$ collisions, this optimistic estimate of the life time of the system could still be smaller than the mean free path of the propagating parton inside the medium. Thus, one does not have to expect a significant effect of parton energy loss on the final hadron spectra at large p_T. Otherwise, it will be difficult to reconcile the absence of parton energy loss with the strong parton interaction which maintains a long-lived partonic system.

Another conclusion one can also make is that the dense hadronic matter which has existed for a period of time in the final stage of heavy-ion collisions does not cause any apparent parton energy loss or jet quenching. A high p_T physical pion from jet fragmentation has a very long formation time. One does not have to worry about its scattering with other soft hadrons in the system which could cause suppression of high p_T pion spectra. We still do not understand the reason why a fragmenting parton does not loss much energy when it propagates through a dense hadronic matter. However, it might be related to the absence of energy loss to the quarks and anti-quarks prior to Drell-Yan hard processes in $p+A$ and $A+A$ collisions. This observation will make jet quenching a better probe of a long-lived partonic matter since one does not have to worry about the complications arising from the hadronic phase of the evolution. If one observes a dramatic suppression of high p_T hadron spectra at the BNL RHIC energy as predicted [3, 15, 18, 19], then it will clearly indicate an initial condition very different from what has been reached at the CERN SPS energy.

Acknowledgments

This work was supported by the Director, Office of Energy Research, Division of Nuclear Physics of the Office of High Energy and Nuclear Physics of the U.S. Department of Energy under Contract Nos. DE-AC03-76SF00098.

References

[1] K.S. Lee, E. Schnedermann, J. Sollfrank, U. Heinz, Nucl.Phys.A525:523c-526c,1991.

[2] M. Gyulassy and M. Plümer, Phys. Lett. **B243**, 432 (1990).

[3] X.-N. Wang and M. Gyulassy, Phys. Rev. Lett. **68**, 1480 (1992).

[4] M. Gyulassy and X.-N. Wang, Nucl. Phys. **B**420, 583 (1994); X.-N. Wang, M. Gyulassy and M. Plümer, Phys. Rev. **D** 51, 3436 (1995).

[5] R. Baier, Yu. L. Dokshitzer, S. Peigné and D. Schiff, Phys. Lett. **B**345, 277 (1995).

[6] R. Baier, Yu. L. Dokshitzer, A. Mueller, S. Peigné and D. Schiff, Nucl. Phys. **B484**, 265 (1997).

[7] X.-N. Wang, Phys. Rev. C **58**, 2321 (1998).

[8] D. C. Hom, *et al.*, Phys. Rev. Lett. **37**, 1374 (1976); D. M. Kaplan, *et al.*, Phys. Rev. Lett. **40**, 435 (1978)

[9] For a review, see J. F. Owens, Rev. Mod. Phys. **59**, 465 (1987).

[10] J. Binnewies, B. A. Kniehl and G. Kramer, Z. Phys. **C**65, 471 (1995).

[11] K. J. Eskola and X.-N. Wang, Int. J. Mod. Phys. A **10**, 3071 (1995).

[12] R. D. Field, *Applications of Perturbative QCD*, Frontiers in Physics Lecture, Vol. 77, Ch. 5.6 (Addison Wesley, 1989).

[13] D. Antreasyan, *et al.*, Phys. Rev. D**19**, 764 (1979).

[14] D. E. Jaffe, *et al.*, Phys. Rev D**40**, 2777 (1989).

[15] X.-N. Wang, nucl-th/9812021

[16] X.-N. Wang and M. Gyulassy, Phys. Rev. D **44**, 3501 (1991); Comp. Phys. Comm. **83**, 307 (1994).

[17] P. B. Straub, *et al.*, Phys. Rev. Lett. **68**, 452 (1992).

[18] X.-N. Wang and Z. Huang, Phys. Rev. C **55**, 3047 (1997).

[19] X.-N. Wang, Z. Huang and I. Sarcevic, Phys. Rev. Lett. **77**, 231 (1996).

[20] R. Albrecht *et al.*, WA80 Collaboration, Eur. Phys. J. C **5**, 255 (1998).

[21] M. M. Aggarwal *et al.*, WA98 Collaboration, nucl-ex/9806004 (to be published).

[22] T. Alber *et al.* (NA49 experiment), Phys. Rev. Lett. **75**, 3814 (1995).

[23] X.-N. Wang, Phys. Rev. Lett. **81**, 2655 (1998).

Chapter 26

THE STAR EMC PROJECT

Gary D. Westfall for the STAR EMC Collaboration
National Superconducting Cyclotron Laboratory and
Department of Physics and Astronomy
Michigan State University
*East Lansing, MI 48824-1321 USA**
westfall@nscl.msu.edu

*Partial funding provided by grant DE-FG02-98ER41010.

Abstract The STAR Electromagnetic Calorimeter (STAR EMC) is a lead-scintillator sampling calorimeter covering $-1 \leq \eta \leq 1$ and $0 \leq \phi \leq 2\pi$ with a segmentation of $(\Delta\eta, \Delta\phi) = (0.05, 0.05)$. The STAR EMC also contains a shower maximum detector at 5 X_o with an effective segmentation of $(\Delta\eta, \Delta\phi) = (0.007, 0.007)$. The STAR EMC will allow STAR to study phenomena in Au-Au collisions at RHIC such as disoriented chiral condensates, jet quenching, and high p_t particles spectra. In p-p and p-Au collisions the STAR EMC will enable the study phenomena such as the gluon structure function of the proton and gluon shadowing. The STAR EMC is essential for the spin physics program in STAR.

Keywords: RHIC, STAR, EMC, calorimeter, quarks, gluons

1. THE STAR EMC

The STAR Electromagnetic Calorimeter (STAR EMC) consists of a barrel calorimeter and an endcap calorimeter.[1] Both the barrel and endcap calorimeters are lead scintillator sampling calorimeters with shower maximum detectors placed at 5 X_o. The barrel calorimeter covers $-1 \leq \eta \leq 1$ and $0 \leq \phi \leq 2\pi$ with a segmentation of $(\Delta\eta, \Delta\phi) = (0.05, 0.05)$. The endcap calorimeter covers $1 \leq \eta \leq 2$ and $0 \leq \phi \leq 2\pi$ and has a segmentation of $(\Delta\eta, \Delta\phi) \leq (0.05, 0.05)$.

The barrel is composed of 120 modules each covering $(\Delta\eta, \Delta\phi) = (1.0, 0.1)$. Each module contains 40 towers each subtending $(\Delta\eta, \Delta\phi) = (0.05, 0.05)$ giving a total of 4800 towers in the barrel calorimeter. The endcap calorimeter will have 720 towers. Schematic drawings of the STAR EMC are shown in figures 26.1 and 26.2.

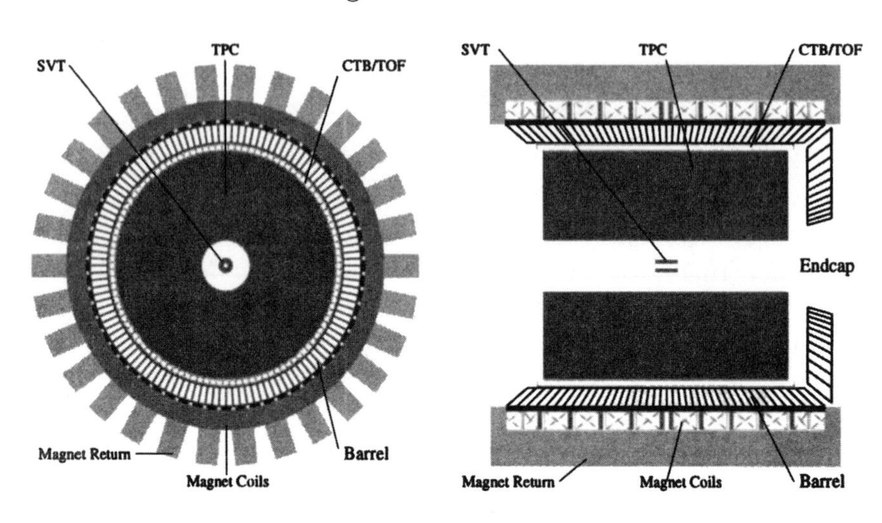

Figure 26.1 Schematic end view of STAR showing the ϕ segmentation of the STAR barrel calorimeter. There are 120 divisions in ϕ each covering $\Delta\phi = 0.05$.

Figure 26.2 Schematic side view of STAR showing the η segmentation of the STAR barrel calorimeter. There are 40 divisions each covering $\Delta\eta = 0.05$.

The towers of the barrel EMC are composed of 21 layers of lead and scintillator corresponding to 19 X_o. The lead is 5 mm thick. The scintillator tiles are 5 mm thick except for the first two tiles which are 6 mm thick. Each tile is read out with a wavelength shifting (WLS) fiber placed in a σ groove. The WLS fibers are coupled to 3.5 m long clear fiber using an optical connector. These clear fibers route the light from the tiles through the STAR magnet structure to the photomultiplier (PMT) boxes located on the magnet return legs. The clear fiber cables are coupled to the PMT box using optical connectors. Inside the PMT box, the fibers from each tower are attached to a PMT tube using a cookie and an air-coupled mixer.

To increase the ability of the STAR EMC to distinguish between electromagnetic particles and hadronic particles, the first two layers of the towers are read out with two fibers. This technique relies on the fact that electromagnetic particle shower earlier in the stack than do hadronic particles. The thickness of the first two tiles is increased from 5 mm to 6 mm to compensate for the loss of light collection efficiency caused by having two fibers in the same groove. A schematic drawing of a tower and its WLS fibers is shown in figure 26.3.

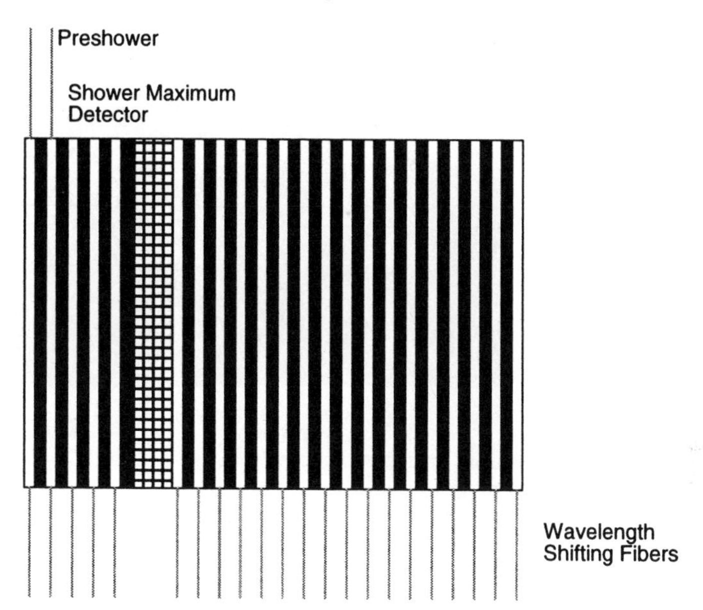

Figure 26.3 Schematic drawing of one tower of the STAR EMC showing the WLS fibers and the read out scheme for the preshower. The shower maximum detector is shown at 5 X_o.

The barrel modules are constructed are constructed using layers of lead and scintillator megatiles. Each scintillator layer consists of two megatiles each making up 20 separate tiles. The megatiles are machined as one piece. Grooves are machined between the tiles and are filled with opaque epoxy. The grooves for the WLS fibers are then machined. The WLS fibers are pushed into the grooves after the layers are stacked. A schematic drawing depicting the megatiles is shown in figure 26.4. The WLS fibers are routed to an optical connector mounted directly to the module. The 3.5 m long clear fibers are connected to the module and routed through the STAR magnet structure to the PMT box.

There are 110,400 tiles in the barrel calorimeter. For each tile, there is a WLS fiber, a 3.5 m long clear fiber, and a 1 m long clear fiber. Thus there are 331,200 separate fibers arranged in 36,000 assemblies.

The optical fiber connectors provide the ability to handle the 1 ton barrel modules without the 100 fiber bundles being connected. The connectors were designed and are used by CDF in the end-plug calorimeter. The connectors hold 10 fibers each. Because there are 23 fibers coming from each tower, 5 connectors are required for each tower pair, 4 con-

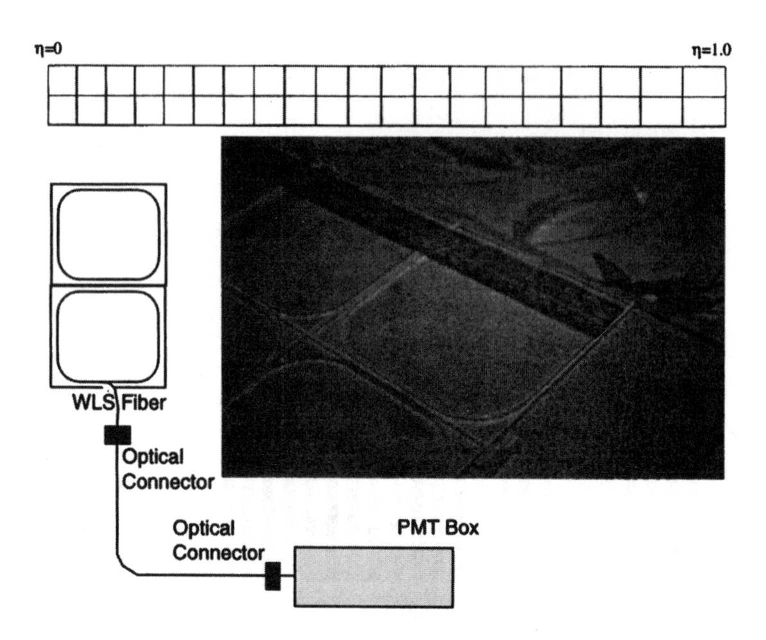

Figure 26.4 Schematic drawing of the last scintillator megatile (top). The photograph shows a machined megatile. On the left and bottom is a schematic drawing of the WLS fibers in the tiles, an optical connector, clear fiber from the WLS fibers to the PMT box, another optical connector, and the clear fiber inside the PMT box.

nectors with 10 fibers each and 1 connector with 6 fibers. Thus there are 100 fiber assemblies for each barrel module. In figure 26.5, the WLS fibers are shown being glued.

The STAR EMC has a shower maximum detector (SMD) installed at 5 X_o. This detector provides high spatial resolution for characterizing showers. The SMD consists of a double sided gas/strip detector. Each module contains one SMD chamber covering $(\Delta\eta, \Delta\phi) = (1.0, 0.1)$. There are 15 wires running the length of the module. A schematic drawing of the SMD is shown in figure 26.6.

One side of the SMD has strips defining the η direction. Each strip covers $\Delta\phi = 0.1$. The strips from $\eta = 0$ to 0.5 are all of the same size. The strips from $\eta = 0.5$ to 1.0 are all the same size. The strips defining the ϕ direction cover $\Delta\eta = 0.1$. There are 15 equal-sized ϕ strips for each η bin. This segmentation leads to 300 strips for each module and 36,000 strips for the entire barrel EMC.

Figure 26.7 shows the first production-style module being assembled at Wayne State University.

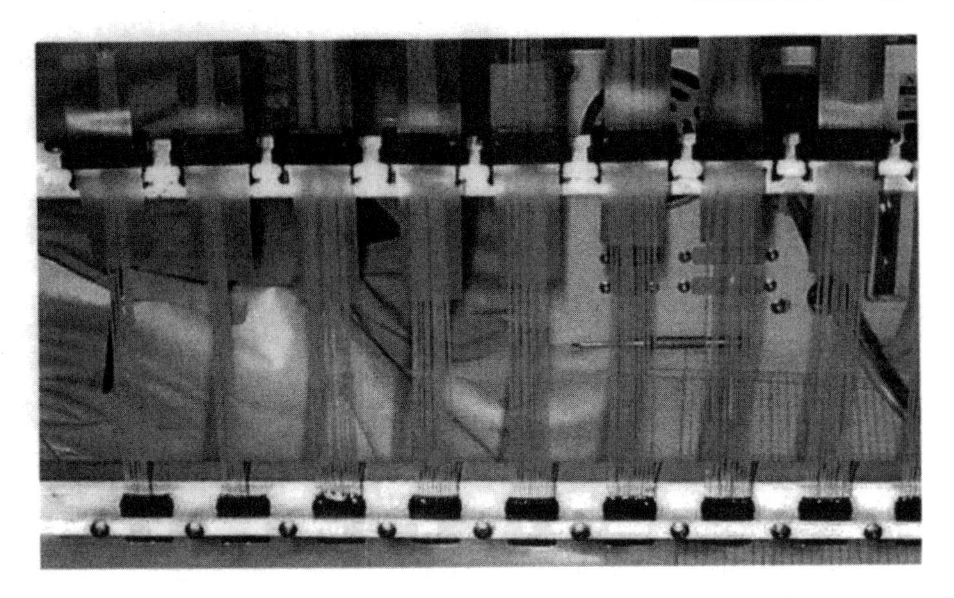

Figure 26.5 Photograph showing the WLS fibers being glued into the optical fiber connectors.

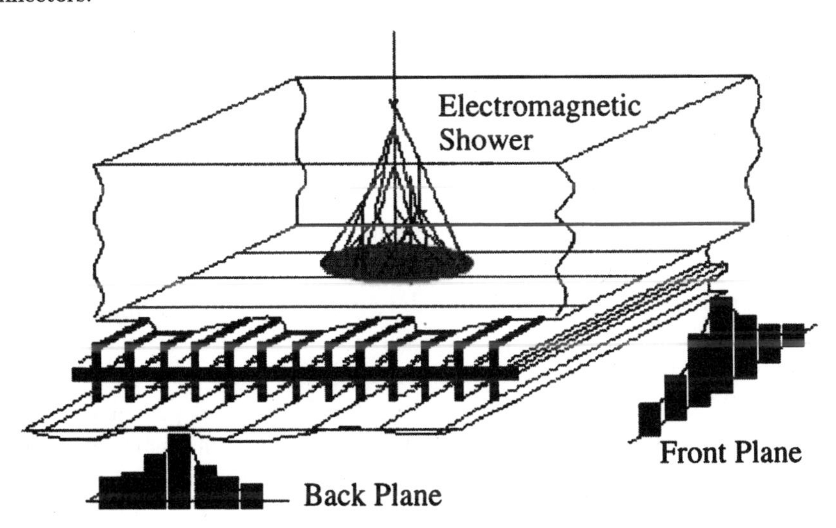

Figure 26.6 Schematic drawing of the STAR EMC shower maximum detector. The detector is installed at 5 X_o. The front and back planes provide high spatial resolution for shower characterization.

2. PHYSICS WITH THE STAR EMC

The STAR EMC provides STAR with a large range of capabilities for Au-Au, p-Au, and p-p interactions. For Au-Au, the EMC provides the

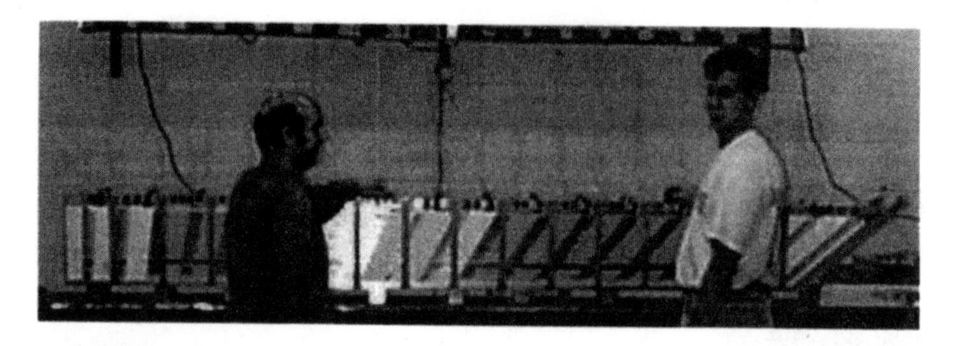

Figure 26.7 Photograph of the first production style STAR barrel EMC module. Visible in the picture are the pseudo-projective towers and shower maximum detector.

ability to study phenomena such as E_t distributions[2], disoriented chiral condensates (DCCs)[3], high p_t π^o spectra, direct photons, and J/ψ suppression. In p-p collisions, the EMC provides access to jets and direct photons through which the quark distributions (structure functions), the gluon distribution functions, and the fragmentation functions can be measured. The spin dependence of these observables will also be measured.

Using the knowledge gained from the p-p measurements, p-Au collisions can be studied and the gluon distributions and gluon shadowing in Au can be studies. Having this basic knowledge in hand, one can then compare with results for moderate to high p_t particle spectra from Au-Au reactions. Thus one can sort out contributions from the superposition of p-p collisions from signals of the quark gluon plasma (QGP).

An example of STAR's ability to detect J/ψ is shown in figure 26.8 in which the mass spectrum of J/ψ's is shown[4]. To detect these particles, the following observables are needed: tracking from the TPC, energy from the EMC, position resolution from the SMD, and preshower information. In addition, the transverse momentum of each electron must be at least 1.5 GeV/c and the transverse momentum of the J/ψ must be greater than 1.2 GeV/c.

3. STAR EMC AS TRIGGER

STAR can record one central Au-Au event per second (more for less central collisions). RHIC will be able to produce 2000 Au-Au reactions per second and 1 million p-p interactions per second. Thus precise triggers are essential for the extraction of any observable at RHIC using STAR. In Au-Au collisions, the STAR EMC can provide triggers for E_t, E_t and isospin fluctuations, high p_t particles, and peripheral collision

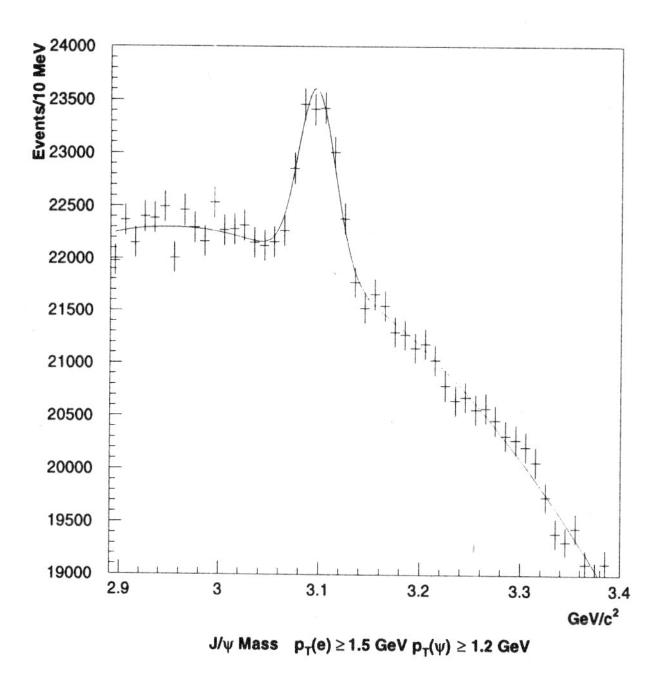

Figure 26.8 Simulated mass spectrum of J/ψ detected using STAR and the STAR EMC.

physics. In p-p and p-Au reactions, the EMC will provide triggers for jets and direct photons.

An example of trigger on isospin fluctuations is shown in figure 26.9 in which the number of charged particles detected by the central trigger barrel (CTB) is plotted against the total energy detected by the EMC[5]. The filled circles represent HIJING model calculations filtered through the STAR acceptance. The open squares represent the same event with 50% of the charged π's changed to π^o's. The open triangles represent a Bose-Einstein condensate in which the momentum of the π's are reduced by 30% and the number of π's is increased to conserve energy. Thus the STAR EMC can easily provide triggers on various types of phenomena related to DCC's and Bose-Einstein condensates.

4. TEST RUN RESULTS

In December, 1998 the first production style STAR EMC module was tested in the Brookhaven B2 Test Beam Facility using electrons, pions, and muons from 0.5 GeV/c to 7.0 GeV/c. Linearity and tile-to-tile variations were checked for 4 towers at $\eta = 0$, 4 towers at $\eta = 0.5$, and

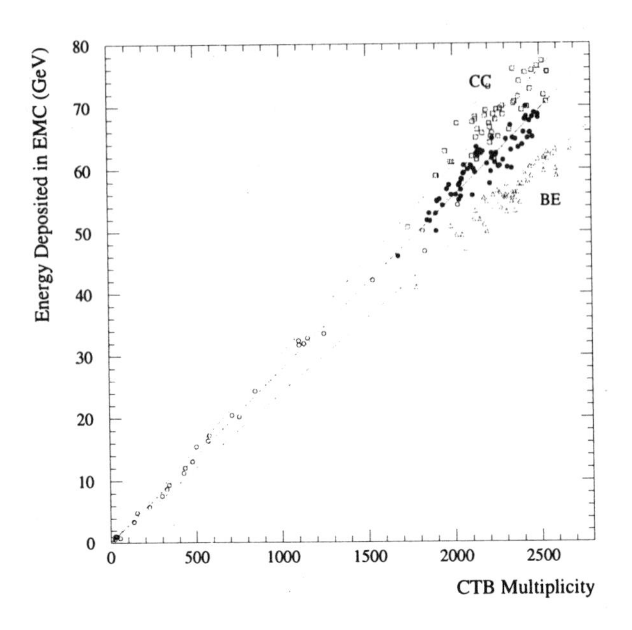

Figure 26.9 Response of STAR using the EMC to plain HIJING events (closed circles). The open squares represent the same events with 50% of the charged pions changed to neutral pions. The open triangles show the same events in which the momenta of the pions is decreased by 30% while the number of pions is increased to conserve energy.

4 towers at $\eta = 1.0$. In addition a production style SMD chamber and its front-end electronics were also successfully tested.

In figure 26.10 the response of tower R1T2 (near $\eta = 0$) is shown for electrons with energies from 0.5 to 7 GeV. The response of the tower is clearly very linear.

Acknowledgments

This work was done in collaboration with the STAR Collaboration and the STAR EMC Collaboration. This work was partially supported by DOE Grant DE-FG02-98ER41070.

References

[1] The STAR EMC Technical Design Report, May, 1998.

[2] T. Alber et al. (NA49), Phys. Rev. Lett. **75**, 3814 (1995).

[3] J.D. Bjorken, Acta Physica Polonica **B23**, 637 (1992).

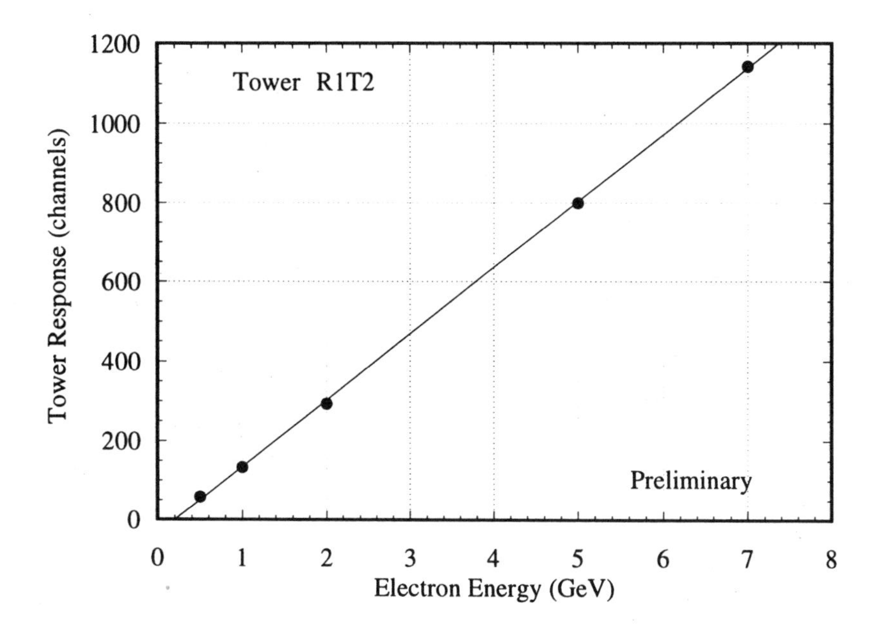

Figure 26.10 Reponse of STAR using the EMC to plain HIJING events (closed circles). The open squares represent the same events with 50% of the charged pions changed to neutral pions. The open triangles show the same events in which the momenta of the pions is decreased by 30% while the number of pions is increased to conserve energy.

[4] Tom LeCompte, STAR Note 368.

[5] B. Hubner, A.M. Vander Molen, and G.D. Westfall, STAR Note 275.

Chapter 27

PION INTERFEROMETRY IN AU+AU COLLISIONS AT THE AGS

J.H. Lee
Physics Department
Brookhaven National Laboratory
Upton, NY 11973 USA
jhlee@bnl.gov

for the E866 Collaboration

Abstract Two-pion Bose-Einstein correlations have been studied using the BNL-E866 Forward Spectrometer in 11.6 A·GeV/c Au + Au collisions. The data were analyzed using three-dimensional correlation parameterizations to study transverse momentum-dependent source parameters. The freeze-out time and the duration of emission were derived from the source radii parameters.

Keywords: AGS, Au+Au, pion correlation

1. INTRODUCTION

Pion interferometry measures an enhancement at small relative momenta from Bose-Einstein symmetrizations of two identical pions, which leads to information on the space-time structure of the particle-emitting source in heavy ion collisions [1]. The dynamical nature of collisions limits the Bose-Einstein Correlation measurements to the extent of space-time "region of homogeneity" from which bosons of similar momenta are emitted. Thus correlation studies as s function of transverse momentum enable one to obtain a more complete picture of the dynamical space-time structure of the source [2].

Advances in Nuclear Dynamics, 5,
Edited by Bauer and Westfall, Kluwer Academic / Plenum Publishers, New York, 1999.

2. EXPERIMENT

The beam of 11.6 A·GeV/c Au ions from the BNL Tandem-AGS complex was incident normally upon a Au target. Data were taken using the

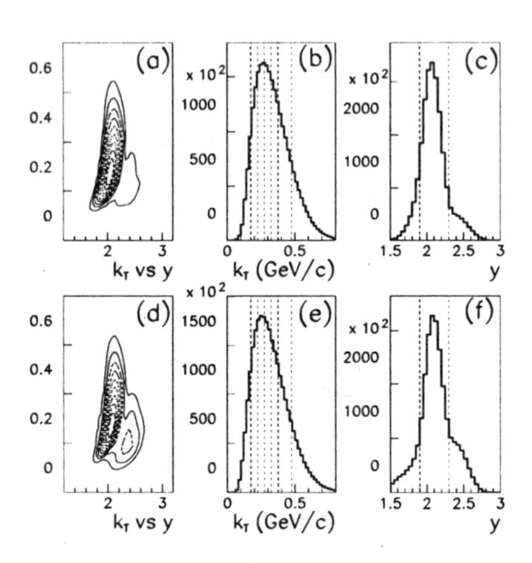

Figure 27.1 The distribution of the average transverse momentum (K_T) vs. the rapidity (y) and their projections for the $\pi^+\pi^+$ (a,b,c) and $\pi^-\pi^-$ (d,e,f) pairs. The dotted lines in figures (b) and (e) represent the K_T bins, and the dotted lines in figures (c) and (f) show the rapidity regions used in the analysis.

E866 Forward Spectrometer [3]. which covers the forward rapidity region from 6 degrees to 28 degrees to the beam by rotation of the spectrometer. Particle momentum measurements were done with two tracking stations, each of which station consists of a TPC and drift chambers, located fore and aft of a dipole magnet. A Time-of-Flight wall was used for particle identification up to a particle momentum of 4 GeV/c for pions.

The total accumulations of analyzed $\pi^+\pi^+$, $\pi^-\pi^-$, and $\pi^+\pi^-$ pairs are 1.5×10^6, 1.5×10^6, and 2.5×10^6, respectively. The 24% of the total interaction cross section events gated by the New Multiplicity Array [4] were selected for the analysis. Fig. 27.1 shows the rapidity, y, and the average transverse momentum, K_T, distributions for $\pi^+\pi^+$ and $\pi^-\pi^-$ pairs. The K_T used for the analysis is from 0.175 GeV/c to 0.475 GeV/c in 5 bins, corresponding to $< K_T > = 0.2$, 0.25, 0.3, 0.35, and 0.425, respectively, as displayed in Fig. 27.1. The rapidity range of pairs was limited to $1.9 \leq y_{\pi\pi} \leq 2.3$ for this analysis.

3. CORRELATION FUNCTION PARAMETERIZATION

The correlation function C_2 is defined as the ratio of the two-particle probability to the product of the single-particle probabilities for particles with momenta $\boldsymbol{p}_i$. Experimentally it is given by the normalized ratio of the number of correlated (signal) pairs S to the number of uncorrelated (background) pairs B:

$$C_2(\boldsymbol{q}, \boldsymbol{K}) = \frac{P(\boldsymbol{p}_1, \boldsymbol{p}_2)}{P(\boldsymbol{p}_1)P(\boldsymbol{p}_2)} = \mathcal{N} \cdot \frac{S}{B \cdot w_{Coulomb}}, \tag{27.1}$$

where $(q_0, \boldsymbol{q}) = (E_1 - E_2, \boldsymbol{p}_1 - \boldsymbol{p}_2)$, $\boldsymbol{K} = \frac{\boldsymbol{p}_1 + \boldsymbol{p}_2}{2}$ denote the relative and average momentum of the pair, respectively, and $\mathcal{N}$ is an overall normalization constant which holds no physical significance. The signal pairs are from the same events and the background pairs are calculated by mixing different events with the same experimental conditions. The background pairs were constructed with 20 times more statistics than the number of correlated pairs, and weighted with the Coulomb factor $w_{Coulomb}$ to remove the correlation created by the Coulomb effect in the correlated pairs.

For the three-dimensional parameterization, the data were fit with the Yano-Koonin-Potgoretskii (YKP) [5, 6] scheme which is a function of the space-time components $(R_\perp, R_\parallel, R_0)$ and the average longitudinal velocity v:

$$C_2(\boldsymbol{q}, \boldsymbol{K}) = \mathcal{N}[1 + \lambda e^{-R_\perp^2(\boldsymbol{K})q_\perp^2 - R_\parallel^2(\boldsymbol{K})(q_\parallel^2 - q_0^2) - (R_0^2(\boldsymbol{K}) + R_\parallel^2(\boldsymbol{K}))(q \cdot U(\boldsymbol{K}))^2}]. \tag{27.2}$$

Here $q_\perp = \sqrt{q_{side}^2 + q_{out}^2}$ and $q_\parallel$ is the longitudinal component. The four-velocity U has only a longitudinal spatial component:

$$U(\boldsymbol{K}) = \gamma(\boldsymbol{K})(1, 0, 0, v(\boldsymbol{K})), \text{ with } \gamma = \frac{1}{\sqrt{1 - v^2}}. \tag{27.3}$$

The data have been also parameterized with the Bertsch-Pratt (BP) Cartesian decomposition [7, 8]:

$$C_2(\boldsymbol{q}, \boldsymbol{K}) = \mathcal{N}[1 + \lambda e^{-R_{long}^2(\boldsymbol{K})q_{long}^2 - R_{side}^2(\boldsymbol{K})q_{side}^2 - R_{out}^2(\boldsymbol{K})q_{out}^2 - 2R_{out,long}^2(\boldsymbol{K})q_{out}q_{long}}] \tag{27.4}$$

where λ is a chaoticity parameter, with λ=0 corresponding to a totally coherent and λ=1 corresponding to a totally chaotic source. The longitudinal component of $\boldsymbol{q}$, q_{long} is parallel to the beam, q_{side} is perpendicular both to the beam and to the average momentum, and q_{out}

is orthogonal to q_{side} and q_{long}. The source radius parameters R_{long}, R_{side}, and R_{out} correspond to three dimensional radius parameters in the (q_{long},q_{side},q_{out}) space. The "out-longitudinal" cross term [9] $R^2_{out,long}$ is included in the function to accommodate the rotation of the q_{out}-q_{long} plane. The correlation functions, Eq. 27.2 and Eq. 27.4, are evaluated in the Longitudinally Co-Moving System (LCMS), which is the longitudinally boosted frame where the average pair momentum $\boldsymbol{K}$ has no longitudinal component, $\beta_L = 0$.

4. ANALYSIS

The final state Coulomb interaction between two pions was obtained by an iterative method which takes into account the pion source size [10] and the momentum resolution. The resolution functions were calculated from the Monte-Carlo simulations of the detector system using GEANT [11] and the data reconstruction chain. Fig. 27.2 shows Coulomb correction factors for two pions with and without the momentum resolution smearing. For comparison, the Gamov function which assumes a point-like source is also displayed in the figure. To test the quality of the Coulomb corrections, the correction functions are applied to $\pi^+\pi^-$ correlations which are dominated by the mutual Coulomb interactions for small q. It is demonstrated in Fig. 27.3 that the $\pi^+\pi^-$ correction function becomes close to unity after applying the Coulomb correction function calculated for a finite source and folded with the experimental momentum resolutions. No correction is applied for Coulomb effects between the pair and the remaining particles.

Fitting the data to the correlation functions was done by minimizing the logarithm of the likelihood function:

$$-\ln \mathcal{L} = \sum_i^n [C_i B_i - S_i \ln(C_i B_i) + \ln(S_i!)], \qquad (27.5)$$

where C_i is the correlation function, S_i is the number of signal counts in the i-th three dimensional $\boldsymbol{q}$ bin, and B_i is the number of background events in the bin weighted by a Coulomb factor. The minimization is done using MINUIT [12].

The resolution factors contribute to an underestimation of the source parameters. The systematic uncertainties of the fitted radius parameter are estimated to be less than 10%. No correction for this effect has been applied to the data.

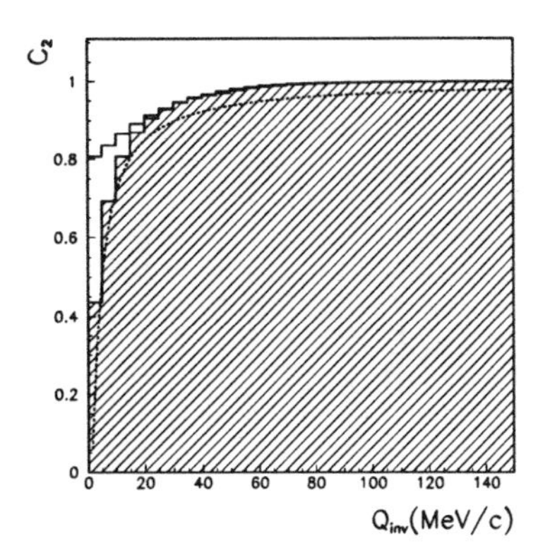

Figure 27.2 Coulomb correction functions without (shaded distribution) and with (unshaded) momentum resolution. The Gamov function is shown by the dotted curve.

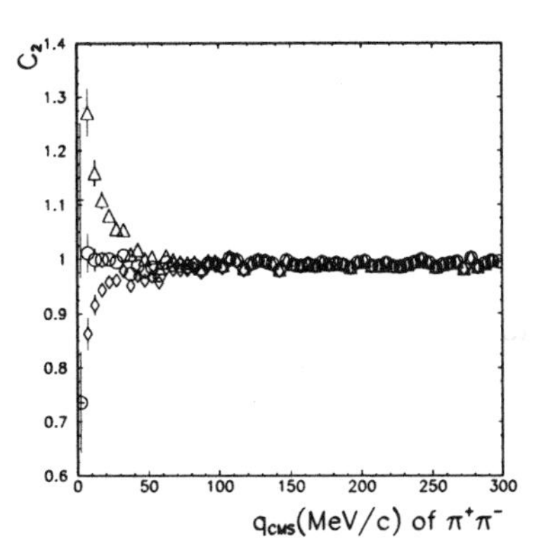

Figure 27.3 $\pi^+\pi^-$ correlation functions C_2 as a function of $|q|$ (in CMS system): The raw correlation ($\triangle$), correlations corrected with the Coulomb function ($\diamond$) and corrected for Coulomb and momentum resolution ($\bigcirc$) are shown.

5. RESULTS

Multi-dimensional source radius parameters for the five K_T bins used in the analysis are shown in Fig. 27.4 for the YKP parameterization, and for the BP parameterization in Fig. 27.5. The regions fitted for each point are indicated by lines in the horizontal direction. The longitudinal component $R_{||}$ (and also R_{long}) features a clear K_T dependence. Since the slope of the decrease is expected to grow with the expansion rate of the source [2, 13, 14, 15], it can be interpreted as an indication of a longitudinal expansion of the source before freeze-out. To quantify this systematic decrease, $R_{||}$ values are fit with $R_{||} \propto 1/m_T^{\alpha}$, where $m_T = \sqrt{K_T^2 + m_\pi^2}$, since the approximate analytic "m_T scaling" ($R_{||} \propto 1/\sqrt{m_T}$) from the hydro-dynamical models [13, 14] has been commonly used as a basis for experimental and theoretical characterization of the dependence of the source radii on the transverse momentum. Fitted values of α for $R_{||}$ in Fig. 27.4 are 0.75 ± 0.04 (with $\chi^2 = 1.4$) for $\pi^+\pi^+$, and 0.57 ± 0.09 ($\chi^2 = 0.29$) for $\pi^-\pi^-$, which indicates that the K_T-dependence of $R_{||}$ is similar to or stronger than the $1/\sqrt{m_T}$ behavior. The values of $R_\perp$ show a weaker and less clear K_T-dependency compared to $R_{||}$. Fitting $R_\perp$ to $1/m_T^{\alpha}$ for a quantitative comparison with $R_{||}$ results in $\alpha = 0.26 \pm 0.07$ ($\chi^2 = 0.7$) for $\pi^+\pi^+$, and 0.15 ± 0.08 ($\chi^2 = 2.0$) for $\pi^-\pi^-$. This may support the interpretation that the

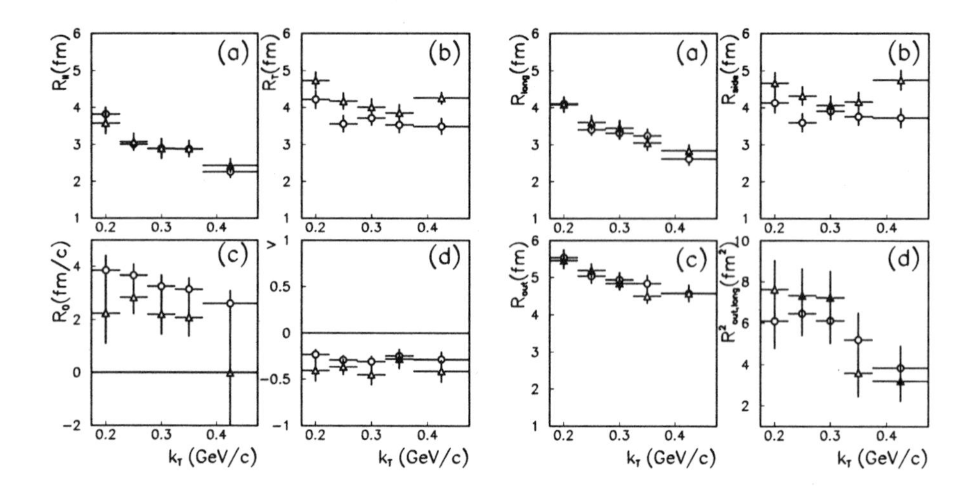

Figure 27.4 Fitted radius parameters for the YKP parameterization for $R_{\|}$ (a), $R_{\perp}$ (b), R_0 (c), and v (d). Circles (○) are shown for $\pi^+\pi^+$ and triangles (△) for $\pi^-\pi^-$. The error bars are statistical only.

Figure 27.5 Fitted radius parameters for the BP parameterization for R_{long} (a), R_{side} (b), R_{out} (c), and $R_{out,long}$ (d). Circles (○) are shown for $\pi^+\pi^+$ and triangles (△) for $\pi^-\pi^-$. The error bars are statistical only.

source undergoes some transverse expansion [15] as well as longitudinal expansion. The geometrical source parameter value $R_{\perp} \approx 4.5$ fm at small K_T corresponds to an r.m.s. radius of $r_{\text{r.m.s.}} = \sqrt{3}R_{\perp} \approx 7.8$ fm to be compared with an r.m.s. charge radius of Au nuclei measured by electron scattering [16] of 5.33 fm. The $r_{\text{r.m.s.}}$ values from the correlation measurement give larger radii than the Au r.m.s. radius. The difference can be understood as a transverse expansion contribution to the geometrical radius $R_{\perp}$. The source size differences between $\pi^+\pi^+$ and $\pi^-\pi^-$ are only apparent for the $R_{\perp}$ dimension, which qualitatively agrees with nuclear Coulomb field effects [17]. The $R_{\perp}$ values show a reasonable overall agreement with R_{side} within systematic uncertainties.

One can derive an estimate of the approximate average freeze-out time τ_0 by assuming a purely longitudinally expanding system without transverse flow from [13, 14]:

$$R_{\|} = \tau_0 \sqrt{\frac{T}{m_T}}, \qquad (27.6)$$

where T is a freeze-out temperature. By assuming [18] $T = 130$ MeV, the freeze-out time is $\tau_0 \approx 5$ fm/c at $< K_T > = 200$ MeV/c, and $\tau_0 \approx 4.5$ fm/c at $< K_T > = 400$ MeV/c.

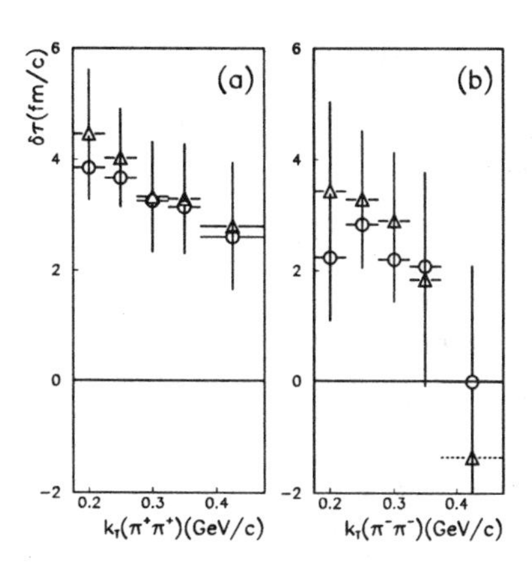

Figure 27.6 $\delta\tau$ calculated from the BP parameterization and YKP parameterizations for $\pi^+\pi^+$ (a) and $\pi^-\pi^-$ (b) as functions of K_T are shown with triangles ($\triangle$). The last point (dotted line) in the (b) panel has $R_{side} > R_{out}$: thus an absolute value was taken for the square root in Eq. 27.7. For a comparison, R_0 values from YKP parameterization are also shown in the figures with circles ($\bigcirc$).

An estimate of the duration of the freeze-out, $\delta\tau$ can be provided by the temporal parameter R_0, or can be obtained [19] from the difference of R^2_{out} and R^2_{side} from the BP parameterization,

$$\delta\tau = \frac{1}{\beta_\perp}\sqrt{R^2_{out} - R^2_{side}}, \tag{27.7}$$

where $\beta_\perp$ is the transverse velocity of the pair. The duration of emission given by both parameterizations are $\delta\tau \approx 2 - 4$ fm/c showing $\delta\tau \lesssim R_\perp$. This result excludes the possibility of a long-lived state created from a phase transition [20] at the AGS energy. It is shown that the $\delta\tau$ values for $\pi^+\pi^+$ are larger than for $\pi^-\pi^-$, and they increase at small K_T for both $\pi^+\pi^+$ and $\pi^-\pi^-$. Those systematic features of the effective lifetime are consistent with both parameterizations within the systematic uncertainties as shown in Fig. 27.6.

6. SUMMARY

Preliminary results on the transverse momentum-dependent source size measurement for $\pi^+\pi^+$ and $\pi^-\pi^-$ in Au+Au collisions using the E866 Forward Spectrometer at the AGS have been presented. Multi-dimensional Bose-Einstein correlation functions parameterized by the Yano-Koonin-Potgoretskii method show a rapid decrease of $R_{||}$ and a slower and less clear decrease of $R_\perp$ as the transverse momentum increases. This may indicate a strong longitudinal and a moderate transverse expansion before freeze-out. An approximate freeze-out time was estimated as $\tau_0 \approx 4.5 - 5$ fm/c with the duration of emission $\delta\tau \approx 2 - 4$ fm/c. These results agree with a Bertsch-Pratt parameterization within the systematic uncertainties.

Acknowledgments

This work was supported by the U.S. Department of Energy under contracts with BNL (DE-AC02-98CH10886), Columbia University (DE-FG02-86-ER40281), LLNL (W-7405-ENG-48), MIT (DE-AC02-76ER03069), UC Riverside (DE-FG03-86ER40271), and by NASA (NGR-05-003-513), under contract with the University of California, and by Ministry of Education and KOSEF in Korea, and by the Ministry of Education, Science, and Culture of Japan.

References

[1] See D.H. Boal, C.G. Gelbke, and B. Jennings, *Rev. Mod. Phys.* **62**, 553 (1990) for a review.

[2] U.A. Wiedemann, P. Scotto, and U. Heinz, *Phys. Rev.* C **53**, 918 (1996).

[3] L. Ahle *et al.*, *Phys. Rev.* C **57**, R466 (1998).

[4] L. Ahle, Ph.D. Thesis, MIT (1997).

[5] F. Yano and S. Koonin, *Phys. Lett.* B **78**, 556 (1978).

[6] S. Chapman, J.R. Nix and U. Heinz, *Phys. Rev.* C **52**, 2694 (1995).

[7] S. Pratt, *Phys. Rev.* D **33**, 1314 (1986).

[8] G. Bertsch, *Nucl. Phys.* A **498**, 173c (1989).

[9] S. Chapman, P. Scotto and U. Heinz, *Nucl. Phys.* B **52**, 2694 (1995).

[10] M. Baker *et al.*, *Nucl. Phys.* A **610**, 213c (1996).

[11] CERN Computing Division, CERN Program Library Long Writeup W5013.

[12] F. James, CERN Program Library Long Writeup D506.

[13] A.N. Makhlin and Yu.M. Sinyukov, *Z. Phys.* C **39**, 69 (1988).

[14] J. Bolz *et al.*, *Phys. Lett.* B **300**, 404 (1993).

[15] Wu Y.-F. *et al.*, *Eur. Phy. J.* C **1**, 599 (1998).

[16] M.A. Preston and R.K. Bhaduri, *Structure of the Nucleus*, p.99, Addison-Wesley, Massachusetts (1975).

[17] H.W. Barz, nucl-th/980827.

[18] J. Stachel, *Nucl. Phys.* A **610**, 509c (1996).

[19] H. Heiselberg, *Phys. Lett.* B **379**, 27 (1996).

[20] D.H. Rischke and M. Gyulassy, *Nucl. Phys.* A **608**, 479 (1996).

Chapter 28

BEAM ENERGY DEPENDENCE OF TWO-PROTON CORRELATIONS AT THE AGS

Sergei Y. Panitkin,[7] for the E895 Collaboration
panitkin@sseos.lbl.gov

N.N. Ajitanand,[12] J. Alexander,[12] M. Anderson,[5] D. Best,[1] F.P. Brady,[5] T. Case,[1] W. Caskey,[5] D. Cebra,[5] J. Chance,[5] P. Chung,[12] B. Cole,[4] K. Crowe,[1] A. Das,[10] J. Draper,[5] M. Gilkes,[11] S. Gushue,[2] M. Heffner,[5] A. Hirsch,[11] E. Hjort,[11] L. Huo,[6] M. Justice,[7] M. Kaplan,[3] D. Keane,[7] J. Kintner,[8] J. Klay,[5] D. Krofcheck,[9] R. Lacey,[12] M. Lisa,[10] H. Liu,[7] Y. Liu,[6] R. McGrath,[12] Z. Milosevich,[3], G. Odyniec,[1] D. Olson,[1] C. Pinkenburg,[12] N. Porile,[11] G. Rai,[1] H.-G. Ritter,[1] J. Romero,[5] R. Scharenberg,[11] L. Schroeder,[1] B. Srivastava,[11] N. Stone,[2] T.J.M. Symons,[1] S. Wang,[7] J. Whitfield,[3] T. Wienold,[1] R. Witt,[7] L. Wood,[5] X. Yang,[4] W. Zhang,[6] Y. Zhang[4]

[1] *Lawrence Berkeley National Laboratory, Berkeley, California 94720*
[2] *Brookhaven National Laboratory, Upton, New York 11973*
[3] *Carnegie Mellon University, Pittsburgh, Pennsylvania 15213*
[4] *Columbia University, New York, New York 10027*
[5] *University of California, Davis, California 95616*
[6] *Harbin Institute of Technology, Harbin 150001, P. R. China*
[7] *Kent State University, Kent, Ohio 44242*
[8] *St. Mary's College of California, Moraga, California 94575*
[9] *University of Auckland, Auckland, New Zealand*
[10] *The Ohio State University, Columbus, Ohio 43210*
[11] *Purdue University, West Lafayette, Indiana 47907*
[12] *State University of New York, Stony Brook, New York 11794*

Abstract First measurements of the beam energy dependence of the two proton correlation function in central Au+Au collisions are performed by the E895 Collaboration at the BNL AGS. No significant changes with beam

Advances in Nuclear Dynamics, 5,
Edited by Bauer and Westfall, Kluwer Academic / Plenum Publishers, New York, 1999.

energy were observed. The imaging technique of Brown-Danielewicz is used in order to extract information about the space-time content of the proton source at freeze-out. Extracted source functions show peculiar enhancement at low relative separation.

Keywords: heavy-ion collisions, two-particle correlations

1. INTRODUCTION

Two-particle correlations are widely considered to be a valuable tool in extracting information about the space-time extent of the system created in the collisions of heavy ions [1, 2, 4, 3]. The complex nature of the heavy-ion reaction requires utilization of different particle species in order to obtain reliable and complete picture of the system created in the collision. The majority of the existing experimental two-particle correlation data in ultra-relativistic heavy-ion collisions was obtained using mesons as a probe. The data on baryon correlations is sparse at best. Since the physics of heavy-ion collisions in the beam energy range between 1 and 11 AGeV is dominated by baryons and baryon resonances, the information related to the space-time extent of the baryon source, obtained via two-proton correlations is clearly very interesting. In this paper we present preliminary results of the first measurement of the beam energy dependence of the two-proton correlation function in the central Au+Au collisions at 2,4,6 and 8 AGeV performed by the E895 Collaboration at the Brookhaven National Lab (BNL) Alternating Gradient Synchrotron (AGS). Preliminary results of the pion correlation analysis were published elsewhere [5].

2. EXPERIMENTAL DETAILS

E895 is a fixed target experiment at the BNL AGS. The goal of E895 is to study multiparticle correlations and particle production with Au beams incident on a variety of targets, over a range of AGS energies. More information about the E895 experimental setup can be found elsewhere [6, 7]. We will describe in the following only detailes relevant to the presented analysis.

Beams of gold ions (^{197}Au) were available at different energies - 2,4,6 and 8 AGeV. They were used to bombard targets of different materials- Be, Cu, Ag and Au. Charged particles produced in the collision were detected with time projection chamber (TPC) [6], positioned in side the MPS magnet, and multi-sampling ionization chamber (MUSIC) [7] located downstream from the magnet. For the presented results only information from the TPC was used. The time projection chamber is filled with P10 gas and has rectangular fiducial volume which is about 150 *cm* long, 75 *cm* high, and 100 *cm* wide. The ionization produced by charged particles in the chamber is detected by a segmented cathode plane at the bottom of the TPC. The cathod plane has 15360(120 by 128) pads. The dimension of the pads are 0.8 *cm* by 1.2 *cm*. The signal from each pad is sampled 140 times at 10 MHz by a 12 bit flash ADC yielding more than 2 millions pixels per event used for track reconstruction. The TPC was capable of detecting and tracking software of reconstructing up to several hundreds tracks per event. The magnetic field of the MPS magnet was typically 0.75 or 1 Tesla. Particle identification was performed via simultaneous measurement of particle momentum and specific ionization in the TPC gas. It was possible to resolve positively charged particles up to charge 6 and obtain reliable identification of protons up to 0.9 GeV/c in momentum.

3. DATA ANALYSIS

Good momentum resolution and good particle identification capabilities together with high reconstructed charged particle multiplicity allowed the performance of two-particle correlation studies. In order to obtain the two-proton correlation function C_2 experimentally, the mixed event technique was used. We employ the following definition of the correlation function

$$C_2(q_{inv}) = \frac{N_{tr}(q_{inv})}{N_{bk}(q_{inv})} \quad , \tag{28.1}$$

where

$$q_{inv} = q = \frac{1}{2}\sqrt{-(p_1^\mu - p_2^\mu)^2} \tag{28.2}$$

is the half relative invariant momentum between the two identical particles with four-momenta p_1^μ and p_2^μ. The quantities N_{tr} and N_{bk} are the "true" and "background" two-particle distributions obtained by selecting particles from the same and different events, respectively. Before calculating the correlation function, several cuts are applied. In order to insure a reliable particle identification and high purity of the proton sample, a cut on proton longitudinal momentum $P_z < 800$ MeV/c is applied. Contamination of the identified proton sample by other particles, in this momentum interval, was estimated to be less than 2%. Event centrality selection is based on a reconstructed charged particles multiplicity. For the present analysis events are selected with a multiplicity cut corresponding to the upper 5% of the inelastic cross section for the Au+Au collisions. Single proton tracks are required to satisfy certain quality cuts. Number of hits belonging to the track should be greater than 20, thus suppressing short tracks from delta electrons and remnants of the split tracks. Track should point into vicinity of the event vertex with distance of closest approach (DCA) less than 2.5 cm. Tracks should be properly reconstructed by the tracking code with the corresponding χ^2 per degree of freedom less than 1.5. Tracks are required to be reconstructed from fairly continuous sequence of hits with no significant hit losses, the fraction of hits assigned to the track should exceed 50% of the theoretically available number for the corresponding trajectory inside the fiducial volume of the TPC. This cut has been shown to be effective in suppression of the track splitting effects in the correlation analysis [5]. In order to suppress effects of track merging, a cut on angular separation of two tracks was imposed. For pairs from both "true" and "background" distributions the angle between two tracks was required to be greater than 3 degrees. Figure 28.1 shows measured two-proton correlation functions for Au+Au central collisions at 2,4,6 and 8 AGeV. Within currently available statistical accuracy no significant changes of the measured correlation functions with beam energy were observed.

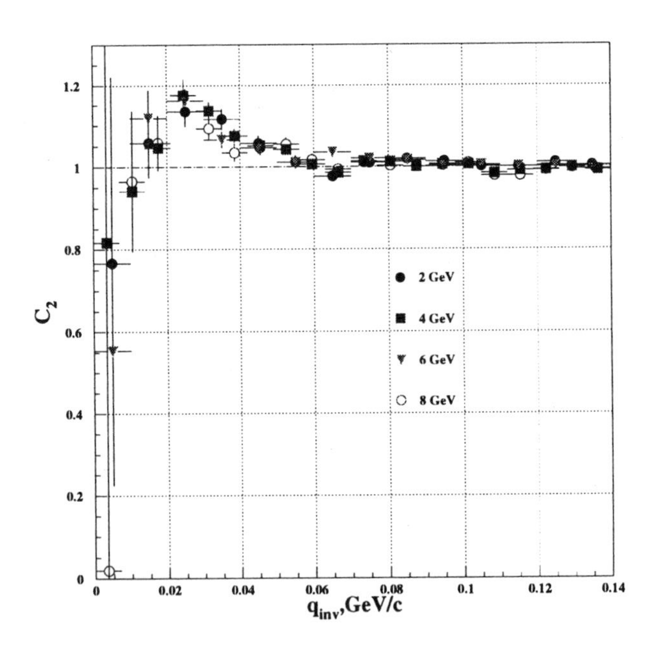

Figure 28.1 Measured two-proton correlation functions for Au+Au central collisions at different beam energies.

4. SOURCE IMAGING

The so called source imaging technique of Brown-Danielewicz was used to extract information about the space-time extent of the proton source. Here we will give just a brief sketch of the method, see Refs [8, 9], for a more detailed description. The two-particle correlation function may be expressed in the following way:

$$C_{\mathbf{P}}(\mathbf{q}) = \frac{dN_2/d\mathbf{p}_1\, d\mathbf{p}_2}{(dN_1/d\mathbf{p}_1)\,(dN_1/d\mathbf{p}_2)} \simeq \int d\mathbf{r}\, |\Phi_{\mathbf{q}}^{(-)}(\mathbf{r})|^2\, S_{\mathbf{P}}(\mathbf{r}). \tag{28.3}$$

$S_{\mathbf{P}}(\mathbf{r})$ is the distribution of relative separation of emission points for the two particles, in their center of mass and $\Phi_{\mathbf{q}}^{(-)}(\mathbf{r})$ is a relative wave function. Using single-particle sources,

$$S_{\mathbf{P}}(\mathbf{r}) = \int d\mathbf{R}\, dt_1\, dt_2\, \overline{D}(0, \mathbf{R} + \mathbf{r}/2, t_1)\, \overline{D}(0, \mathbf{R} - \mathbf{r}/2, t_2)\,. \tag{28.4}$$

where $\overline{D}$ is an averaged distribution of freeze-out points of the particles. In the proton-proton case, the angle and spin averaged relative wave function can be expressed as

$$|\Phi_{\mathbf{q}}^{(-)}(\mathbf{r})|^2 = \frac{1}{2} \sum_{js\ell\ell'} (2j+1)\, \left(g_{js}^{\ell\ell'}(r)\right)^2, \tag{28.5}$$

where $g_{js}^{\ell\ell'}$ is the radial wave function with outgoing asymptotic angular momentum ℓ, which can be calculated numerically given a particular description of the final state interaction. In the present analysis, the proton relative wave functions were calculated by solving the Schrödinger equation with the REID93[10] and Coulomb potentials. The imaging method is concerned with the determination of the relative source function ($S_{\mathbf{P}}(\mathbf{r})$ in Eq. 28.3) knowing $C_{\mathbf{P}}(\mathbf{q})$. Taking into account that the nontrivial part of the correlation function is deviation from unity, one may rewrite Eq. 28.3 in the following way

$$C_{\mathbf{P}}(\mathbf{q}) - 1 = \int d\mathbf{r}\, \left(|\Phi_{\mathbf{q}}^{(-)}(\mathbf{r})|^2 - 1\right)\, S_{\mathbf{P}}(\mathbf{r}) = \int d\mathbf{r}\, K(\mathbf{q}, \mathbf{r})\, S_{\mathbf{P}}(\mathbf{r}), \tag{28.6}$$

where $K = |\Phi_{\mathbf{q}}^{(-)}|^2 - 1$. The problem of imaging then reduces to the more general problem of inversion [11] of K in Eq. 28.6.

Figure 28.2 shows the relative distribution of emission points of protons for central Au+Au collisions at 2,4,6 and 8 AGeV obtained as a result of the application of the imaging technique described above. In order to check the quality of the imaging and numerical stability of the

Table 28.1 Fit parameters for gaussian and exponential parameterizations of the relative source functions for different beam energies. See description in the text.

E_b (AGeV)	λ Exp.	R Exp.(fm)	λ Gauss.	R Gauss.(fm)	R_N (fm)
2.0	1.37	6.85	0.98	9.00	18.7
4.0	1.27	6.95	0.87	9.45	21.2
6.0	0.97	5.61	0.69	7.88	18.7
8.0	1.10	5.66	0.79	7.89	19.2

inversion procedure the two-proton correlation functions are calculated using the relative source functions shown on Figure 28.2 as an input in Equation 28.3. The result of such "double inversion" procedure is shown on Figure 28.3 for the beam energy 4 AGeV. The agreement between the measured and reconstructed correlation function is quite good.

It can be seen from Figure 28.2 that the relative proton source functions have similar shapes at all measured energies. Extracted source functions show enhancement at low relative separation which may be induced by momentum position correlations in the source, possibly due to collective flow. Further investigation and understanding of the origin of this enhancement is clearly needed. Even though the source functions have a non-trivial overall shape, the tail of the relative source function may be described by gaussian(28.7) or exponential (28.8) forms:

$$S(r) = \frac{\lambda}{(2\pi R^2)^{3/2}} exp(-r^2/2R^2) \tag{28.7}$$

$$S(r) = \frac{\lambda}{2R^3} exp(-r/R) \tag{28.8}$$

Parameter λ is a so called generalized chaoticity parameter [8] defined as:

$$\lambda(r_N) = \int_{r<r_N} d\mathbf{r}\, S(\mathbf{r}) \tag{28.9}$$

and has the meaning of an integral of the source over a region ($r < r_N$) where the distribution is significant. Results of the fit of the tail of the correlation function using these two parameterizations are shown at Figure 28.4. Fit parameters are presented in Table 28.1. At the current level of precision of the data both parameterizations provide an adequate description of the relative source function, except for the separations smaller than 2 fm.

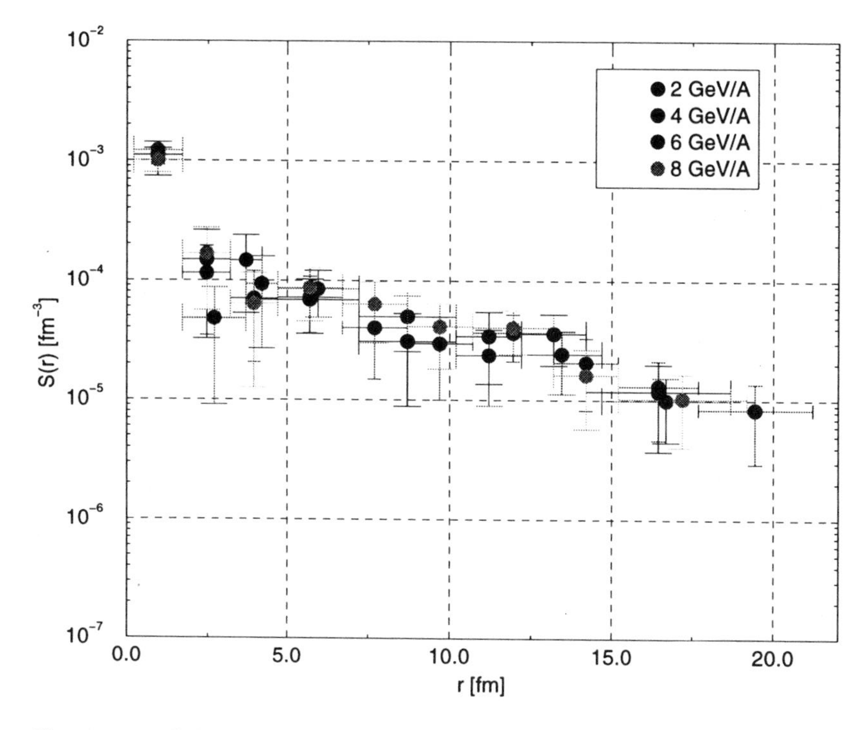

Figure 28.2 Relative source functions extracted from the proton correlation data at 2,4,6 and 8 AGeV.

5. SUMMARY

We reported preliminary results of the analysis of the beam energy dependence of the two-proton correlation function in the target fragmentation region ($P < 800$ MeV/c). The correlation functions were measured for the first time for protons in central Au+Au collisions at beam energies 2,4,6 and 8 AGeV. Within currently available statistical accuracy no significant changes with beam energy were observed. The source imaging technique of Brown-Danielewicz was used to extract information about the space-time extent of the proton source. It was found that the relative proton source functions have similar shapes at all measured energies. Extracted source functions show enhancement at low relative separation which may be induced by momentum-position correlations in the source, possibly due to collective flow. Further investigation of the origin of this enhancement is clearly needed.

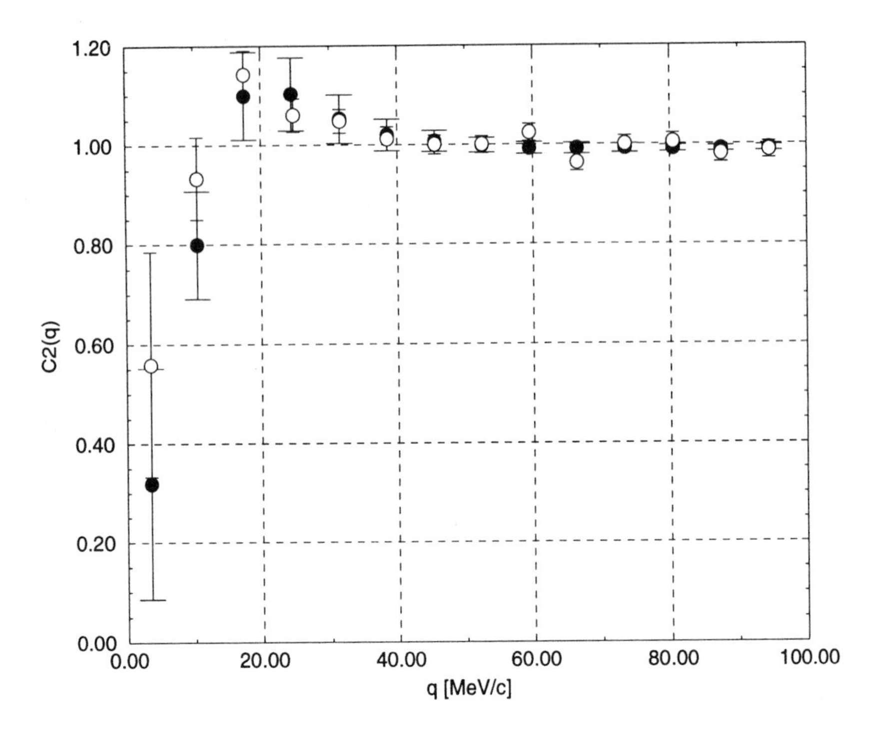

Figure 28.3 Experimentally measured two-proton correlation function (open circles) and correlation function restored from the relative source (filled circles) for beam energy 4 AGeV. See description in the text.

Acknowledgments

Stimulating discussions with Drs. N.Xu and S. Voloshin are gratefully acknowledged. The author wish to thank P. Danielewicz and D. Brown for performing source imaging calculations. This research is supported by the U.S. Department of Energy , the U.S. National Science Foundation and by University of Auckland, New Zealand, Research Committee.

References

[1] Koonin, S. (1977). Phys. Lett. B**70**, 43.

[2] Lednicky, R. and Lyuboshitz, V.L. (1982). Sov. J. Nucl. Phys. **35**, 770.

[3] Pratt, S., Csörgő T. and Zimányi T. (1990). Phys. Rev. C**42**, 2646.

[4] Gelbke, C. and Jennings, B.K. (1990). Rev. Mod. Phys. **62**, 553.

[5] Lisa, M. (1998). in *Advances in Nuclear Dynamics 4.*, edited by W. Bauer and H.-G. Ritter, Plenum Press, New York, 183.

[6] Rai, G., et al. (1990). IEEE Trans. Nucl. Sci. **37**, 56.

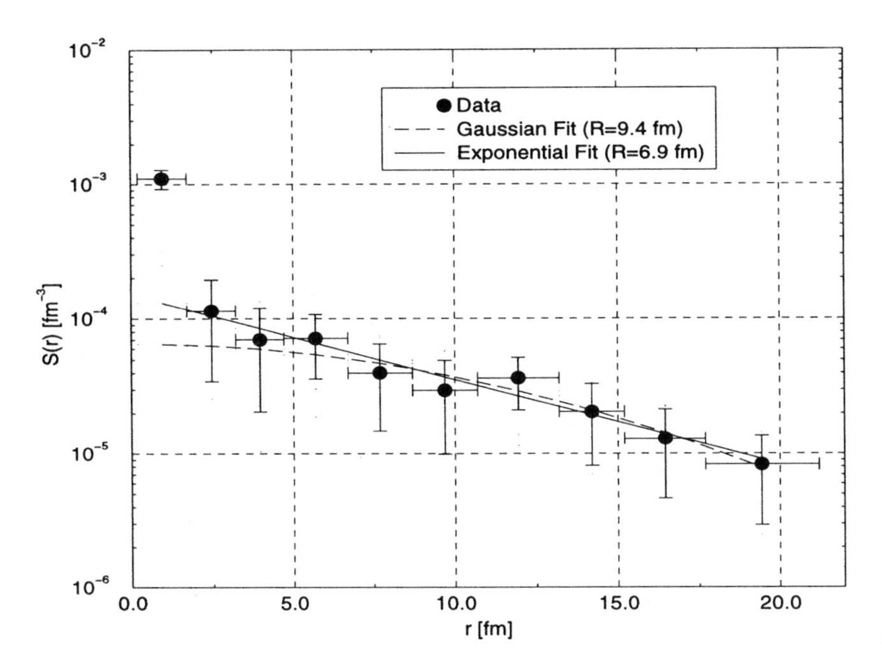

Figure 28.4 Gaussian and exponential fits to the relative source reconstructed for the beam energy 4 AGeV.

[7] Bauer, G. et al. (1997). NIM A**386**, 249

[8] Brown, D.A. and Danielewicz, P. (1997). Phys. Lett. B**398**, 252.

[9] Brown, D.A. and Danielewicz, P. (1998). Phys. Rev. C**57**, 2474.

[10] Stoks W.G.J. et al. (1994). Phys. Rev. C**49**, 2950.

[11] Tarantola, A. (1987). *Inverse Problem Theory*, Elsevier.

Chapter 29

DISAPPEARANCE OF TRANSVERSE FLOW IN AU+AU COLLISIONS

Dan J. Magestro
National Superconducting Cyclotron Laboratory and
Department of Physics and Astronomy
Michigan State University
East Lansing, MI 48824-1321
magestro@msu.edu
Keywords: flow, BUU, heavy-ion collision, Au+Au

1. INTRODUCTION

The study of transverse flow as a means of probing nuclear reactions has progressed greatly since it was first suggested by Danielewicz and Odyniec[1] in 1985. Early experiments (1984-1987) at the Bevalac established the existence of transverse flow at relativistic energies[2], while flow was also experimentally observed (1989) at low energies at the NSCL [3]. At high energies, flow was attributed to repulsive hydrodynamic side-splash, while at low energies, attractive mean-field effects caused the observed flow.

Experimentally, attractive and repulsive scattering cannot be discerned, *i.e.* one cannot tell which side of the target nucleus the projectile nucleus struck. The direction of the measured reaction plane points in the direction of the flow, regardless of the nature of the scattering. However, a minimum in the flow excitation function can be extracted, corresponding to a balance between the attractive and repulsive effects. The incident energy at which this occurs, termed the balance energy, E_{bal}, provides an excellent probe of the nuclear equation of state, the in-medium nucleon-nucleon cross section, and the traits of the nuclear mean field.

The balance energy is a convenient quantity for testing a theoretical model and its parameters. Finite transverse flow is difficult to compare to theory because the theory needs accurate fragment formation. In addition, experimental biases need to be incorporated correctly, such as the dispersion of the true reaction plane and the limited acceptance of the detector system. The robustness of E_{bal} allows models to be tested without knowledge of the experimental apparatus. Studies of the balance energy have revealed a reduction in the in-medium nucleon-nucleon cross section[4], as well as a momentum-dependent mean field[5].

2. MOTIVATION FOR AU+AU EXPERIMENT

Previously, transverse flow has been observed in Au+Au collisions at incident energies as low as 75 MeV/nucleon[6]. Other experiments at Bevalac [7] and SIS [8] attempted to extract values of E_{bal} for Au+Au by fitting flow measurements at higher energies with logarithmic and Fermi functions and extrapolating to zero flow. These experiments obtained balance energies of 47 ± 5 MeV/nucleon and 65 ± 14 MeV/nucleon, respectively, for semi-central events. Using the previously established system mass dependence of E_{bal} [4], a balance energy of 43 MeV/nucleon is expected.

Obtaining E_{bal} for a large system such as Au+Au (A=394) can help to better determine the functional dependence of E_{bal} on the system mass. This in turn allows us to better establish the amount of reduction in the in-medium nucleon-nucleon cross section, as well as the incompressibility of nuclear matter. *i.e.* stiffness of the equation of state. More importantly, Au+Au flow measurements at intermediate energies can be added to data taken at SIS, AGS, SPS, and RHIC, aiding in the development of models which can be applied to the full range of beam energies.

The question of whether the balance energy can be measured at all for Au+Au was posed by Soff *et al* [9], who suggested that Coulomb repulsion at low beam energies would cancel the mean-field attraction. QMD model calculations showed that if the initial pre-contact rotation due to the coulomb interaction is subracted, negative flow is restored.

3. EXPERIMENTAL SETUP

The present measurements were carried out with the 4π Array [10] at the National Superconducting Cyclotron Laboratory (NSCL) using Au beams from the K1200 cyclotron. Data was taken at incident beam energies of 25, 29, 35, 40, 45, 50, 55, and 59 MeV/nucleon. The Au beams were focused directly onto a Au target with a thickness ranging from 2 mg/cm^2 to 6 mg/cm^2. Beam current was approximately 10-100

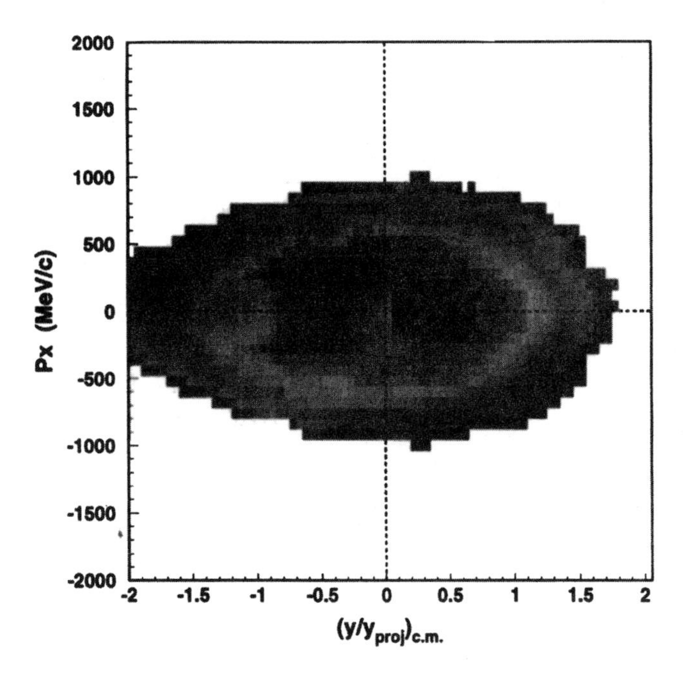

Figure 29.1 The acceptance of the 4π Array for Z=2 fragments in Au+Au collisions at 40 MeV/nucleon, in terms of the transverse momentum in the reaction plane vs. reduced c.m. rapidity. The detection efficiency is weak at backward mid-rapidity; therefore, only forward rapidity particles are used in the present flow analysis.

electrical pA. More than 100 million events were recorded in a 6 day span.

The main ball of the 4π Array is an icosahedron which consists of 55 Bragg curve counters in front of 170 phoswich counters (arranged in hexagonal and pentagonal subarrays) covering the laboratory frame polar angles $18° < \theta_{lab} < 162°$. The Bragg curve counters served as ΔE detectors, allowing the detection of charged fragments from $Z = 1$ to $Z = 12$. In addition, the High Rate Array consists of 45 phoswich detectors covering $3° < \theta_{lab} < 18°$, with detection of charged fragments $1 \leq Z \leq 18$. By using the Bragg counters, the main ball telescopes have lower energy thresholds of approximately 3 MeV/nucleon for ^{7}Li. The High Rate Array phoswiches have corresponding energy thresholds of ~10 MeV/nucleon. The 4π Array is most characterized by its very fast data acquisition system, which is capable of recording thousands of events per second.

4. EVENT CHARACTERIZATION

To extract collective flow in nucleus-nucleus collisions, the impact parameter, b, and the reaction plane first must be determined. Central collisions were selected using total transverse kinetic energy ($E_t = \sum_{i=1,N_c} E_i \sin^2(\theta_i)$). A large value of E_t corresponds to a small impact parameter. In order to compare to previous studies, events were placed into 5 bins, each containing 20% of the events. In the present analysis, central (E5: $b/b_{\mathrm{max}} < 0.39$) and semi-central (E4: $0.39 < b/b_{\mathrm{max}} < 0.56$) collisions are studied. Here, b_{max} is the maximum estimated impact parameter.

To determine the reaction plane, the method of azimuthal correlations was used[11]. This method is useful when flow is weak, *i.e.* near the balance energy. The azimuthal correlation method finds the line which best aligns with the transverse momentum vectors. In order to avoid autocorrelation, the flow particle of interest (POI) is removed from the reaction plane determination. Therefore, an event with multiplicity N can be thought of as having N separate sub-events. Particles in the backward hemisphere of the reaction are weighted with a minus sign in the reaction plane determination. Fourier analysis is applied to correct for the anisotropy of the reaction plane distribution, which results from experimental biases and azimuthal granularity.

Figure 29.3 shows the acceptance of the 4π Array in terms of transverse momentum and rapidity. The assigned (θ,ϕ) of each particle is distributed over the active area of the incident detector, as determined by simulations and the 4π software filter. The acceptance is mostly at forward rapidities at the beam energies studied here. For this reason, only the forward-moving particles are considered in the transverse momentum analysis. Forward/backward symmetry in the collision is assumed, in accordance with previously published work [8].

5. RESULTS

Figure 29.2 shows the transverse momentum for Au+Au projected into the reaction plane, as a function of normalized rapidity. The error bars are statistical, and the solid lines correspond to linear least square fits for the midrapidity region $-0.6 \leq (y/y_{\mathrm{proj}})_{\mathrm{c.m.}} \leq 0.6$. In these plots, $Z = 2$ is the flow particle of interest, and data is for the most central (E5) bin. Forward/backward symmetry is assumed, and the data is reflected about $y_{\mathrm{c.m.}} = 0$. The flow, corresponding to the slope of the plots at mid-rapidity, is clearly maximum at the lowest and highest beam energies.

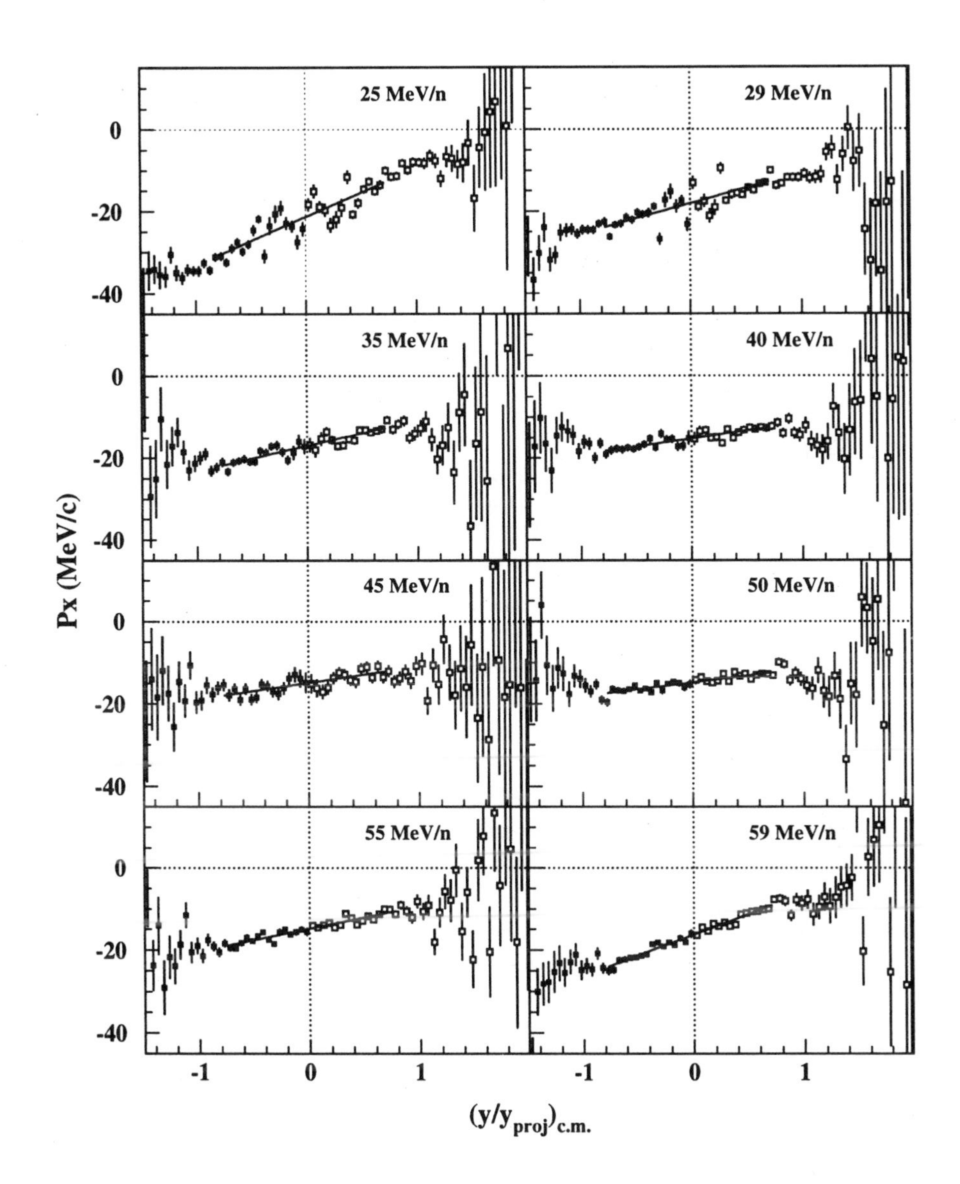

Figure 29.2 Mean transverse momentum in the reaction plane plotted verses the reduced c.m. rapidity for Z=2 fragments from central collisions. Open squares are experimental data, solid squares are reflected about $y = 0$ assuming forward/backward symmetry.

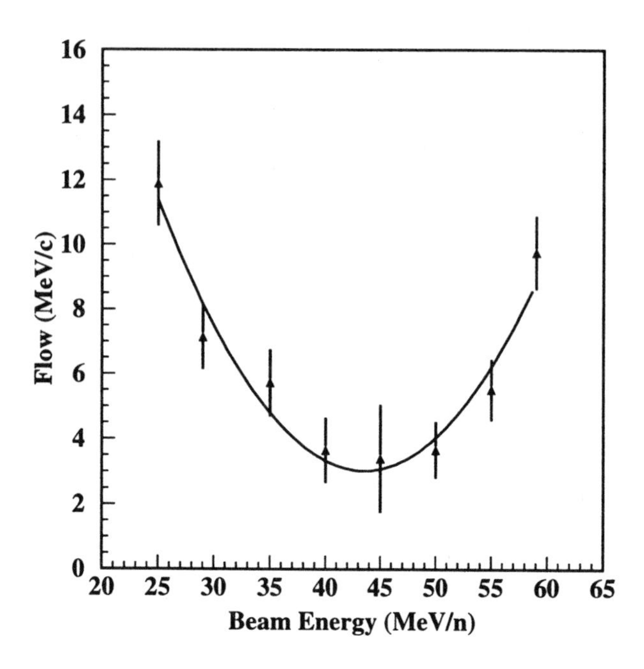

Figure 29.3 Extracted flow vs. incident beam energy for central (E5) collisions, Z=2 is POI. The fit is binomial, and the minimum corresponds to the balance energy for the system.

A question arises as to the source of the negative offset of the data from the origin. The offset has been seen in experimental data for many different systems. Two experimental biases are believed to contribute to this offset. First, since the particle of interest is excluded from the reaction plane determination, there is an inherent lack of conservation of momentum in the event. This means that the extracted reaction plane is shifted from the true reaction plane by an amount at least as large as the momentum of the POI. However, this effect is expected to be small. Since the 4π Array does not successfully detect or identify all of the particles, there is already a lack of momentum conservation in the detected particles, far greater than due to a single particle of interest.

Therefore, the negative offset is mostly attributed to double hits in detectors which are more likely to be hit due to the direction of an event's reaction plane. In other words, if a POI contributes positively to the transverse momentum, it is more likely to be undetected because of an increased chance of being part of a double-hit. Conversely, particles contributing negatively are more likely to be unencumbered. The negative

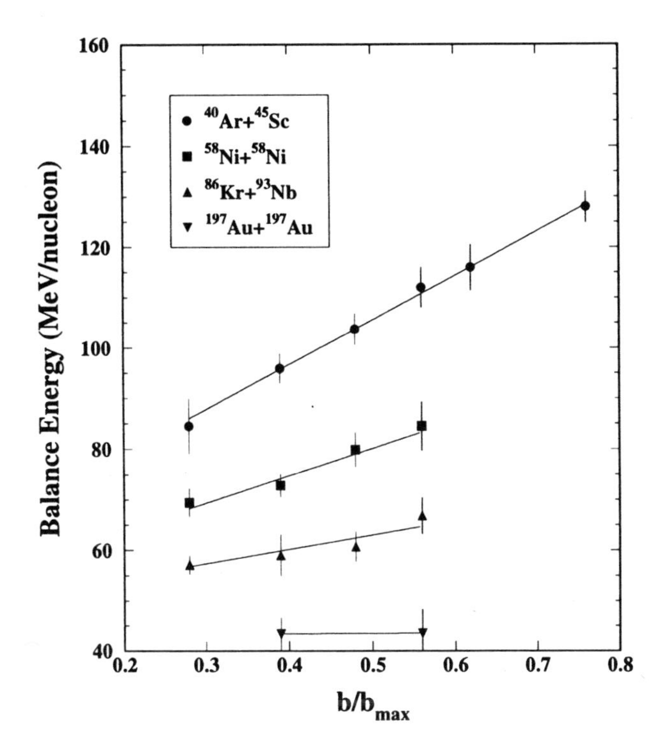

Figure 29.4 The impact-parameter depedence of E_{bal} for several systems. The linear dependence of E_{bal} decreases as a function of system mass.

offset has been observed with the EOS TPC, and it was corrected for by assuming forward/backward symmetry and using knowledge of particles detected at backward rapidities and inefficiency due to two-track resolution. Since our detection efficiency at backward rapidities is poor in the current range of beam energies (25-60 MeV/nucleon), the negative offset cannot be corrected. However, the slope of the transverse momentum at mid-rapidity is the quantity of interest, and an offset is not influential.

In Figure 29.3, the extracted slopes of the transverse momentum at midrapidity are plotted as a function of beam energy for most central (E5) events. The fits are binomial, and the minimum corresponds to the balance energy for the Au+Au system. The balance energy is determined to be 43.4±3.3 MeV/nucleon for the E5 bin, and 43.5±4.8 MeV/nucleon for the E4 bin.

Previously, the balance energy was shown to depend linearly on the impact parameter. However, the relationship was established for a light system (Ar+Sc)[5]. Comparison to QMD models elicited a momentum-dependent mean field.

Figure 29.4 shows the impact-parameter dependence of the balance energy for four systems, measured with the 4π Array. While the dependence is roughly linear for each system, the slope of the fit decreases as the mass of the system increases. In fact, for Au+Au, the impact-parameter dependence seems to vanish. This has not been previously observed and needs to be compared to transport models to understand the effect.

The observed balance energy for Au+Au agrees well with the previously established system mass dependence of $E_{\rm bal}$. This agreement supports the assumption that the low-energy flow values correspond to collisions dominated by attractive mean-field effects, rather than collisions dominated by the repulsive Coulomb force. Of course, as the beam energy drops further, the interaction is completely repulsive, due to the long-range Coulomb interaction.

The directly measured $E_{\rm bal}$ also agrees within error bars with the value inferred using the EOS TPC. However, the directly measured $E_{\rm bal}$ is somewhat lower than the extrapolated measurement by the FOPI group. This is most likely due to small errors in the transverse flow measurements at the higher beam energies, or in over-corrections to the reaction plane dispersion.

6. CONCLUSIONS

The Au+Au system has been studied with a 4π detector at a range of intermediate beam energies. Although the flow signal is weak, the flow excitation function exhibits a minimum, corresponding to a balance between the repulsive and attractive interactions in the collision. The directly measured balance energy agrees well with previously established power-law relationship between $E_{\rm bal}$ and system mass. Also, the impact-parameter dependence of $E_{\rm bal}$ appears to vanish as the system size is increased to heavy systems such as Au+Au. This result needs to be investigated with transport models that can untangle the complicated overlapping of repulsive and attractive interactions as a function of impact parameter.

The discrepancy between the measured value of 43 ± 5 MeV/nucleon and the value of 65 ± 14 MeV/nucleon extrapolated by the FOPI group illustrates the difficulty in assigning finite flow values. However, if experimental biases and dispersion from the true reaction planes are accounted for consistently and effectively, transverse flow can be studied over a wide range of beam energies with a single model as a probe.

This experiment was done in collaboration with Omar Bjarki, Jennifer Crispin, Ed Norbeck, Mike Miller, Marguerite Tonjes, Robert Pak, Andrew Vander Molen, Gary Westfall, and Wolfgang Bauer. The author also wishes to thank the Winter Workshop participants, particularly Dan Cebra and Bill Caskey, for insightful discussions regarding the present analysis. This work has been supported by the U.S. National Science Foundation under Grant No. PHY 95-28844.

References

[1] P. Danielewicz and G. Odyniec, Phys Lett. **157B** 146 (1985).

[2] K.G.R. Doss *et al.*, Phys. Rev. Lett. **59**, 2720 (1987).

[3] C.A. Ogilvie *et al.*, Phys. Rev. C **40**, 654 (1989).

[4] G.D. Westfall *et al.*, Phys. Rev. Lett. **71**, 1986 (1993).

[5] R. Pak *et al.*, Phys. Rev. C **54**, 2457 (1996).

[6] W.M. Zhang *et al.*, Phys. Rev. C **42**, R491 (1990).

[7] M.D. Partlan *et al.*, Phys. Rev. Lett. **75** 2100 (1995).

[8] P. Crochet *et al.*, Nuc. Phys. A **624** 755 (1997).

[9] S. Soff *et al.*, Phys. Rev. C **51** 3320 (1995).

[10] G.D. Westfall *et al.*, Nucl. Instru. and Methods **A238** 347 (1985).

[11] W.K. Wilson *et al.*, Phys. Rev. C **45**, 738 (1992).

Chapter 30

PHOBOS: A STATUS REPORT

Stephen G. Steadman *
For the PHOBOS Collaboration
Argonne National Laboratory, Brookhaven National laboratory, Case-Western Reserve, Institute of Nuclear Physics (Krakow), Jagellonian University (Krakow), Massachusetts Institute of Technology, National Central University (Taiwan), University of Illinois at Chicago, University of Maryland, University of Rochester
Stephen.Steadman@science.doe.gov

*Present address: SC-23, U.S. Department of Energy

Abstract The PHOBOS detector for RHIC is now well underway in assembly, and will be ready to begin taking physics data with the start of RHIC physics running in November, 1999.

Keywords: RHIC, Detector

1. INTRODUCTION

The PHOBOS experiment is one of the four relativistic heavy-ion experiments that will be ready when RHIC begins scheduled physics data taking on November 1, 1999. Although it is one of the two 'smaller' experiments, nevertheless, at a cost of about 8 Million dollars it is still a 'large' experiment that offers substantial physics capabilities. An excellent description of these capabilities was presented at the 1997 Winter Workshop by Rudi Ganz[1]. Since that time we are pleased to report that the National Science foundation has approved funding for time-of-flight walls behind the spectrometers, and these are now well under construction at the University of Rochester. The PHOBOS Collaboration consists of 63 members from 10 institutions.

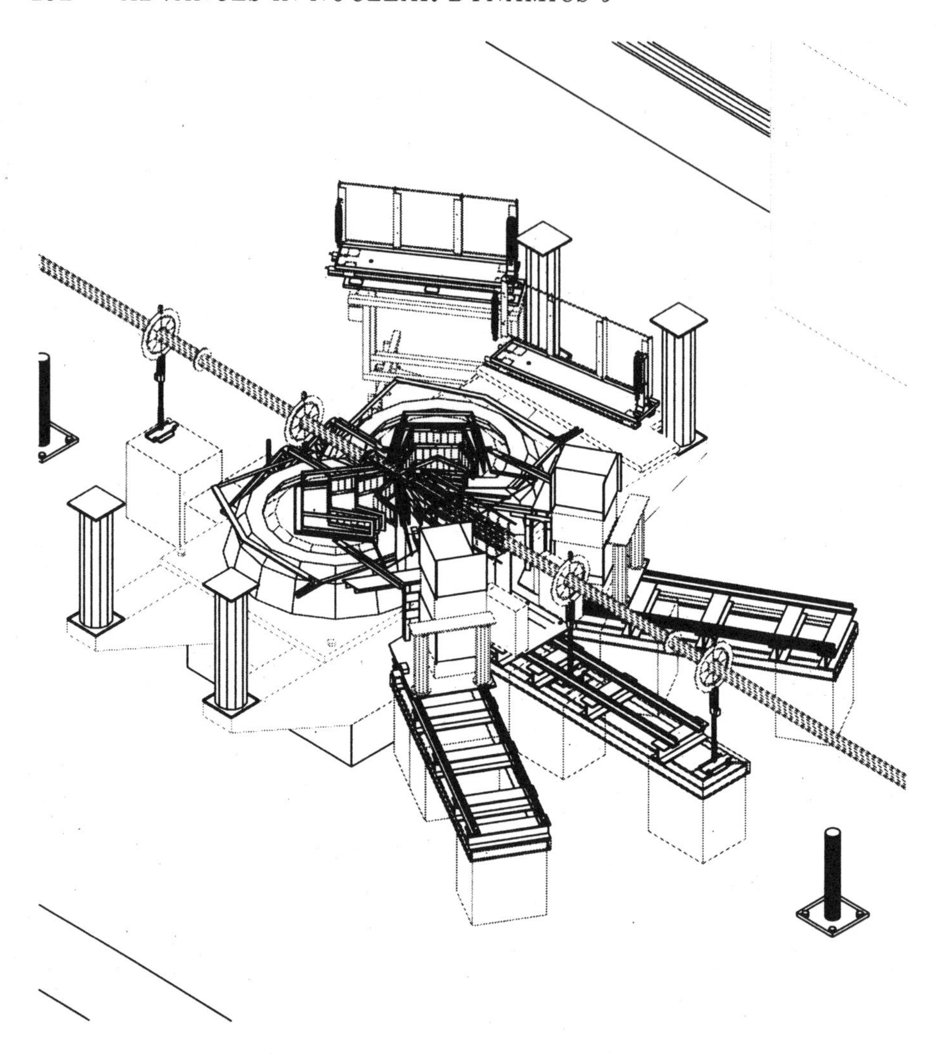

Figure 30.1 PHOBOS detector layout in the RHIC tunnel at the 10 o'clock crossing. The top magnet return yoke has been removed in order to see the spectrometer arms.

2. LAYOUT OF THE DETECTOR

A layout of the detector is shown in Figure 30.1. It consists of two basic parts using components dominantly of a common technology, namely 300 μm silicon pad detectors. The electronic readout has excellent signal/noise performance, so that the energy loss of the charged particles can be well measured (Figure 30.2).

There is a multiplicity detector which consists of an octagon array about the Be beampipe at mid-rapidity with 11,776 elements, as well as six rings with a total of 3,072 elements. The total coverage for the whole

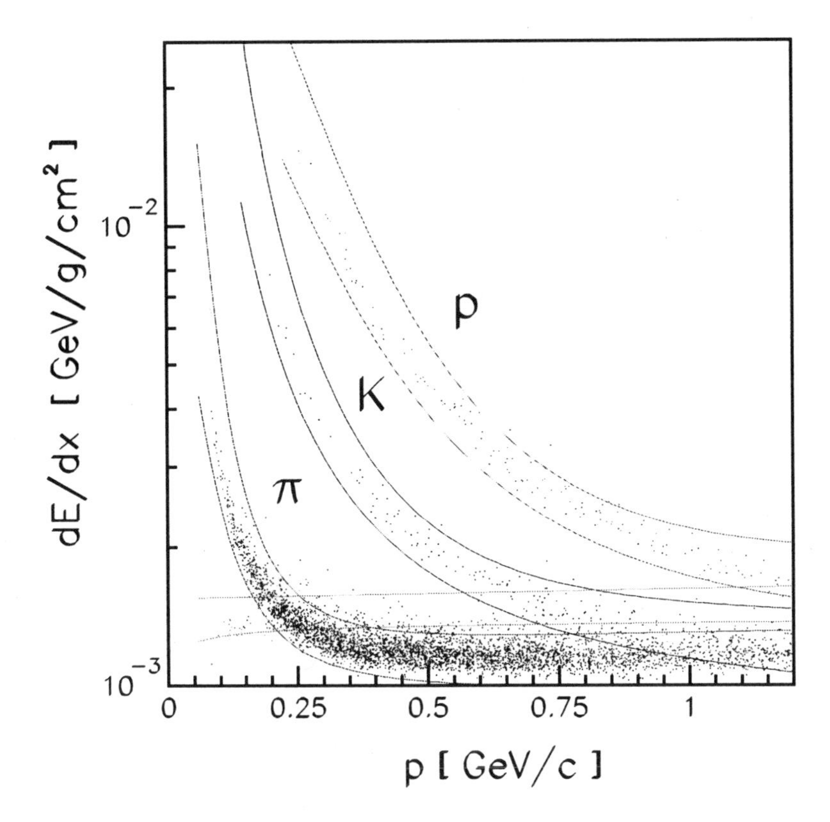

Figure 30.2 Simulated reconstructed dE/dx for the PHOBOS spectrometer arms as a function of particle momentum.

detector is 85% of the 4π available total solid angle ($-5.3 < \eta < 5.3$), which is the largest coverage of any of the four RHIC experiments. Although many pad elements for central Au-Au collisions register multiple hits, the granularity is fine enough that the number of hits can be extracted and the background subtracted on an event-by-event basis through pulse height. A comparison of a simulated 'measured' charged-particle distribution for such a central event is compared with a HIJET generated event in Figure 30.3. This detector will provide opportunities for analyzing the character of events, including multiplicity and reaction-plane determination for each event, but also can be used to select events with (interesting) large multiplicity or large local fluctuations in the rapidity or azimuthal distributions for further study with the two spectrometer arms. About mid-rapidity there is mounted an additional two planes of elements that together can be used as a vertex finder to locate the vertex to within 0.5 mm for central events.

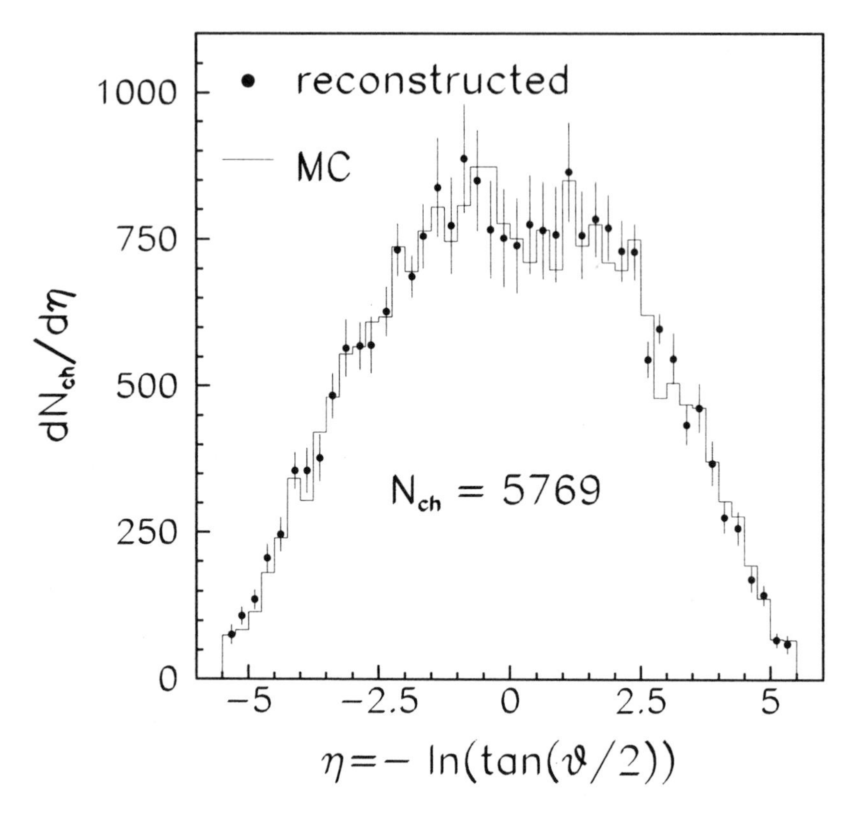

Figure 30.3 Comparison of a simulated reconstructed charged particle pseudo-rapidity distribution for a single central Au-Au collision with its HIJET generated event.

3. SPECTROMETER ARMS

About 1% of the emitted particles are detected in two spectrometer arms, each consisting of 14 planes, with roughly half of the planes located in 2 T magnetic fields, which will be able to measure the momentum and determine the particle identification within about 1 unit of acceptance in pseudo-rapidity. The number of planes is adequate to provide excellent tracking: greater than 85% track reconstruction efficiency, good momentum resolution (less than 1% $\Delta p/p$), and good particle identification using the energy loss information from the Si detectors. The momentum and rapidity acceptance of the spectrometer arms for pions is shown in Figure 30.4 and the quality of the particle identification is shown in Figure 30.2. The long diamond beam crossing region of ± 20 cm allows a larger span of acceptance when spectra are averaged over many events. For the first year of running only one arm will be fully instrumented due to financial limitations. The time-of-flight walls allow a considerable increase in the momentum coverage with good particle identification, as shown in Figure 30.4.

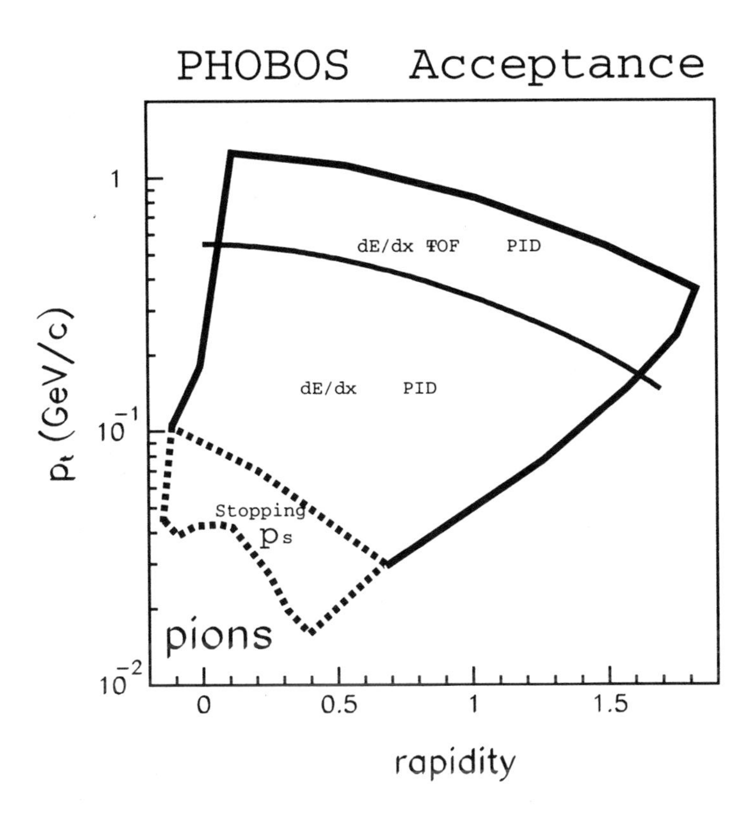

Figure 30.4 Acceptance in rapidity and transverse momentum of the spectrometer arms for pions.

The design of the electronic readout system underwent a major revision in the fall of 1997. It now consists of a multiplexed readout of either 64 or 128-channel Viking amplifier/shaper chips with over an 80 mip dynamic range[2]. The characteristic integration time for these chips of about 1 μs provides superior signal-to-noise performance of typically 16/1 for minimum ionizing particles (1 mip) (Figure 30.5). The multiplexed readout supports event rates of over 200 Hz for central Au-Au collisions (200 kBytes/event readout from the detector). To maximize the troughput of useful data to mass storage, an on-line zero-suppression/common-noise-subtraction system utilizing fast data signal processors in an off-the-shelf system provided by Mercury Computer Systems[3] is employed. This reduces the data rate to mass storage to an estimated 40 kBytes per central Au-Au event.

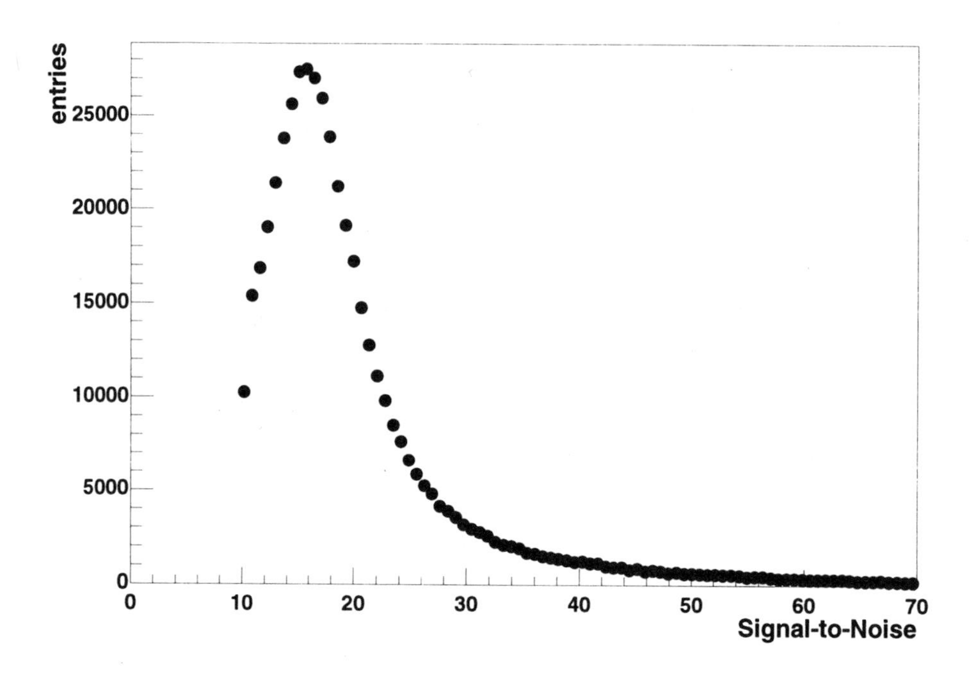

Figure 30.5 Signal-to-noise distribution for a typical spectrometer module.

4. DETECTOR ASSEMBLY

Progress in assembly of the experiment is excellent. The silicon pad detectors employ a double metallized layer, which allows a flexible geometry of different pad sizes to be coupled to a standardardized pitch readout, facilitating ease in readout design. They were fabricated by Miracle Technology, Taiwan, using designs provided by the National Central University group. All the needed silicon sensors have been fabricated and tested (at MIT and UIC). These are mounted on ceramic hybrids (6 different basic types for the spectrometer and 2 types for the vertex detector) or printed circuit boards (for the multiplicity detector). This assembly is about 2/3 complete for one spectrometer arm (at MIT) and 1/2 complete for the multiplicity detector (at UIC and ANL) as of February 1, 1999. The remaining modules will be fully assembled by June.

Spectrometer detector modules have been tested using beams at the AGS during Fall, 1998 and fully meet specifications. Prototype boards for the data readout including digitization, multiplexing, and transmission over optical fiber have been satisfactorily tested in a full string data readout test. One of the two time-of-flight walls is completely assembled, and the second wall will be completed by summer 1999. Most of the mechanical support structure has been fabricated (in Krakow, Poland), and the magnet, built by Danfysik,[4] has been installed at RHIC and its field mapped. Basic trigger counters consisting of rings of Cherenkov and scintillator paddles are being fabricated according to a University of Maryland/BNL design. These trigger counters could be used to select events by centrality, but the experiment is designed for a high rate, so that a highly restrictive trigger would be unneccessary at the design luminosity. Indeed, for the first year, all minimum-bias events at the full collision rate could be easily handled.

The spectrometer data will provide spectral distributions for kaons, pions, and protons, especially to low momentum, 45 MeV/c for pions. This ability to measure to low transverse momentum is especially important for the detection of pion decay from disoriented chiral condensates (DCC's). These arms also have good acceptance for detecting the decay of the ϕ (composed of a strange and anti-strange quark) into a kaon pair, although both arms are needed to obtain the low tansverse momentum ϕ's. The modest acceptance of the spectrometer arms also allows an excellent study of identical particle correlations: $\pi\pi$, KK, and pp. Monte Carlo simulations indicate that source sizes up to 20 fm can be measured with good precision relatively easily - within several days of running at the planned year-one luminosity of 10% of the design goal. Thus, there is a rich program available already in the first year of RHIC operation at the expected reduced luminosity for Au-Au collisions. Physics results should be available shortly after RHIC turnon.

5. SUMMARY

The assembly of the PHOBOS detector is proceeding well, with the detector modules to be completed by June, 1999 and a significant part of the multiplicity detector already planned for installation for the June engineering run, as well as a time-of-flight wall, trigger counters, and the electronic readout system. The remaining elements of the multiplicity/vertex detector and spectrometer arm will be installed and commissioned by November 1. The tracking software is proceeding well beyond the proof-of-principle stage to detailed software modules. We look forward to exciting physics results from the first days of RHIC physics running.

Acknowledgments

The author wishes to thank K. Gulbrandsen, H. Pernegger, and G. Stephans for supplying figures.

References

[1] Ganz, Rudi (1997) *Advances in Nuclear Dynamics*, Proceedings of the XIII Winter Workshop on Nuclear Dynamics, Marathon, FL

[2] IDE AS, Oslo, Norway

[3] Mercury Computer Systems, Chelmsford, MA

[4] Danfysik A/S, Jyllinge 001, Denmark

Chapter 31

PROTON-NUCLEUS DYNAMICS DEFINING, MEASURING, AND STUDYING CENTRALITY DEPENDENCE

Ron A. Soltz for the BNL E910 Collaboration
Lawrence Livermore National Laboratory
*Livermore, CA USA**
soltz@llnl.gov

*Supported by U.S. DOE contract W-7405-ENG-48.

Abstract Experiment E910 has measured slow protons and deuterons from collisions of 18 GeV/c protons with Be, Cu, and Au targets at the BNL AGS. These correspond to the "grey tracks" first observed in emulsion experiments. We report on their momentum and angular distributions and investigate their use in measuring the centrality of a collision, as defined by the mean number of projectile-nucleon interactions. The relation between the measured N_{grey} and the mean number of interactions, $\overline{\nu}(N_{grey})$, is studied using several simple models, one newly proposed, as well as the RQMD event generator. For the Au target, we report a relative systematic error for extracting $\overline{\nu}(N_{grey})$ that lies between 10% and 20% over all N_{grey}.

Keywords: proton-nucleus, p-A, centrality, stopping

1. INTRODUCTION

The use of high-energy collisions of hadrons with nuclear targets to study the space-time development of produced particles was first suggested many years ago [1, 2, 3, 4]. Early experiments indicated that at sufficiently high energies, the projectile will undergo on average a number of inelastic hadron-nucleon scatterings roughly equal to the mean interaction thickness, $\overline{\nu} = A\sigma_{hp}/\sigma_{hA}$, with most particles forming well outside the target nucleus [5, 6, 7, 8]. These data suggest that a single p-

A collision can effectively be modeled by a cascade of ν proton-nucleon interactions, with ~1 fm formation times for produced particles. For reasons given below, any conflicts between such a cascade model and p-A data have yet to be demonstrated.

The differences between a p-A collision and a p-nucleon cascade are especially important to discover and understand in light of recent experiments with relativistic heavy-ion collisions at BNL and CERN. Here the complex hadronic physics processes that we wish to study in p-A form a significant background in the search for a QCD phase transition. The overwhelming complexity of A-A collisions makes it difficult to study these processes directly, whereas p-A collisions are simpler and may provide more insight.

Many previous p-A experiments were limited by their inability to trigger on central collisions, those with small impact parameter and in which ν attains the highest values — essential to studying the effects of multiple interactions. Other experiments which could trigger on centrality were limited by low rates (low statistics) and/or insufficient phase space coverage for identified particles. However, they were able to establish a relationship between ν and a measurable observable, the number of slow singly charged fragments (grey tracks[9]) emitted in the collisions. This relationship is expressed as a conditional probability for detecting N_{grey} grey tracks given a collision in which there were ν interactions, $P(N_{grey}|\nu)$. Given a distribution $\pi(\nu)$ for the number of interactions, the relevant quantity for measuring centrality in p-A collisions is,

$$\overline{\nu}(N_{grey}) = \sum_{\nu} \nu P(N_{grey}|\nu)\pi(\nu). \tag{31.1}$$

Several forms have been proposed for $P(N_{grey}|\nu)$ [10, 11, 12, 13, 14], yet there have been few systematic studies to test the validity of the models' assumptions and assess the accuracy of the extracted values of $\overline{\nu}(N_{grey})$. We will focus on the two models which have been most commonly applied to data: the Geometric Cascade Model (GCM) of Andersson *et al.* [10], and the intra-nuclear cascade calculation of Hegab and Hüfner [13, 12]. We will then present a new model which draws on elements of both.

We present a high statistics analysis of low momenta protons from collisions of 18 GeV/c protons incident on three nuclear targets: Be, Cu, and Au. The data were taken by BNL E910, a large acceptance TPC spectrometer experiment with additional particle identification from time-of-flight (TOF) and Čerenkov (CKOV) detectors. Additional details of the experimental setup are given in [16]. Protons with nominal beam momenta of 6, 12, and 18 GeV/c were normally incident on targets of Be, Cu, Au, and U. Only the 18 GeV/c beam and Be, Cu, and Au targets are included in this analysis. The targets, 4% Be, 3% Cu and 2% Au targets were 3.9, 4.2, and 3.4 gm/cm^2 thick respectively, and were located in the TPC re-entrant window, 10 cm before the TPC active volume. E910 ran in the A1 secondary beam line of the AGS, with a typical intensity of $3 \cdot 10^4 s^{-1}$. For these data the beam momentum was determined to be 17.5 ± 0.2(sys) GeV/c.

2. DATA REDUCTION

All particle tracking and identification of slow protons comes from the TPC analysis. A GEANT simulation of the TPC shows the momentum resolution for the N_{grey} tracks to be dominated by multiple scattering, with a resolution of 15 MeV/c for 1 GeV/c protons.

Particles are identified in the TPC through their ionization energy loss, dE/dx, calculated using a 70% truncated mean. The dE/dx distributions have been fit to the Bethe-Bloch formula with momentum dependent gaussian widths. This analysis does not correct for saturation or non-linearities in the pulse-heights. Particles with dE/dx within 2.25σ of that for a proton and further than 1.5σ from the pion dE/dx are identified as protons. We require that deuterons lie within 2.25σ of dE/dx for a deuteron and further than 2.25σ from the proton and pion bands. Protons are identified up to a momentum of 1.2 GeV/c and deuterons up to 2.4 GeV/c. Two additional cuts are required to limit positron contamination coming from photon conversions in the target. Positive tracks which can be paired with a negative track to yield small transverse momentum are rejected. We also reject positive tracks with dE/dx consistent with that of a positron that are forward of $\cos(\theta_f)$. The value of $\cos(\theta_f)$, given in Table 31.1, was chosen separately for protons and deuterons for each target to minimize the contamination while preserving statistics. From the angular distributions of the paired-positrons we estimate final positron contamination to be less than 5% of the overall N_{grey} sample for all targets. Additional offline interaction cuts are given in [16]. Final N_{grey} statistics are given in Table 31.1. Acceptance corrected momentum and angular distributions for protons are given in Fig. 31.1. Distributions are shown only for $p > 0.1$ GeV/c and $\cos(\theta) > 0.3$, where the acceptance is greater than 10%. The angular distributions for all targets are nearly isotropic in the lowest bin, becoming progressively more forward peaked at higher momenta. The momentum distributions peak near 0.5 GeV/c for Au and at higher momentum for the lighter targets. Based on these distributions, and the previous work in multi-fragmentation, we use a range of $0.25 < p < 1.2$ GeV/c for protons, and $0.5 < p < 2.4$ GeV/c for deuterons for our definition

Table 31.1 Event statistics and forward angle cuts for all targets. Cuts for deuterons are in parentheses.

Target	*Events*	$\cos\theta_f$	protons	deuterons
Au	35520	0.98 (0.97)	56881	10622
Cu	49331	0.98 (0.96)	45784	6224
Be	100609	0.94 (0.94)	30622	3366

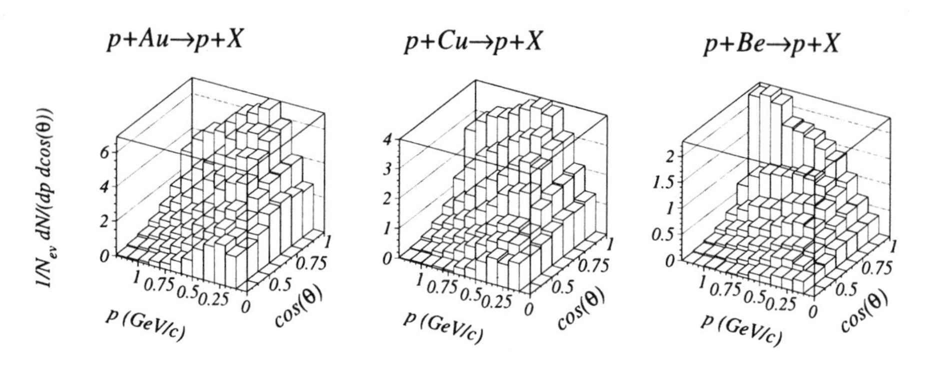

Figure 31.1 Acceptance corrected distribution for protons.

of N_{grey}. The upper bounds reflect the limits of particle identification, 1.2 GeV/c for protons and 2.4 GeV/c for deuterons. The upper limits are higher than for most experiments, while the lower limits are comparable. The distributions are then corrected for contributions from target out and secondary interactions.

3. RESULTS

The GCM [10] assumes a normalized geometric distribution of grey tracks for a single proton-nucleon interaction. Taking μ as the average measured N_{grey} when $\nu = 1$ and convoluting ν independent interactions yields,

$$P(N_{grey}|\nu) = \left(\begin{array}{c} N_{grey} + \nu - 1 \\ \nu - 1 \end{array} \right) (1 - X)^{\nu} X^{N_{grey}}, \tag{31.2}$$

where $X = \frac{\mu}{1+\mu}$. This distribution is recognizable as a negative binomial, where ν is the standard k-parameter, and the mean, $\overline{N_{grey}}(\nu)$ is given by $\nu\mu$. Taking the weighted sum over ν,

$$\overline{N_{grey}} = \sum_{\nu} \pi(\nu)\overline{N_{grey}}(\nu) = \overline{\nu} \cdot \mu. \tag{31.3}$$

Thus, the linear dependence of N_{grey} on $\overline{\nu}$ is a direct consequence of the sum over ν independent distributions. The full distribution is given by,

$$P(N_{grey}) = \sum_{\nu} P(N_{grey}|\nu)\pi(\nu). \tag{31.4}$$

Two calculations for $\pi(\nu)$, Glauber and Hijing, have been used for the three targets. Both calculate ν from an optical model [18] using a value of 30 mb for the p-N cross-section and a Wood-Saxon distribution of the nucleus. The Hijing results come from the HIJING Monte-Carlo event generator which in this context is equivalent to the LUND geometry code. The two distributions are similar. We use the Hijing distribution for all further analysis unless explicitly stated otherwise.

We follow the method of [15] and allow X to be a free parameter in the fit of the N_{grey} distributions. The GCM fits are displayed as the dashed curves in Fig. 31.2. The model tends to fall below the data for low N_{grey}, and above the data for high values. Note that the GCM distribution imposes no maximum on the number of protons that can be emitted from a single nucleus. The mean and dispersion for ν are given by the probability distribution in Eq. 31.2, displayed as the open circles in Fig. 31.3.

The intra-nuclear cascade of [12, 13, 20] takes a very different approach in relating N_{grey} to ν. It assumes that all primary struck nucleons follow the initial projectile trajectory and only secondary nucleons and a fraction of the primary protons contribute to N_{grey}. In this case $\overline{\nu}$ is very nearly proportional to $\sqrt{N_{grey}}$,

$$\overline{\nu}(N_{grey}) = \overline{\nu}\sqrt{N_{grey}/\overline{N_{grey}}}. \tag{31.5}$$

Applying Eq. 31.5 leads to the solid curves in Fig. 31.3, which differ significantly from the predictions of the GCM. Furthermore, the quadratic dependence of N_{grey} on ν is very different from the linear relationship of the GCM.

The contradictory nature of these two models led us to introduce another model, which allows for both a linear and quadratic dependence of N_{grey} on ν, with the relative strengths determined by a fit to the data. The principal assumption is that for a given target, there exists a relation between the mean number of grey tracks detected and the

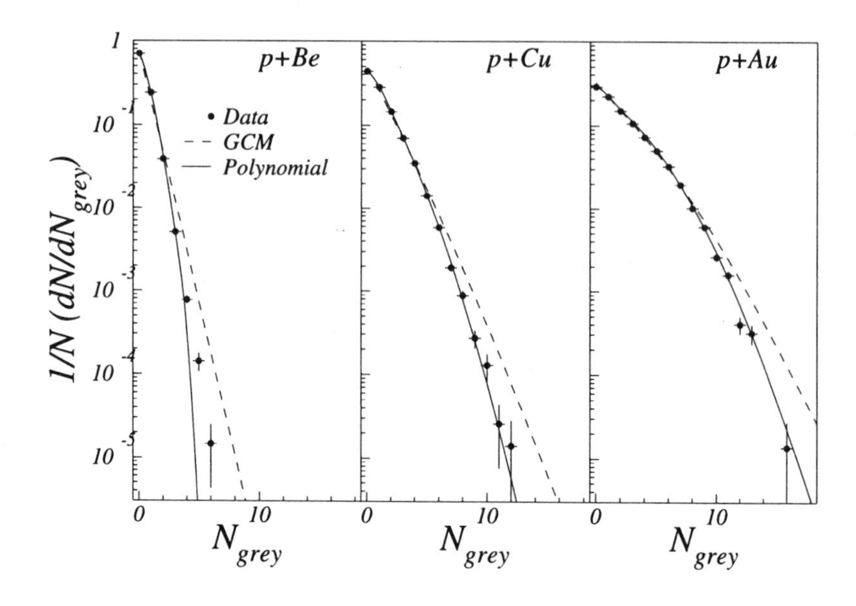

Figure 31.2 Log-likelihood fits to the event normalized N_{grey} distributions for Be, Cu, and Au targets with two models: The Geometric Cascade Model (dashed lines) and the Polynomial Model (solid lines).

number of primary interactions which takes the form of a second degree polynomial,

$$\overline{N_{grey}}(\nu) = c_0 + c_1\nu + c_2\nu^2. \tag{31.6}$$

We furthermore assume that the distribution is governed by binomial statistics; a total of Z target protons exist which can be emitted and detected with probability $\overline{N_{grey}}(\nu)/Z$,

$$P(N_{grey}|\nu) = \binom{Z}{N_{grey}} \left(\frac{\overline{N_{grey}}(\nu)}{Z}\right)^{N_{grey}} \left(1 - \frac{\overline{N_{grey}}(\nu)}{Z}\right)^{Z-N_{grey}} \tag{31.7}$$

The full distribution of $P(N_{grey})$ is again given by a weighted sum over $\pi(\nu)$ of Eq. 31.4, and the coefficients of Eq. 31.6 are derived from a fit to the data. The fitted function for this *polynomial* model is shown as the solid curve in Fig. 31.2, and the coefficients are given in Table 31.2. The quadratic coefficients for both the Au and Cu targets were determined to be zero. For the Be target, the distribution does not extend far enough to allow independent determination of a linear and quadratic coefficient.

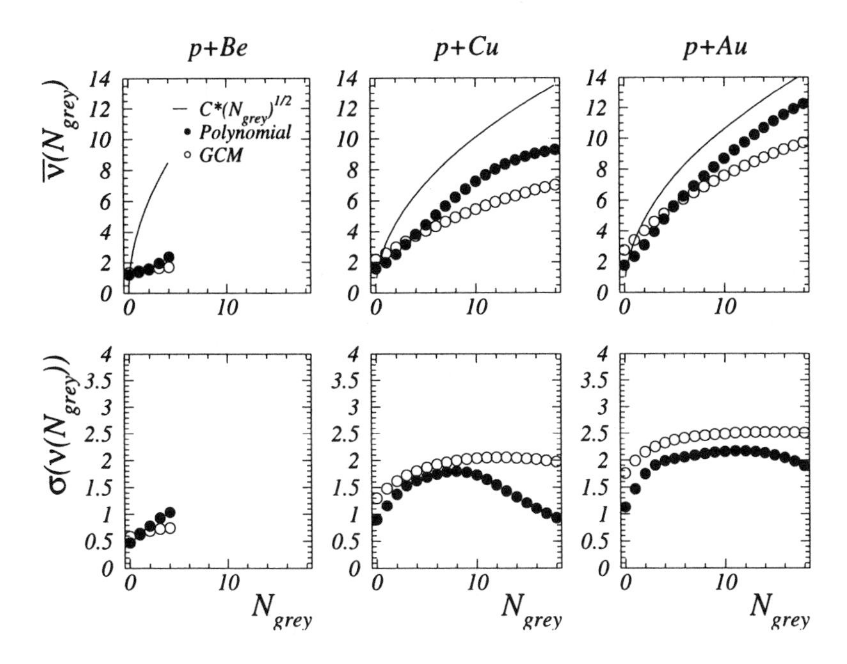

Figure 31.3 $\overline{\nu}(N_{grey})$ and $\sigma(\nu(N_{grey}))$ generated from the Polynomial Model (solid circles) and the GCM (open circles), and $\overline{\nu}(N_{grey})$ according to the $\overline{\nu}^2$ ansatz (solid line) for p+Be, p+Cu and p+Au.

Table 31.2 Coefficients for polynomial fit to N_{grey}.

Target	c_0	c_1	c_2	χ^2/dof
Au	-0.27 ± 0.02	0.63 ± 0.01	-0.0008 ± 0.0012	1639/13
Cu	-0.17 ± 0.02	0.51 ± 0.02	-0.00005 ± 0.00242	15/10
Be	-0.075 ± 0.008	0.306 ± 0.006	—	95/5

Given that in the fits to heavier targets the linear term is dominant and the quadratic term is negligible, we remove the quadratic component for the fits to the Be data. Fig. 31.2 shows that the polynomial model reproduces the data more accurately than the GCM. For a negligible quadratic term, the polynomial model differs from the GCM in only two respects, the presence of a constant term in Eq. 31.6 and the use of binomial statistics. Fig. 31.3 gives $\overline{\nu}(N_{grey})$ for all three models, and the dispersions for the GCM and polynomial models. The polynomial and GCM results are quite similar; they seldom differ by more than 15%, and never more than the dispersion of the GCM. In contrast, the intra-nuclear cascade differs significantly from the other two, with the difference increasing for the lighter targets.

4. MODEL COMPARISONS

A model data set of 200 K p+Au interaction events were generated with RQMD 2.2 running in fast cascade mode in the fireball approximation with all strong decays enforced. The N_{grey} count from our RQMD simulations includes only protons. Presumably some protons that contribute to N_{grey} would bind with neutrons to form deuterons with roughly twice the momentum of the proton. These deuterons would then fall within the N_{grey} momentum range for deuterons, leaving the overall N_{grey} unaltered. The RQMD output was then passed as input to the same GEANT simulation and track reconstruction used to calculate the E910 acceptance. The same momentum cuts were used to define the grey tracks, although proton identification was taken directly from the input. We did not simulate the positron contamination and no forward angle cuts were applied. This data set is labeled "RQMD E910". We also examine the full distribution of N_{grey} (no acceptance cuts), which includes all protons within the momentum range specified for N_{grey}. We refer to this data set as "RQMD 4π". The N_{grey} distributions and comparisons to data are given in [16]. The GCM and polynomial fits were performed for both the RQMD E910 and RQMD 4π N_{grey} distributions. The analysis procedure remains the same as it was for the data; the Hijing distribution for $\pi(\nu)$ is again used in the fit.

The main goal in analyzing RQMD with the GCM and polynomial model is to compare the extracted $\overline{\nu}(N_{grey})$ values with the intrinsic ν of RQMD. For RQMD, the definition of ν requires some explanation. Above a certain energy threshold, cross-sections in RQMD are governed by the Additive Quark Model (AQM). To obtain the appropriate value of ν for comparison, we examine the history file and count all collisions suffered by particles that carry valence quarks of the projectile. Counting for produced particles stops when the formation times elapse, and multiple collisions of valence quark-bearing particles with the same target nucleon are counted only once.

The comparison for $\overline{\nu}(N_{grey})$ and $\sigma(\nu(N_{grey}))$ among the GCM and polynomial analyses of RQMD and their intrinsic values in RQMD are shown in Fig. 31.4. The relative difference between the intrinsic ν of

RQMD and that extracted from the polynomial fit is re-plotted as a relative quantity in Fig. 31.5b. This is compared to the combined relative systematic errors in $\overline{\nu}(N_{grey})$ of Fig. 31.5a. The three changes in the analysis used to estimate systematic errors correspond to an alternate (historical) set of momentum cuts, the Glauber $\pi(\nu)$ distribution, and excluding the N_{grey}=0 bin. See [16] for further details.

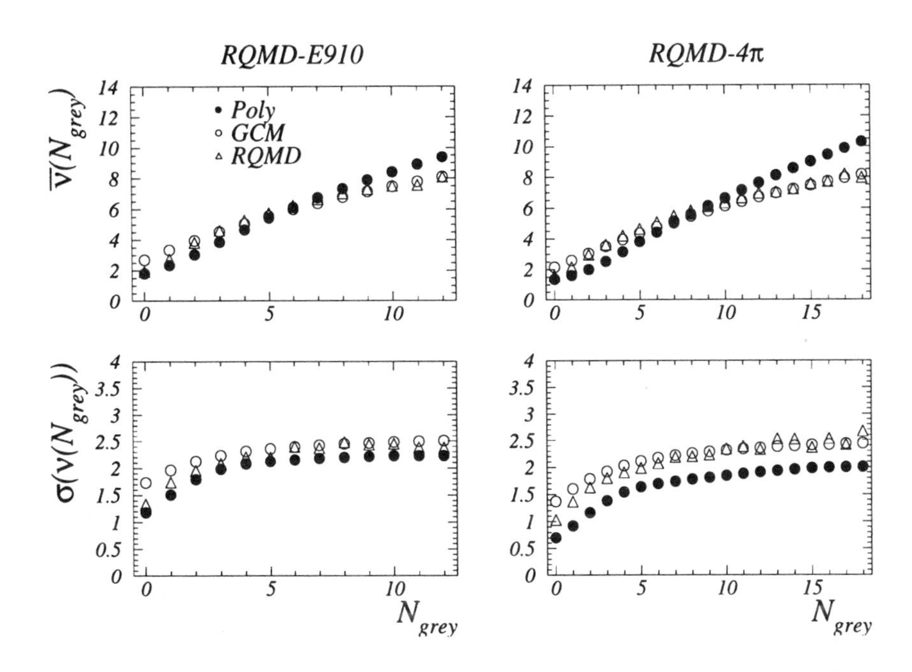

Figure 31.4 Comparison of $\overline{\nu}(N_{grey})$ and $\sigma(\nu(N_{grey}))$ for RQMD values and GCM and Polynomial fits to the RQMD N_{grey} within the E910 acceptance and over 4π.

5. CONCLUSIONS

We have measured the slow proton and deuteron production in 18 GeV/c proton collisions with three targets, Be, Cu, and Au in the momentum range relevant to a determination of collision centrality. RQMD, a full intra-nuclear cascade model, provides reasonable agreement with the N_{grey} distribution for the p-Au data. The simple GCM and Polynomial models are also fit to the data as part of the procedure to extract $\overline{\nu}(N_{grey})$. The GCM imposes no upper bound on the number of protons that can be emitted and therefore over-predicts the N_{grey} distributions for all targets. Though not a perfect fit (χ^2/dof of 1–100), the Polynomial model gives a better description of the data.

Our main result is the determination of centrality for a set of collisions from the measured N_{grey} with two different models. The predictions of the two models differ by less than the predicted dispersions for most N_{grey}. Both models have been checked against a full cascade model, RQMD, and the intrinsic $\overline{\nu}(N_{grey})^{\mathrm{RQMD}}$ lies between the GCM and polynomial results. On the basis of the fits to the data, we ascribe the more accurate measure of $\overline{\nu}(N_{grey})$ to the polynomial model. Finally, we establish a systematic error for this centrality measure that is 10-20% of $\overline{\nu}(N_{grey})$.

We are then able to use $\overline{\nu}(N_{grey})$ to study the centrality dependence of particle production and stopping. We find that stopping, as measured by the rapidity loss of the leading baryon reaches approximately Δy [illegible] after 2-3 collisions, and saturates thereafter. This result is independent of target. In contrast, negative pion production depends strongly on target. For Au and Cu targets, pion production is roughly given by the wounded nucleon model. Saturation for very central collisions appears to result from pions peaking backwards of our acceptance. Negative pion production with the Be target increases faster than predicted by the wounded nucleons model, and is more closely approximated by the binary collision model. These results are given in [21].

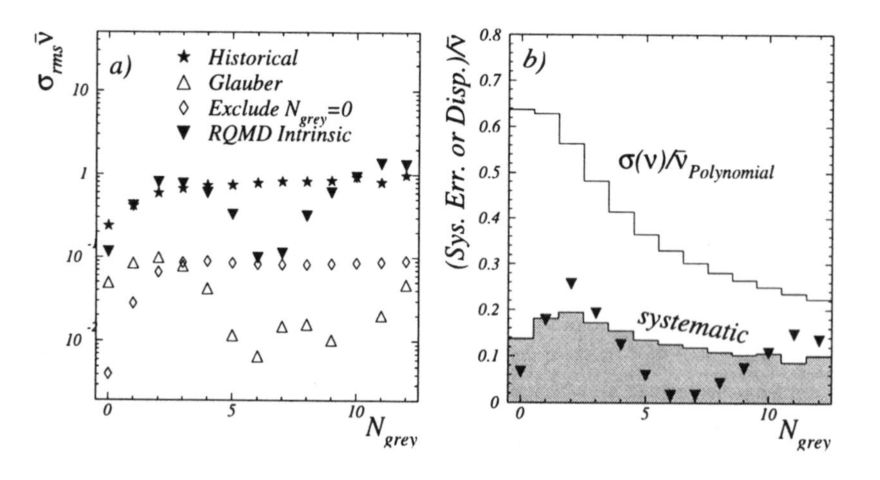

Figure 31.5 a) the rms distribution for $(\overline{\nu}' - \overline{\nu})$ as a function of N_{grey}, where the prime indicates an alternate set of systematic cuts, or the intrinsic value from RQMD. b) the relative systematic error (shaded) for the combined one standard deviation errors from the three re-analyses described in the text. For comparison the RQMD relative intrinsic difference (triangles) and the polynomial model relative dispersion (unshaded) are also shown.

References

[1] E. Feǐnberg, Jour. Exp. Theor. Phys. **23**, 132 (1966).

[2] A. Dar and Vary, Phys. Rev. D **6**, 2412 (1972).

[3] A. Goldhaber, Phys. Rev. D **7**, 765 (1973).

[4] P. Fishbane and J. Trefil, Phys. Lett. B **51**, 139 (1974).

[5] W. Busza *et al.*, Phys. Rev. Lett **34**, 836 (1975).

[6] J. Elias *et al.*, Phys. Rev. D **22**, 13 (1980).

[7] W. Yeager *et al.*, Phys. Rev. D **16**, 1294 (1977).

[8] W. Busza, in *High-Energy Physics and Nuclear Structure*, edited by D. Nagle *et al.* (AIP Conf. Proc., New York, 1975).

[9] This terminology comes from the identification of emulsion tracks.

[10] B. Andersson, I. Otterlund, and E. Stenlund, Phys. Lett. B **73**, 343 (1978).

[11] E. Stenlund and I. Otterlund, Nucl. Phys. **B198**, 407 (1982).

[12] M. K. Hegab and J. Hufner, Phys. Lett. B **105**, 103 (1981).

[13] M. K. Hegab and J. Hufner, Nucl. Phys. **A384**, 353 (1982).

[14] N. Suzuki, Prog. Theor. Phys. **67**, 571 (1982).

[15] R. Albrecht *et al.*, Z. Phys. C **57**, 37 (1993).

[16] I. Chemakin *et al.*, , nucl-ex/9902003, submitted for publication in Phys. Rev. C.

[17] H. Sorge, H. Stoecker, and W. Greiner, Ann. Phys. **192**, 266 (1989).

[18] R. Glauber, in *High Energy Physics and Nuclear Structure*, edited by A. Gideon (North-Holland, Amsterdam, 1967), p. 311.

[19] X. Wang and M. Gyulassy, Phys. Rev. D **44**, 3501 (1991).

[20] W. Chao, Hegab, and Hufner, Nucl. Phys. **A395**, 482 (1983).

[21] I. Chemakin *et al.*, , nucl-ex/9902009, submitted for publication in Phys. Rev. Lett.

Chapter 32

STATUS AND PHYSICS PERSPECTIVES OF THE NEW DIELECTRON SPECTROMETER HADES

J.Stroth, for the HADES collaboration
Institut für Kernphysik
Johann Wolfgang Goethe-Universität
60486 Frankfurt, August-Euler-Strasse 6, Germany
J.Stroth@gsi.de

Abstract HADES is a High Acceptance DiElectron Spectrometer designed to measure dielectrons emitted in central collisions of the heaviest ions at energies in the few GeV/u regime. The spectrometer is being installed at GSI, Germany. Making use of heavy ion and particle (π, p) beams the physics of in-medium modifications of hadrons and electromagnetic transition form factors of mesons in the vacuum will be addressed. This contribution discusses the experimental concept and programme and the status of the project.

Keywords: dielectrons, heavy ion collisions, in-medium modifications, vector mesons, electromagnetic form factors, spectrometer

1. INTRODUCTION

Nuclear fireballs formed in central collisions of heavy ions serve as laboratories for the investigation of nuclear matter under extreme conditions. Such fireballs are transient states with life times of a few 10 fm/c ending in a state of a non (strong) interacting hadron gas. Depending on the size of the collision system and available centre of mass energy the nuclear fireball has a certain initial temperature and density and can even pass through a deconfined state of quarks and gluons, the quark gluon plasma [1]. In contrast to hadronic probes, the kinematical properties of lepton pairs are not governed by the freeze-out conditions. As soon as they are produced they leave the fireball without perturbation and carry the information about their production vertex to the detectors. Hence they are considered a direct signal, emerging from any time during the expansion of the fireball.

A challenging experimental task is the spectroscopy of in-medium properties of hadrons using dileptons. The light neutral vector mesons ρ, ω and ϕ are best suited for this purpose. Carrying the same quantum numbers like the photon, they can directly annihilate by radiating off a virtual gamma materialising as a lepton pair. Thus the twofold production cross section for lepton pairs

$$\mathrm{d}^2\sigma^{l^+l^-}/(\mathrm{d}M\mathrm{d}p),$$

with M the invariant mass of the pair and p its momentum in the fireball rest frame, is directly linked to the in-medium spectral function of the vector mesons. A number of theoretical papers published in the recent past demonstrated the relevance of such information for the understanding of the non-perturbative sector of QCD and the role of partial restoration of chiral symmetry in hot and dense nuclear matter [2].

Until today, dielectron spectra from proton induced reactions on heavy ions (HI) and central HI collisions were measured at BEVALAC and SPS energies [3, 4]. From the interpretation of the experimental data it became evident that a certain fraction of the dilepton yield does not origin from hadrons after freeze-out. The origin of that extra yield can only be understood assuming "medium-effects"[1]. However, an unambiguous identification of the sources and their properties is not possible at the moment. Limited mass resolution of the experiments (around 10 %), the small signal to background ratio and partially the limited statistics are the main experimental reasons.

The HADES collaborations has constituted to set up a second generation dielectron spectrometer at the SIS (few GeV/u) at GSI[2]. The experimental programme comprises dielectron spectroscopy in HI collisions and p and π induced reactions. In the following section we will briefly explore the experimental concept of the HADES spectrometer. The physics programme of HADES with emphasis on the particular aspects of the fireball dynamics at SIS energies and its impact on the dilepton signal is discussed in section 3. The status of the project is outlined in section 4.

2. THE EXPERIMENTAL CONCEPT

The HADES spectrometer is designed to measure electron pairs (e^+, e^-) produced in central collisions of very heavy ions at typical beam energies of 1-2 GeV/u, thus covering exactly the beam energy range of the two-arm dilepton spectrometer DLS operated at the BEVALAC. It is

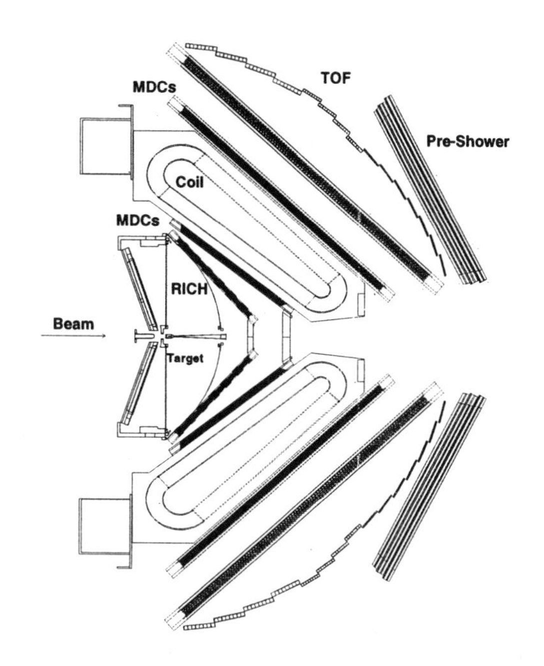

Figure 32.1 Schematic cross section through the symmetry lines of two opposite sectors of the HADES spectrometer. Two coils of the super conducting toroid, which are positioned at the boarders of the sectors, are projected into that plane. The beam enters the set-up from left, the target position is in the interior of the RICH.

therefore considered a second generation experiment aiming at considerably improved experimental capabilities. Two design parameters are essential in that respect: (a) An invariant mass resolution for electron pairs of 1% (σ) in the region of light vector mesons. This will allow to distinguish between contributions to the invariant mass spectrum from the ω and the ρ decay. (b) Sufficient statistics to accumulate invariant mass spectra for various cuts in the observed parameter space e.g. centrality, transverse momentum and rapidity of the lepton pair. Besides that, low Z material was used in the construction and emphasis was put on a good two-track resolution, particularly of the inner two drift chambers, to keep combinatorial background low.

The HADES spectrometer is composed out of six identical sectors each of trapezoidal shape and covering 60° of the full azimuth. The sectors are arranged in a frustum-like geometry around the beam pipe with polar acceptance between $18^o < \Theta < 85^o$. Fig. 32.1 shows a schematic cross section through the spectrometer. The detector systems include (going from the target outwards) a fast Ring Imaging Cherenkov detector (RICH), four sets of planar Multi-wire Drift Chambers (MDC), a Time-Of-Flight wall (TOF) and a Pre-Shower detector for polar angles $\Theta < 45^o$. A momentum kick to the charged particles is given by a toroidal field produced by six super-conducting coils positioned at the border of the sectors. With this setup, a flat electron pair acceptance of $\simeq 40\%$

over a wide rapidity and transverse momentum range is achieved for pairs emitted from the fireball. The detector systems, read-out electronics and data acquisition is designed to cope with interaction rates as high as 10^6 HI collisions/s. To allow that, the spectrometer is operated with a multi-level trigger system relying on fast lepton identification and track generation.

2.1 LEPTON IDENTIFICATION

Typically in about 1 out of 10^5 central collisions of e.g. $Au + Au$ at 1 GeV/u an electron pair due to an ω decay is emitted from the fireball. The fast identification of events containing lepton pairs is therefore crucial. This task is mainly accomplished by the RICH. Its radiator volume is filled with C_4F_{10} gas which has a $\gamma_{th} \simeq 18$, high enough to be blind with respect to hadrons (in particular pions) emitted at moderate velocities at these beam energies. Cherenkov light radiated off by electrons and positrons is reflected backward by a thin spherical UV-mirror. The rings are imaged on a fast MWPC with a solid CsI photocathode, which is positioned upstream the target where direct irradiation by charged particles is suppressed. The rings have a constant diameter of 5 cm and are formed of typically > 10 detected photons ($N_0 = 117$). The overall thickness seen by a particle traversing the RICH is $x/X_0 = 2\,10^{-2}$.

At the outer part of the spectrometer, electron candidates are discriminated a second time making use of time-of-flight information, and, at small polar angles, where pions can also reach velocities close to c, by their characteristic showers induced in the Pre-Shower detector. The latter consists of a set of three pad chambers interspersed by two lead converters. Electrons are identified comparing the detector response in the first chamber (where the particles have not yet traversed a converter) with the one in chamber 2 and 3, where the electromagnetic shower process in the lead converter leads to an irradiation of many surrounding pads.

2.2 HIGH RESOLUTION TRACKING

For tracking of the charged particles a novel type of low-mass drift chambers is used [5]. In total 24 planar chamber modules are grouped in four planes, two in front of and two behind the magnetic field. Each chamber has six drift cell layers with stereo angles $\pm 40^o, \pm 20^o$ and 0°, optimized for resolution in the direction of the momentum kick (90°). In order to achieve a 1% invariant mass resolution electrons with momenta larger than 0.1 GeV/c must be tracked with a position resolution better than 80 μm and multiple scattering must be kept on the same order of magnitude. Cell sizes increase from 5 to 14 mm, thus keeping a constant granularity that is able to handle multiplicities of up to 25 charged particles per sector. The contribution of multiple scattering to the momentum resolution, which is dominant below 0.4 GeV/c, is limited using low mass materials in the chamber (Aluminium field and sense wires, 60/40 He/i-butane gas mixture) and by using Helium bags in the magnetic field region. The total radiation thickness of the tracking system amounts to $x/X_0 \simeq 2.5\,10^{-3}$.

2.3 THE TRIGGER SYSTEM

In a typical heavy ion run, the detector system is triggered for central collisions (10% of the highest multiplicities in the TOF array) at trigger rates of up to 10^5/s. For each of these first level (LVL1) trigger, the complete detector information is digitised and stored in front-end pipeline memory. To cope with this primary data rate, which amounts to 1-2 GByte/s, events with true dielectron signature are searched in two subsequent trigger levels. Only for those events, which have fulfilled the trigger conditions, the data is finally written to tape, achieving reduction factors of $> 10^{-3}$. The LVL2 trigger uses combined information from image processors searching for rings in the RICH photon detector and identifying electromagnetic shower candidates in the Pre-Shower detector. The result of a fast tracking in the MDC, not using the drift time information but cell numbers of hit cells only, is compared to track candidates from the LVL2 in the LVL3 decision. The remaining events are finally collected using an ATM network.

3. THE PHYSICS PROGRAMME

The high demands upon the spectrometer are dictated by the particle multiplicity of a central collision of very heavy ions. However, the physics programme of HADES also includes dielectron production in hadron induced heavy ion reactions and elementary reactions. For that purpose, a secondary beam available at GSI can be directed to the experimental site of HADES and the segmented solid target can be exchanged by a liquid Hydrogen target. The expected intensities of secondary pion beams are shown in Fig. 32.2.

3.1 EXPERIMENTS USING HEAVY ION BEAMS

Medium modifications of hadrons, i.e. the change of the spectral function of a hadronic states as a function of the baryon density and tem-

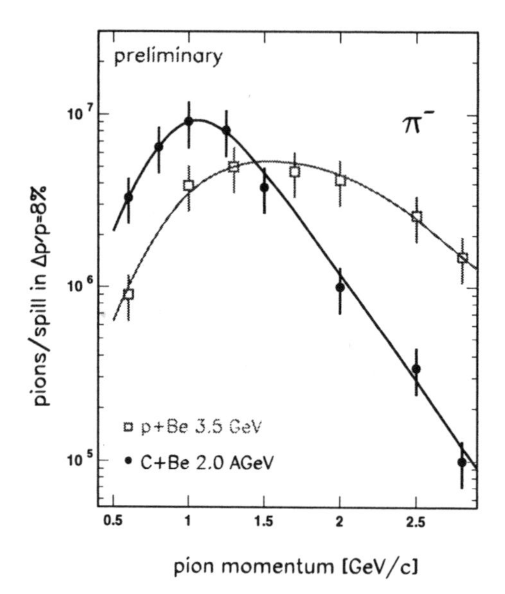

Figure 32.2 Maximum beam intensities of the secondary pion beam at GSI for primary $p + Be$ reactions (squares) and $C + Be$ collisions (dots) [6]. The yields plotted are extrapolations under the assumption that the primary beam intensity is as high as constraint by the space charge limit of the SIS accelerator.

perature of the surrounding nuclear matter, is a prime issue of heavy ion physics at relativistic energies. However, the macroscopic quantities temperature and density characterising the fireball can not easily be experimentally controlled. Even more, the assumption of an adiabatically evolving thermalised fireball is at least in the energy regime of a few GeV/u questionable. On the other hand, possible medium effects are best observed for well-defined (constant) properties of the fireball, which is in contradiction to the fact that the fireball is a highly transient state. A way to tackle this conceptional problem is sub-threshold particle production. The free production threshold for a ρ meson in pp collisions (fixed target) is at 1.85 GeV proton energy. In heavy ion collision with energies per nucleon below this threshold vector mesons are produced through coherent processes. A typical mechanism is the creation of a Δ resonance in a collision of two nucleons where a ρ is produced in a subsequent collision of the Δ with a third nucleon. Through the strong density dependence of such two-step processes, they occur predominantly in the high density phase of the collision.

The time evolution of the central density in HI collisions with different centre of mass energies was investigated by Friman and coworkers [8]. They calculated the time evolution of the chiral condensate expectation value, which is the leading order parameter for chiral symmetry restoration. Loosely spoken, the depletion of the chiral condensate in the interior of the fireball formed in the collision goes with the number

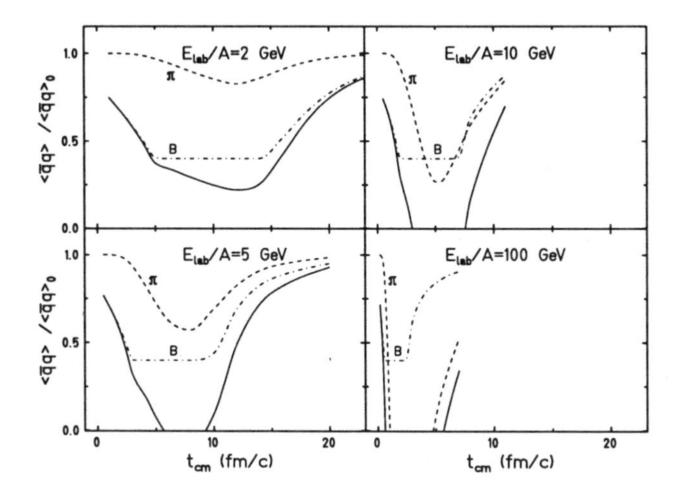

Figure 32.3 Time evolution of the chiral condensate in the central region of a fireball formed in central $Au+Au$ collisions for 4 different beam energies [8]. The depletion is normalized to the vacuum value (full line) and the effect of baryon and pion abundance is indicated as dashed-dotted and dashed lines, respectively.

density of baryons and pions. Their results demonstrate (see Fig. 32.3) that at BEVALAC/GSI energies, where only moderate temperatures are reached, the depletion is driven by the baryon density. At SPS energies, temperature is high and pions dominate over baryons. Here the depletion is caused by the pions. It is worthwhile to note that the moderate temperatures at BEVALAC/GSI energies create less pressure and consequently the fireball stays longer at enhanced baryon densities. Consequently, beam energies around a few GeV/u are well suited to study in-medium modification of vector mesons. The vector mesons production is confined to the high density phase of the collision and the system stays for a reasonably long time at densities of 2-3 ρ_0.

A second promising method to study in-medium modification in a clean environment is recoilless production of vector mesons using a pion beam impinging on heavy targets. At selected pion energies of $\simeq$ 1.3 GeV ω mesons with small laboratory momenta can be produced in collisions with protons [7]. Although omega mesons have a lifetime in vacuum of 23 fm/c, a measurable fraction of omega mesons will decay inside the medium due to the low relative velocities with respect to the target nucleus. In addition, the decay width of the omega meson increases drastically in the medium, basically due to coupling to baryons. The high production cross section and the good signal to background in these collisions overcome the disadvantageous fact that this experiment can obviously be preformed at nuclear matter ground state density only.

3.2 ELEMENTARY REACTIONS

A substantial fraction of electron pairs with invariant masses around the ρ mass and below stems from $\pi\pi$ annihilation. The pion electromagnetic form factor in the time-like region is well described by vector meson dominance. Hence the decay of two pions proceeds through p-wave coupling to a single virtual photon. Two pions can couple in the σ (f_0) channel (s-wave) and then decay by emission of two photons. A process one could think of in loose analogy as a Dalitz-decay of the σ. The role of these two processes to the dielectron yield could be investigated in more detail in πp reactions.

In future, combining HADES with an electromagnetic calorimeter, the spectrometer could be used to study transition form factors of mesons. An interesting case is the transition form factor in the ω-Dalitz decay

$$\omega \rightarrow \pi^0 + e^+ + e^-,$$

which can be investigated through the reaction

$$\pi^- + p \rightarrow \gamma + \gamma + e^+ + e^-.$$

In this reaction, the intermediate ω can be observed by the combined invariant mass of the electron pair and the neutral pion detected in the calorimeter through its gamma decay. An other option would be to detect the outgoing neutron by a hadronic calorimeter (like e.g. the Large Area Neutron Detector LAND [9]) and requiring the missing mass of the pion.

4. STATUS OF THE PROJECT

The R&D of the detector system was successfully completed in 1997. In test measurements with prototypes using proton and Uranium beams electron tracks could be identified. Minimum ionizing particles were tracked with a single drift cell resolution of 70 μm (σ). The mass production of detectors has started and first modules were installed at the experimental site and tested with beam in December 1998. Production is in progress and most of the system will be installed during summer 1999. The assembly of the outer tracking chambers has just started and completion is expected for the end of 2001. However, the physics programme will start, after commissioning runs in this year, early 2000 with a somewhat reduced invariant mass resolution (5-6% σ) using the granular TOF system for particle tracking behind the magnetic field. Among the first experiments addressed are a measurement of the dielectron spectrum for light symmetric systems (i.e. $C+C$ at 2 GeV/u). The search for a narrow ω or ϕ signal in light HI collisions or elementary reactions starts as soon as two sectors with outer drift chambers are installed.

Acknowledgments

Clarifying discussions with B. Friman and M. Lutz are gratefully acknowledged. We thank B. Friman for providing us with Fig. 32.3.
Supported by EC (HTM, TMR programme), ASCR, BMBF, DAAD, DFG, DLR (WTZ), GSI, IN2P3, INFN, INTAS, NATO and the local Institutes.

Notes

1. Quotes are used to indicate that the word is used in its most general sense.
2. At SPS energies the experiments will be continued by the CERES collaboration with an improved apparatus, see contribution to this conference.

References

[1] for recent reviews see:
Harris, J.W. et al. (1996) Annu. Rev. Nucl. Part. Sci. **46**, 71
Shuryak, E.V. (1993) Rev. Mod. Phys. **65**, 1
Bass, S.A. et al. (1999) J.Phys. G **25**, R1

[2] for a recent review see:
Cassing, W., Bratkovskaya, E.L. (1999) Phys. Rep. **308**, 65

[3] DLS collaboration:
Porter, R.J. et al. (1997) Phys. Rev. Lett. **79**, 1229
Wilson, W.K. et al. (1998) Phys. Rev. C **57,4**, 1865

[4] CERES collaboration:
Agakishiev, G. et al. (1995) Phys. Rev. Lett. **75**, 1272
Agakishiev, G. et al. (1995) Phys. Lett. B **422**, 405

[5] Garabatos, C. et al. (1998) Nucl. Inst. Meth. A **412**, 38

[6] Schicker, R. et al. (1999) Spring Meeting of the German Physical Society, Freiburg. Unpublished.

[7] Schoen, W. et al. (1996) Act. Phys. Polon. B **27**, 1908

[8] Friman, B. et al. (1998) Eur. Phys. J. A **3**, 165

[9] LAND collaboration (1992) Nucl. Inst. Meth. A **314**, 136

Chapter 33

OVERVIEW AND STATUS OF THE STAR DETECTOR AT RHIC

W.B. Christie - for the STAR Collaboration
Brookhaven National Laboratory
Upton N.Y., USA
christie@bnl.gov

Abstract Presented here is the current status of the STAR Detector. STAR is one of the four detectors being constructed at the RHIC collider facility. The STAR detector is scheduled to have its first engineering run with the RHIC beams about six months from the date of this conference. The STAR project is on schedule and expects to be complete on time.

Keywords: RHIC, quarks, gluons, hadrons, STAR, QGP

1. INTRODUCTION

The STAR (Solenoidal Tracker At RHIC) detector is one of the four detectors presently under construction at the RHIC (Relativistic Heavy Ion Collider) Facility. In this document the current status of all of the major subsystems of the STAR detector is presented. The baseline STAR detector systems are the Time Projection Chamber (TPC) and its readout electronics, the Central Trigger Barrel (CTB) and the associated trigger electronics, the Solenoid magnet, and the Data Acquisition System (DAQ). Also included in the STAR baseline systems are the Slow Control system, the Online computing system, and the offline event reconstruction and analysis software.

STAR also has a number of additional detectors in its future. These detectors include the Barrel Electromagnetic Calorimeter (BEMC), the Endcap Electromagnetic Calorimeter (EEMC), the Silicon Vertex Tracker (SVT), the Forward Time Projection Chambers (FTPCs), a Time of Flight System (TOF), and a Ring Imaging Cherenkov Detector (RICH).

In what follows I will give a brief overview of the various STAR detector subsystems and give the current status of each system.

2. THE TIME PROJECTION CHAMBER

The TPC is a large acceptance charged particle tracking detector. It has an inner radius of 50 cm, an outer radius of 200 cm, and extends 210 cm in each direction from the central cathode membrane. The acceptance of the TPC covers 2 π in azimuth, and $|\eta| < 1$ in pseudorapidity for tracks that traverse to the outer radius of the device before reaching the anode planes at +- 210 cm. With decreasing momentum resolution the acceptance of the device extends out to $|\eta| < 2$. Tracks that traverse the entire radius of the TPC leave hits on up to 45 padrows. Using the energy loss in the gas (P-10) for the reconstructed charged particle tracks one can extract particle identification for pions and kaons out to transverse momenta (p_T) of about 600 MeV/c, and identify protons out to p_T of about 1.3 GeV/c.

The TPC was designed and constructed at Lawrence Berkeley National Lab (LBNL). It was tested with cosmic rays while still at LBNL in the Fall of 1997. In November of 1998 the TPC was flown out to Brookhaven National Lab (BNL) on a large (C5) military transport plane. The device was then setup in the STAR assembly building (AB). After installing and commissioning the TPC gas system and various temporary support systems (e.g. power supplies, cooling water, high voltage, cosmic ray trigger system, etc.) the TPC was again used to collect cosmic ray data. Only one sector of the TPC was outfitted with electronics at a time for this testing. By moving the electronics around, data was collected for three of the twelve sectors on each end of the TPC. By analyzing the residuals for the reconstructed tracks, a study was done to look for distortions in the drift of the ionization electrons in the TPC. Using information gathered through the analysis of the cosmic ray tracks a few shorts were located in the field cage and some tuning of the voltage potentials at the ends of the field cage was performed. Some additional testing that has taken place for the TPC includes:

- Anode, Cathode, and Ground planes have all passed integration tests.
- DC testing of the gating grids have been completed on all sectors.
- The automatic drift velocity control system has been tested.
- A proof of principal of the laser calibration system has been performed, generating "real" calibration data from the TPC.

3. TPC FRONT END ELECTRONICS

The Front End Electronics (FEE) for the TPC consist of what are known as FEE boards and Readout boards. There are 4,344 Front End Electronics (FEE) boards used for the TPC readout. The present status is that all of the FEE boards are in hand. More than 4,500 FEE boards have been produced and tested. The readout of the TPC is subdivided into twenty-four sectors. Twenty-two of the twenty-four sectors have all of their front end boards installed. There are a total of 144 readout boards required to read out the TPC. 160 of these readout boards have been produced and are currently being tested. Six pre-production prototype readout boards were used in the cosmic ray tests.

The FEE are powered by low voltage power supplies which reside on the three story platform that is attached to the South side of the STAR magnet. All of these low voltage supplies have been designed, procured, and installed. Approximately 80% of the cabling for the TPC has been installed.

4. STAR SOLENOID MAGNET

The STAR solenoid magnet has an inner diameter of 5.26 m, a length of 6.2 m, and a nominal operating field of 0.5 T. The construction of the STAR magnet has been completed. All of the power supplies (five in total) used to power the STAR magnet have been designed, procured, and installed. The water cooling system for the magnet, which supplies 1100 gpm of water to take away the 3.5 MW of heat generated when the magnet is in operation, has been installed and brought online.

The magnetic field for the STAR magnet was mapped in November of 1998. The mapping device that was used was borrowed from CERN. High precision, complete field maps were made at half field (0.25 T) and full field (0.5 T). The maps confirmed that the magnetic field uniformity that had been specified for the design of the magnet, necessary to constrain the azimuthal displacement of the drifting electrons throughout the volume of the TPC, had been achieved.

5. DATA ACQUISITION SYSTEM

The data acquisition system (DAQ) is the only electronic system for STAR that does not reside either directly on the detector itself, or on the three story platforms attached to either side of the detector. The DAQ system for the TPC consists of 144 "receiver" boards, one board for each of the 144 readout boards in the TPC FEE system. The data from the TPC, digitized and multiplexed, is carried from each readout board to a receiver board through 144 optical fibers. Each of the receiver boards has three daughter or mezzanine boards attached to it. In the baseline configuration for STAR these 144 receiver boards reside in twelve VME crates in our DAQ room. Each of these twelve VME crates also contain "sector brokers" and networking hardware. Finally, there is an "event builder" which collects all of the data from the TPC crates, as well as data from the other STAR detectors, and builds the whole event.

The present status of the DAQ system is that 10 of the eventual 177 receiver boards are at BNL and being tested. Twenty out of the 530 mezzanine boards are in house at BNL. All of the receiver boards should be at BNL by May of this year. All twelve of the sector brokers have been procured and tested at BNL. The software for these devices has been developed and tested. The event builder, which utilizes a commercial CPU, has been procured. The alpha version of the software for the device has been written and is currently being tested. Events have been run through parts of the DAQ system and delivered to tape and to disk. The writing to disk is a placeholder for the eventual readout scheme that has the STAR DAQ writing out its events to what's known as a buffer box, which then passes the data to the RHIC computer facility via a dedicated optical fiber link.

6. TRIGGER

The baseline trigger system for STAR consists of the Central Trigger Barrel (CTB), and an electronics system that provides for multiple stage trigger decisions. The Central trigger barrel consists of 120 "trays" that attach to the outer field cage of the TPC. Each tray covers 6 degrees in azimuth and one unit of pseudorapidity (0 to +1 or 0 to -1). Inside of each tray are two scintillator slats, covering one half a unit of pseudorapidity, and their readout photomultiplier tubes.

The present status of this system is that all of the trays have been produced. Tray installation has begun onto the TPC, and will be completed in February. The software interface between the trigger and the STAR online computing system has been defined and is being tested. A production prototype of the "Trigger control unit" (a component of the trigger electronics) has been produced and tested. A prototype Data Storage and Manipulation board (another piece of the electronics) has also been built and tested. Work is ongoing on the design of electronics which will enable the anode wires on the TPC readout to be readout and used in the trigger. This will extend the pseudorapidity coverage of the trigger detectors to four units ($|\eta| < 2$).

7. ONLINE SOFTWARE

The online computing system for STAR ties together the DAQ, trigger, and slow controls systems and provides the user interface for STAR run control. The requirements for this system have been written and have passed their review within STAR. A resource loaded schedule for the online system is complete. The essential architecture for the system has been defined, and the interfaces to the trigger and slow controls systems have been worked out. The interface to the DAQ system is in progress. A prototype of the online system will be ready for testing with the DAQ and trigger systems soon.

8. OFFLINE SOFTWARE AND EVENT RECONSTRUCTION

All areas of the STAR software effort are accelerating rapidly. STAF executables and KUIP macros for TPC analysis have been developed, documented, and made available to users within the collaboration. The software that processes the raw TPC pixel data has been significantly improved (more than an order of magnitude in speed). This includes the raw data reformater, the TPC cluster finder, and the space point reconstruction software.

Software packages exist and are being tuned for fast TPC space point simulation as well as slow TPC raw pixel (raw data) simulation, for tracking, for extracting energy loss for the tracks (dE/dx), and for evaluation of the Monte Carlo data. The data structure definitions for the software have been drafted and are being iterated and refined.

Silicon drift detectors of the same design as will be used in the STAR Silicon Vertex Tracker were used in a heavy ion experiment at the AGS (E896). The data that was collected is being analyzed using STAR SVT software. The data that has been collected during electronics system tests, and cosmic ray and laser calibration tests of the TPC is being analyzed and provides the valuable ability to run the developing software on real data well before the STAR detector sees its first beam-beam collisions at RHIC. There has been one Mock Data Challenge, which is an effort to exercise the detector software and RHIC computing center hardware on simulated data, completed. It proved to be a very useful exercise, pointing out the areas that needed more effort. A second Mock Data challenge is planned for March.

9. ADDITIONAL STAR DETECTORS

The strength of a collaboration and a detector effort can be seen not only by its current status, but also by its plans and prospects for the future.

The Silicon Vertex Tracker, a high precision charged particle tracking device that resides between the beam pipe and the inner radius of the TPC, is funded and well along in its construction. Approximately half of the total number of silicon drift detector wafers necessary for the complete SVT have been produced. The integrated circuit that contains the preamplifier and shaping amplifier (PASA) for the SVT have completed the design stage and are in production. The switched capacitor array (SCA) chip has also been designed and is in production. The Hybrid board that holds the various FEE chips has been designed and is in production, and final prototypes for the Readout components (similar to TPC readout boards) have been produced. Mechanically, the Beryllium support structures that hold the Silicon Drift Detectors (SDD) have been machined, and our French collaborators have delivered the support cone, that holds the SVT in place at the center of STAR, to BNL. The present schedule calls for the SVT to be installed during the summer shutdown of 2000. A fifteen layer array of STAR SDDs was used in E896 at the AGS. Please see the contribution to these proceedings from Rene Bellweid for more details.

The design of the Barrel Electromagnetic Calorimeter (BEMC) for STAR is essentially complete and funding for the construction of the detector has started. A full scale prototype of one of the 120 modules that comprise the BEMC was built and successfully tested this Fall in a test beam at the AGS. The schedule calls for somewhere between 6 to 12 of the BEMC modules to be produced and installed into the STAR detector this summer, so that they are available for use in the first physics run, which starts in November. The remainder of the BEMC modules will be produced and installed over the next three years. Please see the contribution to these proceeding from Gary Westfall for more details about this device.

The Forward Time Projection Chambers (FTPCs) have completed the design stage and are well along in their production. The FTPCs will provide the STAR detector with tracking for the pseudorapidity range of 2.4 to 4.0 on each side of the interaction. The schedule calls for the FTPCs to be installed during the summer 2000 shutdown.

A French group has recently joined the STAR collaboration and are building a layer of silicon strip detectors that will reside in the region between the outer radius of the SVT and the inner radius of the TPC.

This device will enhance the ability to reconstruct short lived particles which decay before they reach the TPC.

There is a funded effort by the STAR collaborators at Frankfurt Germany to supply hardware (primarily CPUs), and software manpower to advance the third level trigger for STAR. The primary feature of this third level trigger is that it quickly reconstructs the tracks in the TPC and allows one to use the tracking information in trigger decisions before the data is written out for archival storage.

There is an approved proposal to install an existing RICH detector into STAR from a group comprised primarily of STAR collaborators from Yale University, CERN personnel, and personnel from Bari(Italy). This device will significantly extend the STAR particle identification information as a function of p_T for a limited solid angle. The schedule calls for this device to be installed this summer. Please see the contribution to these proceedings from Gerd Kunde for more details about this device.

There is a proposal that has been submitted to the NSF, led by a recent addition to the STAR collaboration from Indiana University, to procure funding to build an Endcap Electromagnetic Calorimeter for STAR. The addition of this device will extend the pseudorapidity coverage for electromagnetic calorimetry in STAR from the +-1 that we acheive with the BEMC to -1 to 2. The addition of this detector will significantly extend the ability of the STAR detector to measure both the polarized gluon distributions in pp running and the perhaps modified gluon distribution in nucleons bound in nuclei. The proposed schedule for the EEMC has it being installed into STAR in the summer 2002 shutdown.

Finally, there is a proposal to add a patch of Time of Flight (TOF) coverage to STAR, and an effort to put together a proposal to add Photon Multiplicity Detectors (PMDs) to measure very forward electromagnetic energy.

10. SUMMARY

The facility work for the STAR detector is largely finished, with the remaining work on schedule. The magnet is finished and mapped. Some further power supply tests are planned for this March. The TPC structure is finished and has been well tested outside of the STAR magnet. Cosmic ray data is being analyzed, and approximately 90% of the FEE has been installed. The TPC was installed into the magnet in early December and the final electronics installation and cabling is expected to be complete by mid-March. The water and gas hookups are progressing and should also be completed in March. The DAQ system is in production. A portion of the complete system is up and running and all hardware is scheduled to be on site by May. The trigger electronics is in preproduction testing. The CTB trays are complete and will be fully installed by late February. The STAR software is up and running. MDC2 is the next milestone for the software effort, and then the final polish of the system and the code will take place in preparation for data taking in June. We are actively into planning and working on the upcoming transition in the STAR effort from the construction phase to operations.

The STAR project has every expectation of being ready to take data when the first collisions take place at RHIC this Spring, starting an exciting new era in Heavy Ion physics.

Index